State-of-the-Art and Progress in Metal-Hydrogen Systems

State-of-the-Art and Progress in Metal-Hydrogen Systems

Editors

Terry D. Humphries
Craig E. Buckley
Mark Paskevicius
Torben R. Jensen

Basel • Beijing • Wuhan • Barcelona • Belgrade • Novi Sad • Cluj • Manchester

Editors

Terry D. Humphries
Curtin University
Perth
Australia

Craig E. Buckley
Curtin University
Perth
Australia

Mark Paskevicius
Curtin University
Perth
Australia

Torben R. Jensen
Aarhus University
Aarhus
Denmark

Editorial Office
MDPI
St. Alban-Anlage 66
4052 Basel, Switzerland

This is a reprint of articles from the Special Issue published online in the open access journal *Inorganics* (ISSN 2304-6740) (available at: https://www.mdpi.com/journal/inorganics/special_issues/0745584385).

For citation purposes, cite each article independently as indicated on the article page online and as indicated below:

Lastname, A.A.; Lastname, B.B. Article Title. *Journal Name* **Year**, *Volume Number*, Page Range.

ISBN 978-3-7258-0357-6 (Hbk)
ISBN 978-3-7258-0358-3 (PDF)
doi.org/10.3390/books978-3-7258-0358-3

© 2024 by the authors. Articles in this book are Open Access and distributed under the Creative Commons Attribution (CC BY) license. The book as a whole is distributed by MDPI under the terms and conditions of the Creative Commons Attribution-NonCommercial-NoDerivs (CC BY-NC-ND) license.

Contents

About the Editors vii

Preface ix

Terry D. Humphries, Craig E. Buckley, Mark Paskevicius and Torben R. Jensen
State-of-the-Art and Progress in Metal-Hydrogen Systems
Reprinted from: *Inorganics* 2023, *11*, 476, doi:10.3390/inorganics11120476 1

Ting Zhang, Timothy Steenhaut, Michel Devillers and Yaroslav Filinchuk
Release of Pure H_2 from $Na[BH_3(CH_3NH)BH_2(CH_3NH)BH_3]$ by Introduction of Methyl Substituents
Reprinted from: *Inorganics* 2023, *11*, 202, doi:10.3390/inorganics11050202 9

Filippo Peru, Seyedhosein Payandeh, Torben R. Jensen, Georgia Charalambopoulou and Theodore Steriotis
Destabilization of the $LiBH_4$–$NaBH_4$ Eutectic Mixture through Pore Confinement for Hydrogen Storage
Reprinted from: *Inorganics* 2023, *11*, .128, doi:10.3390/inorganics11030128 20

Xiantun Huang, Chenglin Lu, Yun Li, Haimei Tang, Xingqing Duan, Kuikui Wang and Haizhen Liu
Hydrogen Release and Uptake of MgH_2 Modified by Ti_3CN MXene
Reprinted from: *Inorganics* 2023, *11*, 243, doi:10.3390/inorganics11060243 35

Zhi-Kang Qin, Li-Qing He, Xiao-Li Ding, Ting-Zhi Si, Ping Cui, Hai-Wen Li and Yong-Tao Li
Liquid Channels Built-In Solid Magnesium Hydrides for Boosting Hydrogen Sorption
Reprinted from: *Inorganics* 2023, *11*, 216, doi:10.3390/inorganics11050216 46

Sabrina Sartori, Matteo Amati, Luca Gregoratti, Emil Høj Jensen, Natalia Kudriashova and Jacques Huot
Study of Phase Composition in TiFe + 4 wt.% Zr Alloys by Scanning Photoemission Microscopy
Reprinted from: *Inorganics* 2023, *11*, 26, doi:10.3390/inorganics11010026 55

Lasse G. Kristensen, Mads B. Amdisen, Mie Andersen and Torben R. Jensen
Synthesis, Structure and Mg^{2+} Ionic Conductivity of Isopropylamine Magnesium Borohydride
Reprinted from: *Inorganics* 2023, *11*, 17, doi:10.3390/inorganics11010017 66

John J. Vajo, Jasim Uddin, Son-Jong Hwang and Jason Graetz
Electrolytes in Multiple-Phase Hydrogen Storage Reactions
Reprinted from: *Inorganics* 2023, *11*, 267, doi:10.3390/inorganics11070267 78

Naruki Endo, Yumi Kaneko, Norikazu Dezawa, Yasuhiro Komo and Masanobu Higuchi
Collectable Single Pure-Pd Metal Membrane with High Strength and Flexibility Prepared through Electroplating for Hydrogen Purification
Reprinted from: *Inorganics* 2023, *11*, 111, doi:10.3390/inorganics11030111 90

Xu Zhang, Yu-Yuan Zhao, Bao-Quan Li, Mikhail Prokhorenkov, Elshad Movlaev, Jin Xu, Wei Xiong and et al.
Hydrogen Compression Materials with Output Hydrogen Pressure in a Wide Range of Pressures Using a Low-Potential Heat-Transfer Agent
Reprinted from: *Inorganics* 2023, *11*, 180, doi:10.3390/inorganics11050180 98

Xin Zheng, Hanyang Kong, Desheng Chu, Faping Hu, Yao Wang, Yigang Yan and Chaoling Wu
Stress Reduction of a V-Based BCC Metal Hydride Bed Using Silicone Oil as a Glidant
Reprinted from: *Inorganics* 2022, *10*, 167, doi:10.3390/ inorganics10100167 116

Liang Liu, Alexander Ilyushechkin, Daniel Liang, Ashleigh Cousins, Wendy Tian, Cherry Chen and et al.
Metal Hydride Composite Structures for Improved Heat Transfer and Stability for Hydrogen Storage and Compression Applications
Reprinted from: *Inorganics* 2023, *11*, 181, doi:10.3390/inorganics11050181 126

Isaac Jacob, Dotan Babai, Matvey Bereznitsky and Roni Z. Shneck
The Role of Bulk Stiffening in Reducing the Critical Temperature of the Metal-to-Hydride Phase Transition and the Hydride Stability: The Case of $Zr(Mo_xFe_{1-x})_2$-H_2
Reprinted from: *Inorganics* 2023, *11*, 228, doi:10.3390/inorganics11060228 143

Yiqing Pan, Minh Tam Hoang, Sanaa Mansoor and Maria Alexandra Gomez
Exploring Proton Pair Motion Away from the Global Proton–Tuple Energy Minimum in Yttrium-Doped Barium Zirconate
Reprinted from: *Inorganics* 2023, *11*, 160, doi:10.3390/inorganics11040160 153

Jintao Wang, Qijun Pei, Yang Yu, Jirong Cui, Shangshang Wang, Khai Chen Tan, Jiaquan Guo and et al.
Investigation on the Formation of Rare-Earth Metal Phenoxides via Metathesis
Reprinted from: *Inorganics* 2023, *11*, 115, doi:10.3390/inorganics11030115 163

About the Editors

Terry D. Humphries

Terry D. Humphries is a Senior Research Fellow of the Hydrogen Storage Research Group at Curtin University. He has been developing hydrogen and energy storage materials for 18 years since commencing his Ph.D. studies. Terry was awarded his Ph.D. in Chemistry in 2011 at the University of New Brunswick (UNB), Canada, under the supervision of Prof. G. S. McGrady. He started his first post-doc position at the Institute for Energy Technology (IFE), Norway. During his three years at IFE, Terry was provided with the opportunity to be a guest researcher on two occasions at the University of Hawaii, USA, and one at the University of Aarhus, Denmark. In 2013, Terry moved to the Advanced Institute for Materials Research (WPI-AIMR), Tohoku University, Sendai, Japan, before moving to Australia for a position at Curtin University, where he has worked since 2015. During his career, Terry's research interests have comprised the development of materials for hydrogen storage, thermal energy storage, hydrogen export, and solid-state batteries. Terry is adept in the synthesis and analysis of air sensitive materials and specialises in the development of inert atmosphere and in situ analysis techniques. His research endeavours have led to the publication of over 80 research articles with over 150 global collaborators.

Craig E. Buckley

Craig E. Buckley is a John Curtin Distinguished Professor in Physics and leads the Hydrogen Storage Research Group at Curtin University, Australia. He was awarded a Ph.D. in Physics (Thesis title: Absorption/Desorption Hysteresis in Metal Hydrides) from Griffith University in 1994; since then, he has held various research positions in the UK, USA, and Australia prior to working at Curtin University. He leads Program 2, Hydrogen Exports, and Value Chains in the Future Energy Exports Co-operative Research Centre (CRC). He is a Fellow of the Australian Institute of Physics and a member of several national and international science and advisory committees. He is an Australian executive committee member for the International Energy Agency (IEA) Hydrogen Technology Collaboration Program (TCP), and an Australian expert on the IEA Hydrogen TCP Task 40: Energy Storage and Conversion based on Hydrogen. His research focusses on hydrogen storage materials and the investigation of the structural properties of a wide range of materials. Craig has been a lead/co-investigator on over USD 100 M of research funding and has published >240 peer-reviewed papers.

Mark Paskevicius

Mark Paskevicius is an Associate Professor of the Department of Physics and Astronomy at Curtin University in Australia. His research is focussed on the storage of renewable energy. He is dedicated to developing new materials for solid-state hydrogen storage to store energy for automotive, stationery, and export applications. He is also leading research into new solid-state ion conductors for battery applications, focussing on boron-rich materials. He is the Hydrogen Theme Leader for the Institute for Energy Transition at Curtin and has supervised 28 PhD students and 30+ BSc students in the field of energy storage. He has published over 125 peer-reviewed research publications with an h-index over 35. Mark has been the recipient of the Australian Institute of Nuclear Science and Engineering (AINSE) gold medal for research excellence, multiple Curtinnovation Awards for commercialization of research, along with numerous fellowships, including a 3-year fellowship at Aarhus University, Denmark.

Torben R. Jensen

Torben R. Jensen (TRJ) obtained a PhD degree in Materials Chemistry at the University of Southern Denmark, Odense, in 1999. In October 1998, he joined Risø Natl. Lab as a post-doc, where he changed research topic to Biophysics and conducted surface diffraction and reflectivity of lipids and enzymes. Much of his research was conducted at the synchrotron Hasylab, Hamburg. After 2 years, Oct. 2000, he became an Assistant Prof. at the Department of Chemistry, Aarhus University, and became associated with the Interdisciplinary Nanoscience Center (iNANO). Since then, he has assembled a unique and productive independent research group within energy materials science. His research achievements led to a Doctor of Science (D.Sc.) degree in 2014 and a promotion to Professor in 2016. Over the years, T.R. Jensen developed several new courses in chemistry and nanoscience and has educated 16 post-docs, 19 PhDs, 44 MScs, 61 BSc students and has hosted about 45 guest professors and researchers at all levels. His research group has synthesised a multitude of new metal hydrides by combining solvent-based methods, mechanochemistry, and solid–gas reactions, etc. They also infiltrated materials in nanoporous scaffolds and conducted systematic studies of properties as a function of pore size and surface area to investigate their nanoeffects. This group now focuses on cationic conductivity in the solid state and on the development of novel battery materials. This group is a frequent user of large synchrotron X-ray and neutron facilities and has developed knowhow and new versatile in situ powder X-ray diffraction (PXD) sample environments and gas control systems for the investigation of solid–gas and solid–liquid reactions, also at extreme p,T conditions of up to $p(H_2) = 700$ bar. T. R. Jensen has published >310 peer-reviewed papers.

Preface

Hydrogen is destined to be a future global energy carrier for a clean, innovative, safe, and competitive energy industry with great expectations in the time to come. Today, metal hydrides are explored for a range of applications, including hydrogen export, remote area power systems, solid-state batteries, thermochemical energy storage, and hydrogen diffusion. As such, there is increasing interest from industries to integrate hydrogen technology into their energy portfolio and supply chains. The aim of this Special Issue of Inorganics, entitled 'State-of-the-Art and Progress in Metal-Hydrogen Systems', is to inspire continued research within this important class of materials, particularly for energy-related applications. This Special Issue also serves as a collection of contributions presented at the International Symposium on Metal-Hydrogen Systems, held in Perth, Western Australia, 30 October – 4 November 2022. This meeting, MH2022, was the 17th meeting in a distinguished series of conferences dating back to 1968. Our hope is to convey inspiration for continued research to create a 'greener' future based on renewable energy. This reprint and the special issue are edited by Terry D. Humphries, Craig E. Buckley, Mark Paskevicius and Torben R. Jensen.

Terry D. Humphries, Craig E. Buckley, Mark Paskevicius, and Torben R. Jensen
Editors

Editorial

State-of-the-Art and Progress in Metal-Hydrogen Systems

Terry D. Humphries [1,*], Craig E. Buckley [1], Mark Paskevicius [1] and Torben R. Jensen [2]

1. Physics and Astronomy, Curtin University, GPO Box U1987, Perth, WA 6845, Australia; c.buckley@curtin.edu.au (C.E.B.); mark.paskevicius@gmail.com (M.P.)
2. Interdisciplinary Nanoscience Center and Department of Chemistry, Aarhus University, DK-8000 Aarhus C, Denmark; trj@chem.au.dk
* Correspondence: terry_humphries81@hotmail.com

Citation: Humphries, T.D.; Buckley, C.E.; Paskevicius, M.; Jensen, T.R. State-of-the-Art and Progress in Metal-Hydrogen Systems. *Inorganics* **2023**, *11*, 476. https://doi.org/10.3390/inorganics11120476

Received: 14 November 2023
Accepted: 4 December 2023
Published: 11 December 2023

Copyright: © 2023 by the authors. Licensee MDPI, Basel, Switzerland. This article is an open access article distributed under the terms and conditions of the Creative Commons Attribution (CC BY) license (https://creativecommons.org/licenses/by/4.0/).

1. Introduction

Hydrogen is heralded as a future global energy carrier [1–9]. The development of hydrogen as a clean energy solution is gaining unprecedented global attention, as many countries are now recognising its potential in various applications and including hydrogen in its energy portfolio. This includes stationary, portable, and transport use today, including hydrogen fuel cell vehicles, trains, and ferries. However, hydrogen storage remains a critical challenge in expanding infrastructure for hydrogen-based transportation and industrial utilisation.

Metal hydrides have received much interest over the past several decades, which is evident from a previous related Special Issue published in Inorganics: "Functional Materials Based on Metal Hydrides" [10]. Reversible solid-state hydrogen storage at ambient conditions with moderate energy exchanges with the surroundings is the ultimate challenge to realise a hydrogen-based society. Varieties of novel hydrogen-rich materials have been investigated in the past decades, which has provided many novel compositions, fascinating structures, and functionalities [8,11–14]. Today, metal hydrides are explored for a range of applications, including hydrogen exports, remote area power systems, solid-state batteries, thermochemical energy storage, and hydrogen diffusion [15–23]. Indeed, the International Energy Agency (IEA) is involved in the development of hydrogen and hydrogen storage materials, where Task 40 "Energy Storage and conversion based on hydrogen" has recently released a number of review articles on the research being conducted in this area [24–30].

The aim of this Special Issue of Inorganics, entitled "State-of-the-Art and Progress in Metal-Hydrogen Systems", is to inspire continued research within this important class of materials, in particular for energy-related applications. This Special Issue also serves as a collection of contributions presented at the International Symposium on Metal-Hydrogen Systems, held in Perth, Western Australia, 30 October–3 November 2022 [31]. This meeting, MH2022, is the 17th meeting in a distinguished series of conferences dating back to 1968. This conference was due to be held in 2020; however, COVID forced it to be postponed to 2022. The conference Chair was Prof. Craig Buckley, and the Vice-Chair was Assoc. Prof. Mark Paskevicius, both from Curtin University, Australia. MH2022 had a total of 159 participants from 22 countries. Five plenary talks were presented by Prof. Ping Chen (Dalian Institute of Chemical Physics, Chinese Academy of Sciences), Prof. Petra E de Jongh (Utrecht University, The Netherlands), Prof. Evan Gray (Griffith University, Australia), Dr Patrick Hartley (CSIRO, Australia), and Dr Michael Hirscher (Max Planck Institute for Intelligent Systems, Germany). There were 40 invited and 87 contributed talks and 30 posters on the topic of the fundamentals and applications of metal–hydrogen systems. The contributions covered a wide range of materials topics, including complex hydrides, metal hydrides, and chemical, organic, and nanoporous materials.

The Guest Editors would like to sincerely thank the Scientific Program Committee members for MH2022, the Local Organizing Committee, and the MH series International

Steering Committee for their help in planning and successfully organizing the symposium. We gratefully acknowledge the sponsors and supporting partners of MH2022: Curtin University, The Australian Renewable Energy Agency (ARENA), Toyota, and the exhibitors Anton Paar, Hidden Analytical, Hidden Isochema, John Morris Group, and Suzuki Shokan. Furthermore, the Guest Editors would like to thank all the reviewers who spent their valuable time thoroughly reviewing and improving the articles published in this Special Edition.

2. An Overview of Published Articles

As expressed above, the field of metal hydrides is diverse, and this is exemplified by the varied topics covered in the 14 articles published in this Special Edition. This section provides a brief overview of the included manuscripts, of which they are divided into discreet, albeit cross-cutting, subsections that include: Hydrogen Release and Uptake, Electrolytes, Physical Properties, and Metallic Alloys.

2.1. Hydrogen Release and Uptake

Varieties of novel hydrogen containing compounds with fascinating compositions and structures are continually discovered [1,13,32–37], well-illustrated by the $Na[BH_3(CH_3NH)BH_2(CH_3NH)BH_3]$ material presented by Zhang et al. (in contribution 1). This compound releases 4.6 wt.% of pure hydrogen below 150 °C contrary to the parent counterpart $Na[BH_3NH_2BH_2NH_2BH_3]$. This work illustrates a new route to suppress the release of unwanted gaseous by-products by the introduction of organic methyl groups bonded to nitrogen.

Lithium and sodium borohydride, $LiBH_4$ and $NaBH_4$, are known to possess high gravimetric and volumetric hydrogen densities of 18.5 and 10.6 wt.% and 121 and 113 kg·m^{-3}, respectively [37–43]. However, their high thermal stabilities hamper utilisation as hydrogen storage materials for practical applications. Interestingly, F. Peru et al. investigated the composite $0.71LiBH_4–0.29NaBH_4$ with a eutectic melting point of 219 °C, which allows convenient melt infiltration into a nanoporous carbon material (CMK-3) with a pore size of ~5 nm (contribution 2) [44]. This material reveals an uptake of ~3.5 wt.% H_2 after five hydrogenation/dehydrogenation cycles facilitated by carbon–hydride surface interactions and possibly enhanced heat transfer.

Magnesium hydride, MgH_2, has attracted much attention during the last few decades owing to a high hydrogen content of 7.6 wt.% and possesses good reversibility [23,24,45–48]. However, hydrogen release and uptake occur at temperatures higher than 300 °C and with slow kinetics. The third and fourth paper contributed to this collection reveals new strategies for improving both kinetic and thermodynamic properties of magnesium hydride. Huang et al. (contribution 3) demonstrated that the operating temperature for pristine MgH_2 can be decreased from 322 to 214 °C through the employment of Ti_3CN. Furthermore, the hydrogen-release kinetics of MgH_2 are also improved, as observed by a decrease in activation energy from 121 to 80 kJmol^{-1} for the release of hydrogen. Another strategy, developed by Qin et al. (contribution 4), is to use hydrogen-containing additives such as lithium borohydride in magnesium hydride. A new composite of MgH_2-$LiBH_4$ (5 wt%) released ~7.1 wt.% H_2 within 40 min at 300 °C, where a reference sample of MgH_2 only released <0.7 wt.% H_2. Importantly, the kinetics of hydrogen release are improved by a factor of ten (×10) compared to that of the pure MgH_2 material. This effect is suggested to be due to the creation of ion transfer channels, which can inhibit the agglomeration of MgH_2 particles.

TiFe is a well-established hydrogen storage material due to its low cost, elemental abundance, and near-ambient temperature and mild pressure for hydrogen absorption [15,49]. This makes it perfect for stationary applications [50]. The main hindrance is the difficulty to activate the material for the absorption of hydrogen in the first cycle, where high pressures of ~50 bar and high temperatures of ~400 °C are often required. Contribution five by Sartori, Amati, and Gregoratti et al. studied the influence of adding 4 wt.% Zr to the material and found that the kinetics of the first hydrogenation are greatly improved. The authors employed scanning photoemission microscopy to investigate the composition and chemical state of the various phases present in this alloy and how they change upon hydrogenation/dehydrogenation and found the presence of different oxide phases that were not seen by conventional SEM investigation.

2.2. Electrolytes

Recently, a variety of new complex hydride-based electrolytes were discovered [51–56]. They deviate from classical ionic compounds by offering high cationic conductivity at moderate temperatures, also with divalent cations. The high cationic conductivity is often assigned to weak interactions, such as dihydrogen bonds, providing structural flexibility [26,57]. A new example is magnesium borohydride, templated by isopropylamine, $Mg(BH_4)_2 \cdot (CH_3)_2CHNH_2$, which also has hydrophobic domains, presented by Kristensen et al. (contribution 6). Aluminium oxide nanoparticles and heat treatment produce a highly conductive composite, $\sigma(Mg^{2+}) = 2.86 \times 10^{-7}$ and 2.85×10^{-5} S cm^{-1} at -10 and $40\,°C$, respectively, with a low activation energy, $E_a = 0.65$ eV. Nanoparticles stabilise the partially eutectic molten state and prevent recrystallisation even at low temperatures and provide a high mechanical stability of the composite.

Electrochemical reactions taking place in a solid-state battery cell can be considered multi-phase reactions taking place at interfaces, involving cationic diffusion in the solid state [26,58]. Hydrogen storage systems based on complex hydrides, such as metal borohydrides or metal alanates, reveal similarities. In both cases, multiple solid phases nucleate and grow or are consumed as hydrogen is released or stored or electrochemical reactions take place in a battery. These reactions are limited by the kinetics of atoms or ion diffusion at grain boundaries in the solid state and also by solid state transformations. Thus, Vajo and co-workers (contribution 7) proposed to explore these challenges by investigating combined electrolytes and hydrogen storage systems. This includes the use of eutectic molten composites. An increase in reaction rates was observed in several cases, but the kinetics may become more complex. A reduction in reaction temperature may be observed, but the results also suggest that electrolytes improve intraparticle transport phenomena.

2.3. Metallic Hydrides for Hydrogen Purification and Compression

If the world is to move towards a hydrogen economy, hydrogen must be produced in high purity to avoid any poisoning of catalysts during consumption, e.g., in a fuel cell [59]. Palladium has been determined to be extremely effective towards the application of hydrogen-purification/separation membranes. In contribution 8, Endo et al. discusses the preparation of high-purity palladium (Pd) films using the electroplating method, which is considered a simple and cost-effective technique. However, electroplating can result in stress accumulation in the film, making it challenging to obtain a dense single Pd film. This study successfully addressed this issue by optimising the electroplating process, resulting in a high-purity Pd film with unique surface characteristics. Notably, the plated film exhibited superior mechanical properties compared to rolled Pd films, including twice the displacement and four times the breaking point strength. As such, this research demonstrates the practical feasibility of using electroplating to produce Pd-based membranes for hydrogen purification applications.

For practical applications of hydrogen gas, convenient compression using low amounts of energy is crucial. Metallic hydrides have proven useful for the development of metal hydride–hydrogen compressors [15,30,60]. This is successfully demonstrated by X. Zhang et al. (contribution 9) where the rare earth series AB_5 and Ti/Zr-based AB_2 hydrogen storage materials are investigated. A four-stage compressor is developed with output pressures of 8.90, 25.04, 42.97, and 84.73 MPa operating at 363 K. The first pressure stage was achieved by a $CaCu_5$-type hexagonal structure, the others by $TiCr_2$ and $ZrFe_2$ type alloys.

There are many aspects to overcome to enable metal hydrides to be the ideal hydrogen storage medium, but several researchers are utilising a variety of methods to overcome these problems. Large volume expansions within these powders are observed upon cycling of hydrogen, and this can lead to the occurrence of stress on the walls of the container and the possibility of catastrophic failure [15,30,49]. Zheng et al. (contribution 10) used silicone oil as a glidant to promote stress-free conditions for V-based BCC metal hydrides and have reported that the addition of 5 wt.% silicone oil slightly reduced the initial hydrogen storage capacity of $V_{40}Ti_{26}Cr_{26}Fe_8$ (particle size ~325 µm) but improved the absorption reversibility, regardless of the oil viscosity. The maximum strain on the surface of the hydrogen storage container decreased by $\geq 22.5\%$ after adding 5 wt.% silicone oil.

During absorption and desorption of hydrogen, metal hydrides interact with hydrogen via multiple bonding interactions, including covalent, ionic, and van der Waals forces [61]. This inherently produces heat during absorption and absorbs heat during desorption [22,27]. This heat must be dissipated if a steady reaction rate is to occur and avoid physical degradation, including agglomeration or side reactions that are more favourable at high temperatures. Liu et al. (contribution 11) submitted a review on improving heat transfer and stability for hydrogen storage and compression applications. Currently, several researchers have adopted the method of forming composites of alloy powders with high thermal conductivity materials, such as exfoliated natural graphite (ENG), but there are many options, including liquid-based methods, polymers, and metal foams. This article highlights the state of the art in this field of research.

2.4. Physical Properties

Fundamental research into the physical properties of metal hydrides is required if technological progress is to continue in the future [1,10,14,25,27]. Jacob, Babai, and Bereznitsky et al. (contribution 12) have concentrated on determining the elastic moduli of $Zr(Mo_xFe_{1-x})_2$, $x = 0, 0.5, 1$, as well as hydrogen absorption in $ZrMo_2$, in an attempt to shed light on the unusual trend in hydride stabilities in this system. In general, the stabilities of these hydrides exhibit a unique trend, with stability increasing from $x = 0$ to 0.5, but destabilisation is observed for the end member $ZrMo_2$ hydride. They found that the bulk modulus (B) significantly increases from 148.2 GPa in $ZrFe_2$ to 200.4 GPa in $ZrMo_2$, which is in contrast to the moderate variation in the shear modulus (G). The increase in B suggests a bulk stiffening in $ZrMo_2$.

Pan et al. (contribution 13) provided a study detailing how protons move within yttrium-doped barium zirconate, which is known to be one of the fastest solid-state proton conductors. This study used density functional theory with the Perdew–Burke–Ernzerhof functional to calculate the total electronic energy for each proton pair in an effort to catalogue and understand their motion, especially when the protons are in close proximity to one another. Overall, it was determined that protons are in close proximity to each other and the dopant in the lowest-energy configuration, significantly affecting the backbone structure. This knowledge is important for developing superior materials for proton-conduction applications.

The search for highly tuneable hydrogen storage materials is an ongoing field, especially as the addition of functional groups can produce novel materials with interesting properties such as superconductivity, photoluminescence, etc. Alkali organometallic complexes are known to have favourable thermodynamics and hydrogen capacities, but not many transition metal–organic complexes have been reported. The contribution by Wang, Pei, Yu, and Cui et al. (contribution 14) have demonstrated the formation of yttrium phenoxide and lanthanum phenoxide via metathesis of sodium phenoxide with YCl_3 and $LaCl_3$, respectively. Their properties were elucidated using theoretical calculations, quasi in situ NMR, and UV-vis spectroscopies. Although the hydrogenation of these rare-earth phenoxides was not successful, further research on these materials may lead to progress.

3. Conclusions

The necessity for reliable methods to store renewable energy is driving the development of hydrogen production, utilisation, distribution, and storage. The most-efficient method of storing hydrogen is by using metal hydrides, whereas, simultaneously, many of these metal hydrides are finding applications in a variety of functions, including solid-state batteries, thermal energy storage, compression, etc.

The presentations, posters, and manuscripts that have arisen from MH2022 in Australia are a testament to the world-class research being undertaken in this area. The Editors of this Special Edition "State-of-the-Art and Progress in Metal-Hydrogen Systems" welcome you to read the contributed articles in the issue.

Conflicts of Interest: The authors declare no conflict of interest.

List of Contributions

1. Zhang, T.; Steenhaut, T.; Devillers, M.; Filinchuk, Y. Release of Pure H_2 from $Na[BH_3(CH_3NH)BH_2(CH_3NH)BH_3]$ by Introduction of Methyl Substituents. *Inorganics* **2023**, *11*, 202. https://10.3390/inorganics11050202.
2. Peru, F.; Payandeh, S.; Jensen, T.R.; Charalambopoulou, G.; Steriotis, T. Destabilization of the $LiBH_4$-$NaBH_4$ Eutectic Mixture through Pore Confinement for Hydrogen Storage. *Inorganics* **2023**, *11*, 128. https://10.3390/inorganics11030128.
3. Huang, X.T.; Lu, C.L.; Li, Y.; Tang, H.M.; Duan, X.Q.; Wang, K.K.; Liu, H.Z. Hydrogen Release and Uptake of MgH_2 Modified by Ti_3CN MXene. *Inorganics* **2023**, *11*, 243. https://10.3390/inorganics11060243.
4. Qin, Z.K.; He, L.Q.; Ding, X.L.; Si, T.Z.; Cui, P.; Li, H.W.; Li, Y.T. Liquid Channels Built-In Solid Magnesium Hydrides for Boosting Hydrogen Sorption. *Inorganics* **2023**, *11*, 216. https://10.3390/inorganics11050216.
5. Sartori, S.; Amati, M.; Gregoratti, L.; Jensen, E.H.; Kudriashova, N.; Huot, J. Study of Phase Composition in TiFe + 4 wt.% Zr Alloys by Scanning Photoemission Microscopy. *Inorganics* **2023**, *11*, 26.
6. Kristensen, L.G.; Amdisen, M.B.; Andersen, M.; Jensen, T.R. Synthesis, Structure and Mg^{2+} Ionic Conductivity of Isopropylamine Magnesium Borohydride. *Inorganics* **2023**, *11*, 17. https://10.3390/inorganics11010017.
7. Vajo, J.J.; Uddin, J.; Hwang, S.J.; Graetz, J. Electrolytes in Multiple-Phase Hydrogen Storage Reactions. *Inorganics* **2023**, *11*, 267. https://10.3390/inorganics11070267.
8. Endo, N.; Kaneko, Y.; Dezawa, N.; Komo, Y.; Higuchi, M. Collectable Single Pure-Pd Metal Membrane with High Strength and Flexibility Prepared through Electroplating for Hydrogen Purification. *Inorganics* **2023**, *11*, 111. https://10.3390/inorganics11030111.
9. Zhang, X.; Zhao, Y.Y.; Li, B.Q.; Prokhorenkov, M.; Movlaev, E.; Xu, J.; Xiong, W.; Yan, H.Z.; Mitrokhin, S. Hydrogen Compression Materials with Output Hydrogen Pressure in a Wide Range of Pressures Using a Low-Potential Heat-Transfer Agent. *Inorganics* **2023**, *11*, 180. https://10.3390/inorganics11050180.
10. Zheng, X.; Kong, H.Y.; Chu, D.S.; Hu, F.P.; Wang, Y.; Yan, Y.G.; Wu, C.L. Stress Reduction of a V-Based BCC Metal Hydride Bed Using Silicone Oil as a Glidant. *Inorganics* **2022**, *10*, 167. https://10.3390/inorganics10100167.
11. Liu, L.; Ilyushechkin, A.; Liang, D.; Cousins, A.; Tian, W.; Chen, C.; Yin, J.; Schoeman, L. Metal Hydride Composite Structures for Improved Heat Transfer and Stability for Hydrogen Storage and Compression Applications. *Inorganics* **2023**, *11*, 181. https://10.3390/inorganics11050181.
12. Jacob, I.; Babai, D.; Bereznitsky, M.; Shneck, R.Z. The Role of Bulk Stiffening in Reducing the Critical Temperature of the Metal-to-Hydride Phase Transition and the Hydride Stability: The Case of $Zr(Mo_xFe_{1-x})_2$-H_2. *Inorganics* **2023**, *11*, 228. https://10.3390/inorganics11060228.
13. Pan, Y.Q.; Hoang, M.T.; Mansoor, S.; Gomez, M.A. Exploring Proton Pair Motion Away from the Global Proton-Tuple Energy Minimum in Yttrium-Doped Barium Zirconate. *Inorganics* **2023**, *11*, 160. https://10.3390/inorganics11040160.

14. Wang, J.T.; Pei, Q.J.; Yu, Y.; Cui, J.R.; Wang, S.S.; Tan, K.C.; Guo, J.Q.; He, T.; Chen, P. Investigation on the Formation of Rare-Earth Metal Phenoxides via Metathesis. *Inorganics* **2023**, *11*, 115. https://10.3390/inorganics11030115.

References

1. Hirscher, M.; Yartys, V.A.; Baricco, M.; Bellosta von Colbe, J.; Blanchard, D.; Bowman, R.C.; Broom, D.P.; Buckley, C.E.; Chang, F.; Chen, P.; et al. Materials for hydrogen-based energy storage—Past, recent progress and future outlook. *J. Alloys Compd.* **2020**, *827*, 153548. [CrossRef]
2. Orimo, S.; Nakamori, Y.; Eliseo, J.R.; Zuttel, A.; Jensen, C.M. Complex hydrides for hydrogen storage. *Chem. Rev.* **2007**, *107*, 4111–4132. [CrossRef] [PubMed]
3. van den Berg, A.W.C.; Arean, C.O. Materials for hydrogen storage: Current research trends and perspectives. *Chem. Commun.* **2008**, *6*, 668–681. [CrossRef] [PubMed]
4. Felderhoff, M.; Weidenthaler, C.; von Helmolt, R.; Eberle, U. Hydrogen storage: The remaining scientific and technological challenges. *Phys. Chem. Chem. Phys.* **2007**, *9*, 2643–2653. [CrossRef] [PubMed]
5. Hagemann, H. Boron Hydrogen Compounds: Hydrogen Storage and Battery Applications. *Molecules* **2021**, *26*, 7425. [CrossRef]
6. Crabtree, G.W.; Dresselhaus, M.S.; Buchanan, M.V. The hydrogen economy. *Phys. Today* **2004**, *57*, 39–44. [CrossRef]
7. Abbott, D. Hydrogen Without Tears: Addressing the Global Energy Crisis via a Solar to Hydrogen Pathway [Point of View]. *Proc. IEEE* **2009**, *97*, 1931–1934. [CrossRef]
8. He, T.; Pachfule, P.; Wu, H.; Xu, Q.; Chen, P. Hydrogen carriers. *Nat. Rev. Mater.* **2016**, *1*, 16059. [CrossRef]
9. Egeland-Eriksen, T.; Hajizadeh, A.; Sartori, S. Hydrogen-based systems for integration of renewable energy in power systems: Achievements and perspectives. *Int. J. Hydrogen Energy* **2021**, *46*, 31963–31983. [CrossRef]
10. Li, H.W.; Zhu, M.; Buckley, C.; Jensen, T.R. Functional Materials Based on Metal Hydrides. *Inorganics* **2018**, *6*, 91. [CrossRef]
11. Webb, C.J. A review of catalyst-enhanced magnesium hydride as a hydrogen storage material. *J. Phys. Chem. Solids* **2015**, *84*, 96–106. [CrossRef]
12. Li, C.; Peng, P.; Zhou, D.W.; Wan, L. Research progress in LiBH$_4$ for hydrogen storage: A review. *Int. J. Hydrogen Energy* **2011**, *36*, 14512–14526. [CrossRef]
13. Dovgaliuk, I.; Filinchuk, Y. Aluminium complexes of B- and N-based hydrides: Synthesis, structures and hydrogen storage properties. *Int. J. Hydrogen Energy* **2016**, *41*, 15489–15504. [CrossRef]
14. Callini, E.; Atakli, Z.Ö.K.; Hauback, B.C.; Orimo, S.-i.; Jensen, C.; Dornheim, M.; Grant, D.; Cho, Y.W.; Chen, P.; Hjörvarsson, B. Complex and liquid hydrides for energy storage. *Appl. Phys. A* **2016**, *122*, 353. [CrossRef]
15. Lototskyy, M.V.; Yartys, V.A.; Pollet, B.G.; Bowman, R.C. Metal hydride hydrogen compressors: A review. *Int. J. Hydrogen Energy* **2014**, *39*, 5818–5851. [CrossRef]
16. Pasini, J.M.; Corgnale, C.; van Hassel, B.A.; Motyka, T.; Kumar, S.; Simmons, K.L. Metal hydride material requirements for automotive hydrogen storage systems. *Int. J. Hydrogen Energy* **2013**, *38*, 9755–9765. [CrossRef]
17. de Jongh, P.E.; Allendorf, M.; Vajo, J.J.; Zlotea, C. Nanoconfined light metal hydrides for reversible hydrogen storage. *MRS Bull.* **2013**, *38*, 488–494. [CrossRef]
18. Song, Y. New perspectives on potential hydrogen storage materials using high pressure. *Phys. Chem. Chem. Phys.* **2013**, *15*, 14524–14547. [CrossRef]
19. Mohtadi, R.; Orimo, S.-i. The renaissance of hydrides as energy materials. *Nat. Rev. Mater.* **2016**, *2*, 16091. [CrossRef]
20. Yartys, V.; Noreus, D.; Latroche, M. Metal hydrides as negative electrode materials for Ni–MH batteries. *Appl. Phys. A* **2016**, *122*, 43. [CrossRef]
21. Schouwink, P.; Didelot, E.; Lee, Y.-S.; Mazet, T.; Černý, R. Structural and magnetocaloric properties of novel gadolinium borohydrides. *J. Alloys Compd.* **2016**, *664*, 378–384. [CrossRef]
22. Manickam, K.; Mistry, P.; Walker, G.; Grant, D.; Buckley, C.E.; Humphries, T.D.; Paskevicius, M.; Jensen, T.; Albert, R.; Peinecke, K.; et al. Future perspectives of thermal energy storage with metal hydrides. *Int. J. Hydrogen Energy* **2019**, *44*, 7738–7745. [CrossRef]
23. Yartys, V.A.; Lototskyy, M.V.; Akiba, E.; Albert, R.; Antonov, V.E.; Ares, J.R.; Baricco, M.; Bourgeois, N.; Buckley, C.E.; Bellosta von Colbe, J.M.; et al. Magnesium based materials for hydrogen based energy storage: Past, present and future. *Int. J. Hydrogen Energy* **2019**, *44*, 7809–7859. [CrossRef]
24. Pasquini, L.; Sakaki, K.; Akiba, E.; Allendorf, M.D.; Alvares, E.; Ares, J.R.; Babai, D.; Baricco, M.; Bellosta von Colbe, J.; Bereznitsky, M.; et al. Magnesium- and intermetallic alloys-based hydrides for energy storage: Modelling, synthesis and properties. *Prog. Energy* **2022**, *4*, 032007. [CrossRef]
25. Dematteis, E.M.; Amdisen, M.B.; Autrey, T.; Barale, J.; Bowden, M.E.; Buckley, C.E.; Cho, Y.W.; Deledda, S.; Dornheim, M.; de Jongh, P.; et al. Hydrogen storage in complex hydrides: Past activities and new trends. *Prog. Energy* **2022**, *4*, 032009. [CrossRef]
26. Cuevas, F.; Amdisen, M.B.; Baricco, M.; Buckley, C.E.; Cho, Y.W.; de Jongh, P.; de Kort, L.M.; Grinderslev, J.B.; Gulino, V.; Hauback, B.C.; et al. Metallic and complex hydride-based electrochemical storage of energy. *Prog. Energy* **2022**, *4*, 032001. [CrossRef]
27. Adams, M.; Buckley, C.E.; Busch, M.; Bunzel, R.; Felderhoff, M.; Heo, T.W.; Humphries, T.D.; Jensen, T.R.; Klug, J.; Klug, K.H.; et al. Hydride-based thermal energy storage. *Prog. Energy* **2022**, *4*, 032008. [CrossRef]
28. Ulucan, T.H.; Akhade, S.A.; Ambalakatte, A.; Autrey, T.; Cairns, A.; Chen, P.; Cho, Y.W.; Gallucci, F.; Gao, W.; Grinderslev, J.B.; et al. Hydrogen storage in liquid hydrogen carriers: Recent activities and new trends. *Prog. Energy* **2023**, *5*, 012004. [CrossRef]

29. Zhang, L.; Allendorf, M.D.; Balderas-Xicohténcatl, R.; Broom, D.P.; Fanourgakis, G.S.; Froudakis, G.E.; Gennett, T.; Hurst, K.E.; Ling, S.; Milanese, C.; et al. Fundamentals of hydrogen storage in nanoporous materials. *Prog. Energy* **2022**, *4*, 042013. [CrossRef]
30. Dornheim, M.; Baetcke, L.; Akiba, E.; Ares, J.-R.; Autrey, T.; Barale, J.; Baricco, M.; Brooks, K.; Chalkiadakis, N.; Charbonnier, V.; et al. Research and development of hydrogen carrier based solutions for hydrogen compression and storage. *Prog. Energy* **2022**, *4*, 042005. [CrossRef]
31. 17th International Symposium on Metal-Hydrogen Systems. Available online: https://www.metal-hydrogen2022.com/ (accessed on 31 October 2023).
32. Dovgaliuk, I.; Le Duff, C.S.; Robeyns, K.; Devillers, M.; Filinchuk, Y. Mild Dehydrogenation of Ammonia Borane Complexed with Aluminum Borohydride. *Chem. Mater.* **2015**, *27*, 768–777. [CrossRef]
33. Chua, Y.S.; Chen, P.; Wu, G.; Xiong, Z. Development of amidoboranes for hydrogen storage. *Chem. Commun.* **2011**, *47*, 5116–5129. [CrossRef]
34. Chua, Y.S.; Wu, G.T.; Xiong, Z.T.; Karkamkar, A.; Guo, J.P.; Jian, M.X.; Wong, M.W.; Autrey, T.; Chen, P. Synthesis, structure and dehydrogenation of magnesium amidoborane monoammoniate. *Chem. Commun.* **2010**, *46*, 5752–5754. [CrossRef] [PubMed]
35. Brinks, H.W.; Hauback, B.C.; Jensen, C.M.; Zidan, R. Synthesis and crystal structure of Na_2LiAlD_6. *J. Alloys Compd.* **2005**, *392*, 27–30. [CrossRef]
36. Brinks, H.W.; Istad-Lem, A.; Hauback, B.C. Mechanochemical synthesis and crystal structure of alpha'-AlD_3 and alpha-AlD_3. *J. Phys. Chem. B* **2006**, *110*, 25833–25837. [CrossRef] [PubMed]
37. Filinchuk, Y.; Chernyshov, D.; Dmitriev, V. Light metal borohydrides: Crystal structures and beyond. *Z. Kristallogr.* **2008**, *223*, 649–659. [CrossRef]
38. Paskevicius, M.; Jepsen, L.H.; Schouwink, P.; Cerny, R.; Ravnsbaek, D.B.; Filinchuk, Y.; Dornheim, M.; Besenbacher, F.; Jensen, T.R. Metal borohydrides and derivatives—Synthesis, structure and properties. *Chem. Soc. Rev.* **2017**, *46*, 1565–1634. [CrossRef] [PubMed]
39. Remhof, A.; Mauron, P.; Züttel, A.; Embs, J.P.; Łodziana, Z.; Ramirez-Cuesta, A.J.; Ngene, P.; de Jongh, P. Hydrogen Dynamics in Nanoconfined Lithiumborohydride. *J. Phys. Chem. C* **2013**, *117*, 3789–3798. [CrossRef]
40. Liu, X.; Peaslee, D.; Jost, C.Z.; Baumann, T.F.; Majzoub, E.H. Systematic Pore-Size Effects of Nanoconfinement of $LiBH_4$: Elimination of Diborane Release and Tunable Behavior for Hydrogen Storage Applications. *Chem. Mater.* **2011**, *23*, 1331–1336. [CrossRef]
41. Sartori, S.; Knudsen, K.D.; Hage, F.S.; Heyn, R.H.; Bardaji, E.G.; Zhao-Karger, Z.; Fichtner, M.; Hauback, B.C. Influence of nanoconfinement on morphology and dehydrogenation of the $Li^{11}BD_4$–$Mg(^{11}BD_4)_2$ system. *Nanotechnology* **2012**, *23*, 255704. [CrossRef]
42. Chen, W.; Ouyang, L.Z.; Liu, J.W.; Yao, X.D.; Wang, H.; Liu, Z.W.; Zhu, M. Hydrolysis and regeneration of sodium borohydride ($NaBH_4$)—A combination of hydrogen production and storage. *J. Power Sources* **2017**, *359*, 400–407. [CrossRef]
43. Demirci, U.B.; Akdim, O.; Andrieux, J.; Hannauer, J.; Chamoun, R.; Miele, P. Sodium Borohydride Hydrolysis as Hydrogen Generator: Issues, State of the Art and Applicability Upstream from a Fuel Cell. *Fuel Cells* **2010**, *10*, 335–350. [CrossRef]
44. Schneemann, A.; White, J.L.; Kang, S.; Jeong, S.; Wan, L.F.; Cho, E.S.; Heo, T.W.; Prendergast, D.; Urban, J.J.; Wood, B.C.; et al. Nanostructured Metal Hydrides for Hydrogen Storage. *Chem. Rev.* **2018**, *118*, 10775–10839. [CrossRef] [PubMed]
45. de Jongh, P.E.; Wagemans, R.W.P.; Eggenhuisen, T.M.; Dauvillier, B.S.; Radstake, P.B.; Meeldijk, J.D.; Geus, J.W.; de Jong, K.P. The Preparation of Carbon-Supported Magnesium Nanoparticles using Melt Infiltration. *Chem. Mater.* **2007**, *19*, 6052–6057. [CrossRef]
46. Gross, A.F.; Ahn, C.C.; Van Atta, S.L.; Liu, P.; Vajo, J.J. Fabrication and hydrogen sorption behaviour of nanoparticulate MgH_2 incorporated in a porous carbon host. *Nanotechnology* **2009**, *20*, 204005. [CrossRef]
47. Ares, J.R.; Aguey-Zinsou, K.F.; Klassen, T.; Bormann, R. Influence of impurities on the milling process of MgH_2. *J. Alloys Compd.* **2007**, *434–435*, 729–733. [CrossRef]
48. Oelerich, W.; Klassen, T.; Bormann, R. Metal oxides as catalysts for improved hydrogen sorption in nanocrystalline Mg-based materials. *J. Alloys Compd.* **2001**, *315*, 237–242. [CrossRef]
49. Dematteis, E.M.; Berti, N.; Cuevas, F.; Latroche, M.; Baricco, M. Substitutional effects in TiFe for hydrogen storage: A comprehensive review. *Mater. Adv.* **2021**, *2*, 2524–2560. [CrossRef]
50. Modi, P.; Aguey-Zinsou, K.-F. Room Temperature Metal Hydrides for Stationary and Heat Storage Applications: A Review. *Front. Energy Res.* **2021**, *9*, 616115. [CrossRef]
51. Roedern, E.; Kühnel, R.S.; Remhof, A.; Battaglia, C. Magnesium Ethylenediamine Borohydride as Solid-State Electrolyte for Magnesium Batteries. *Sci. Rep.* **2017**, *7*, 46189. [CrossRef]
52. Choi, Y.S.; Lee, Y.-S.; Choi, D.-J.; Chae, K.H.; Oh, K.H.; Cho, Y.W. Enhanced Li Ion Conductivity in $LiBH_4$–Al_2O_3 Mixture via Interface Engineering. *J. Phys. Chem. C* **2017**, *121*, 26209–26215. [CrossRef]
53. Tang, W.S.; Matsuo, M.; Wu, H.; Stavila, V.; Unemoto, A.; Orimo, S.-i.; Udovic, T.J. Stabilizing lithium and sodium fast-ion conduction in solid polyhedral-borate salts at device-relevant temperatures. *Energy Storage Mater.* **2016**, *4*, 79–83. [CrossRef]
54. Matsuo, M.; Orimo, S.-i. Lithium Fast-Ionic Conduction in Complex Hydrides: Review and Prospects. *Adv. Energy Mater.* **2011**, *1*, 161–172. [CrossRef]
55. Tutusaus, O.; Mohtadi, R.; Arthur, T.S.; Mizuno, F.; Nelson, E.G.; Sevryugina, Y.V. An Efficient Halogen-Free Electrolyte for Use in Rechargeable Magnesium Batteries. *Angew. Chem. Int. Ed.* **2015**, *54*, 7900–7904. [CrossRef] [PubMed]

56. Kisu, K.; Kim, S.; Inukai, M.; Oguchi, H.; Takagi, S.; Orimo, S.-i. Magnesium Borohydride Ammonia Borane as a Magnesium Ionic Conductor. *ACS Appl. Energy Mater.* **2020**, *3*, 3174–3179. [CrossRef]
57. Yan, Y.; Dononelli, W.; Jørgensen, M.; Grinderslev, J.B.; Lee, Y.-S.; Cho, Y.W.; Černý, R.; Hammer, B.; Jensen, T.R. The mechanism of Mg^{2+} conduction in ammine magnesium borohydride promoted by a neutral molecule. *Phys. Chem. Chem. Phys.* **2020**, *22*, 9204–9209. [CrossRef] [PubMed]
58. Hadjixenophontos, E.; Dematteis, E.M.; Berti, N.; Wołczyk, A.R.; Huen, P.; Brighi, M.; Le, T.T.; Santoru, A.; Payandeh, S.; Peru, F.; et al. A Review of the MSCA ITN ECOSTORE—Novel Complex Metal Hydrides for Efficient and Compact Storage of Renewable Energy as Hydrogen and Electricity. *Inorganics* **2020**, *8*, 17. [CrossRef]
59. Faye, O.; Szpunar, J.; Eduok, U. A critical review on the current technologies for the generation, storage, and transportation of hydrogen. *Int. J. Hydrogen Energy* **2022**, *47*, 13771–13802. [CrossRef]
60. Gray, E.M.; Webb, C.J. Metal-hydride hydrogen compressors for laboratory use. *J. Phys. Energy* **2020**, *2*, 034004. [CrossRef]
61. Züttel, A.; Remhof, A.; Borgschulte, A.; Friedrichs, O. Hydrogen: The future energy carrier. *Phil. Trans. R. Soc. A* **2010**, *368*, 3329–3342. [CrossRef]

Disclaimer/Publisher's Note: The statements, opinions and data contained in all publications are solely those of the individual author(s) and contributor(s) and not of MDPI and/or the editor(s). MDPI and/or the editor(s) disclaim responsibility for any injury to people or property resulting from any ideas, methods, instructions or products referred to in the content.

Article

Release of Pure H$_2$ from Na[BH$_3$(CH$_3$NH)BH$_2$(CH$_3$NH)BH$_3$] by Introduction of Methyl Substituents

Ting Zhang, Timothy Steenhaut, Michel Devillers and Yaroslav Filinchuk *

Institute of Condensed Matter and Nanosciences, Université Catholique de Louvain, Place Louis Pasteur 1, 1348 Louvain-la-Neuve, Belgium; ting.zhang@uclouvain.be (T.Z.)
* Correspondence: yaroslav.filinchuk@uclouvain.be

Abstract: Over the last 10 years, hydrogen-rich compounds based on five-membered boron–nitrogen chain anions have attracted attention as potential hydrogen storage candidates. In this work, we synthesized Na[BH$_3$(CH$_3$NH)BH$_2$(CH$_3$NH)BH$_3$] through a simple mechanochemical approach. The structure of this compound, obtained through synchrotron powder X-ray diffraction, is presented here for the first time. Its hydrogen release properties were studied by thermogravimetric analysis and mass spectrometry. It is shown here that Na[BH$_3$(CH$_3$NH)BH$_2$(CH$_3$NH)BH$_3$], on the contrary of its parent counterpart, Na[BH$_3$NH$_2$BH$_2$NH$_2$BH$_3$], is able to release up to 4.6 wt.% of pure hydrogen below 150 °C. These results demonstrate that the introduction of a methyl group on nitrogen atom may be a good strategy to efficiently suppress the release of commonly encountered undesired gaseous by-products during the thermal dehydrogenation of B-N-H compounds.

Keywords: five-membered chain anions; B-N-H compounds; mechanochemical synthesis; thermal dehydrogenation; hydrogen release

Citation: Zhang, T.; Steenhaut, T.; Devillers, M.; Filinchuk, Y. Release of Pure H$_2$ from Na[BH$_3$(CH$_3$NH) BH$_2$(CH$_3$NH)BH$_3$] by Introduction of Methyl Substituents. *Inorganics* **2023**, *11*, 202. https://doi.org/10.3390/inorganics11050202

Academic Editor: Maurizio Peruzzini

Received: 10 April 2023
Revised: 2 May 2023
Accepted: 5 May 2023
Published: 7 May 2023

Copyright: © 2023 by the authors. Licensee MDPI, Basel, Switzerland. This article is an open access article distributed under the terms and conditions of the Creative Commons Attribution (CC BY) license (https:// creativecommons.org/licenses/by/ 4.0/).

1. Introduction

In the field of chemical hydrogen storage [1–3], boron–nitrogen–hydrogen (B-N-H) compounds have emerged as promising candidates owing to the light weight of boron and nitrogen and to their ability of bearing multiple hydrogens. Additionally, B–H and N–H bonds tend to be hydridic and protic, respectively, resulting in normally facile hydrogen release [4–10]. A typical representative of B-N-H materials is ammonia borane (NH$_3$BH$_3$, or AB), which contains three hydridic and protic hydrogens on the N and B atoms, respectively. Ammonia borane has attracted consideration attention for hydrogen storage due to its high gravimetric storage density (up to 19.6 wt.%), high stability under ambient conditions, low toxicity, and high solubility in common solvents [11–15]. However, one of the drawbacks of NH$_3$BH$_3$ for hydrogen storage is the decomposition temperature. It starts releasing the first equivalent of hydrogen at about 120 °C, and a second hydrogen elimination step occurs at approximately 145 °C; the remaining amount of hydrogen is not released until more than 500 °C. Moreover, its decomposition is exothermic and thus irreversible, and it releases multiple volatile byproducts such as NH$_3$, N$_3$B$_3$H$_6$, and B$_2$H$_6$, making the chemical hydrogenation process more challenging. In addition, the thermal decomposition of AB is furthermore paired with severe foaming and volume expansion [14,16,17]. To overcome these disadvantages, several strategies have been employed, including nanoconfinement using nanoscaffolds, catalytic effects, ionic liquid assistance, the hydrolysis reaction, and chemical modification of NH$_3$BH$_3$ through replacing one of the H atoms in the –NH$_3$ group of NH$_3$BH$_3$ by a metal, forming metal amidoboranes (MABs) [7,15,18–25]. Among these strategies, the formation of metal amidoboranes as a popular option show a number of advantages over neutral NH$_3$BH$_3$: (i) lower hydrogen release temperatures than that of pristine NH$_3$BH$_3$ [26]; (ii) generally the released hydrogen is not contaminated with undesirable borazine by-products [19,27,28]; (iii) the de-hydrogenation process is much less exothermic, about 3 to 5 kJ/mol [26,29], vs. 22.5 kJ/mol for NH$_3$BH$_3$ [17,30]. Furthermore, the introduction of metals increases the diversity of hydrogen storage candidates based on B-N-H compounds. Recently, five-membered chain anions having the general formula [BH$_3$NH$_2$BH$_2$NH$_2$BH$_3$]$^-$, also known under the abbreviation [B$_3$N$_2$]$^-$, have emerged as a novel group of ammonia borane derivatives [31–38]. M[B$_3$N$_2$] compounds have a higher hydrogen content than MABs, and the Li and Na [B$_3$N$_2$]$^-$ derivatives are stable at room temperature, on the contrary of their respective MABs [33]. However, the interest in M[B$_3$N$_2$] is much more recent than for MAB, and therefore there are only few reports about their synthesis, structure, characterization, and hydrogen storage properties. In 2011, the salt of Verkade's base (2,8,9-triisobutyl-2,5,8,9-tetraaza-1-phosphabicyclo[3.3.3]undecane, C$_{18}$H$_{39}$N$_4$P, VB, chemical formula see Figure S1) with [B$_3$N$_2$] was synthesized, and its structure was characterized, with the aim of studying the activating effect of VB on the rate and extent of H$_2$ release from NH$_3$BH$_3$ [31]. Two years later, the same authors reported the synthesis of the sodium salt and two substituted Na[BH$_3$N(R)HBH$_2$N(R)HBH$_3$] salts (R = H, Me, and benzyl), to further study the growth of aminoborane oligomers through the de-hydrocoupling reactions of NH$_3$BH$_3$ [32]. Interestingly, since 2014, the salts of the [BH$_3$NH$_2$BH$_2$NH$_2$BH$_3$]$^-$ anion with different cations have been synthesized with a focus on the study of their hydrogen storage properties (see Table S1). Generally, these kinds of complexes that are studied for their hydrogen storage properties can be classified into three types, based on their cations: ionic liquids, ammonium, and alkali metal salts. A total of four [B$_3$N$_2$]$^-$ ionic liquids have been described: [Bu$_4$N][B$_3$N$_2$], [Et$_4$N][B$_3$N$_2$], [C(N$_3$H$_6$)][B$_3$N$_2$], and [C(N$_3$H$_5$CH$_3$)][B$_3$N$_2$] [35]. Among them, [Bu$_4$N][B$_3$N$_2$] and [Et$_4$N][B$_3$N$_2$] release pure hydrogen below 160 °C [35]. [NH$_4$][B$_3$N$_2$] was reported this year, as the minor component of a 1:3 mixture with NH$_3$BH$_2$NH$_2$BH$_2$NH$_2$BH$_3$. Despite its impressive hydrogen content, this system releases H$_2$ with substantial contamination by borazine and traces of ammonia and diborane [38]. Among alkali metal (Li–Cs) salts of [B$_3$N$_2$]$^-$, only Li[B$_3$N$_2$] was shown to release pure hydrogen during thermal decomposition [33,34]. However, it is the sodium salt, Na[B$_3$N$_2$], that was studied the most in the literature until now, with five synthesis approaches (four wet chemical and one dry mechanochemical) reported between 2013 and 2021. In 2013, Sneddon and co-workers reported that Na[B$_3$N$_2$] could be obtained from NaN(SiMe$_3$)$_2$–3 NH$_3$BH$_3$ in fluorobenzene at 50 °C for 24 h [32]. Grochala et al. synthesized the same compound from NaH-3NH$_3$BH$_3$ in THF at room temperature for 24 h and obtained Li[B$_3$N$_2$] by a similar approach [33]. The same authors later used a metathesis method to obtain Na[B$_3$N$_2$] from VBH[B$_3$N$_2$] and M[Al{OC(CF$_3$)$_3$}$_4$] (M = Na) in CH$_2$Cl$_2$ at room temperature for 1 h and were able to obtain the related K, Rb, and Cs salts by this method as well [34]. Although the metathesis is fast, the two precursors involved in this kind of reaction need to be synthesized first, adding a second step to the preparation of the salt. In 2021, Chen et al. reported a facile synthetic method to obtain Na[B$_3$N$_2$], based on the reaction of NaNH$_2$BH$_3$ with NiBr$_2$ or CoCl$_2$ as a catalyst [36]. Results showed that the reaction with 0.05 equiv. of NiBr$_2$ in THF at 0 °C could produce the final Na[B$_3$N$_2$] after 10 h, with a yield of 60%. The main advantages of the dry mechanochemical synthesis are that it avoids the use of solvent and usually simplifies the drying process [39,40]. However, the reported procedures for the synthesis of Li[B$_3$N$_2$] and Na[B$_3$N$_2$] require two stages of milling at room temperature, followed by a removal of the by-products (NH$_3$) upon heating [33,34]. Moreover, long time reaction times and/or complicated operating processes are usually needed for the synthesis of the alkali metal [B$_3$N$_2$] compounds.

With its 12.7 wt.% hydrogen content, Na[B$_3$N$_2$] has potential for hydrogen storage. However, the hydrogen released when heating this compound is contaminated by unwanted by-products, including NH$_3$, B$_2$H$_6$, and larger fragments detected by mass spectrometry [33,34]. In one of our previous studies, we found that, compared to NH$_3$BH$_3$, the reaction of CH$_3$NH$_2$BH$_3$ with NaAlH$_4$ leads to a product with a completely different thermal behavior. This is likely due to the space hindrance and the electronic effect caused by the introduction of the methyl group [41]. Similar introduction of a methyl group on the N atoms of [BH$_3$NH$_2$BH$_2$NH$_2$BH$_3$]$^-$ in Na[B$_3$N$_2$] would affect the geometry of the B-N-B-N-B skeleton and change the inter-anion dihydrogen bonds, potentially positively affecting the hydrogen properties of the compound. With this in mind, we synthesized Na[BH$_3$(CH$_3$NH)BH$_2$(CH$_3$NH)BH$_3$] (abbreviated here as Na[B$_3$(MeN)$_2$]) through a new convenient mechanochemical synthesis method from easily accessible NaH and CH$_3$NH$_2$BH$_3$. We also report that its structure, solved from synchrotron powder X-ray diffraction (PXRD), enables a better understanding of the structure–properties relationships. Its thermal dehydrogenation was also investigated, by thermogravimetric analysis (TGA) and mass spectrometry, revealing a release of pure hydrogen and thus confirming our hypothesis. The purity of hydrogen released from Na[BH$_3$NH$_2$BH$_2$NH$_2$BH$_3$] was enhanced by the introduction of methyl groups on N atoms. This achievement represents the first successful suppression of the unwanted by-product release through the introduction of -CH$_3$ groups on the nitrogen atoms of [B$_3$N$_2$]$^-$. Furthermore, the structure of Na[BH$_3$(CH$_3$NH)BH$_2$(CH$_3$NH)BH$_3$] was analyzed for the first time helping to understand the potential reasons behind the improved hydrogen purity. This study provides valuable insights into the relationship between the hydrogen release properties and the structure of B-N-H compounds. The introduction of methyl groups on the nitrogen atoms of Na[B$_3$N$_2$] to enhance hydrogen purity could potentially be extended to the other M[B$_3$N$_2$] or even other M-B-N-H compound. It could even be expanded to include the use of other small alkyl groups or small electron-donating groups instead of the methyl group.

2. Results and Discussion

The reaction between NaH and NH$_3$BH$_3$ in various molar ratios was reported to produce different hydrogen rich B-N compounds, i.e., NaNH$_2$BH$_3$ (1:1), NaBH$_3$NH$_2$BH$_3$ (1:2), NaBH$_3$NH$_2$BH$_2$NH$_2$BH$_3$ (1:3), NaBH$_3$NH$_2$BH$_2$NH$_2$BH$_2$NH$_2$BH$_3$, and NaBH$_3$NH$_2$BH(NH$_2$BH$_3$)$_2$ (1:4) [42]. All of those have potential for application in hydrogen storage due to their high H content (see Table S2). Despite the potential of this system, there is only one report of the reaction between the methyl-substituted CH$_3$NH$_2$BH$_3$ and NaH, in a 1:3 molar ratio [32]. We thus investigated the reaction of NaH and CH$_3$NH$_2$BH$_3$ by mechanochemistry (Figure 1A), to avoid an incorporation of or a reaction with solvents. Upon milling NaH-CH$_3$NH$_2$BH$_3$ systems in different molar ratios, new peaks appeared on the PXRD pattern of all the tested ratios, along with some unreacted NaH for the 1:1 system (Figure 1B). The ^{11}B NMR spectra of the resulting products furthermore showed the appearance of a new quadruplet signal located between -14.45 and -16.07 ppm, which likely belongs to the BH$_3$ unit of Na[CH$_3$NHBH$_3$]. When the NaH:CH$_3$NH$_2$BH$_3$ ratio was increased to 1:2, the PXRD pattern of the obtained product shows peaks corresponding to crystalline Na[BH$_3$(CH$_3$NH)BH$_2$(CH$_3$NH)BH$_3$]. Although the ^{11}B NMR spectrum of the product displays a triplet signal of BH$_2$ (-2.24 ppm) and quadruplet signal of BH$_3$ (-16.39 ppm), expected for the [BH$_3$(CH$_3$NH)BH$_2$(CH$_3$NH)BH$_3$]$^-$ anion, other signals on the spectrum reveal the presence of unknown non-crystalline by-products (Figure 1C). Further increasing the ratio to 1:3 leads to the appearance of only the signals of BH$_2$ and BH$_3$ from the [BH$_3$(CH$_3$NH)BH$_2$(CH$_3$NH)BH$_3$]$^-$ anion on the ^{11}B NMR spectra. The phase purity of the compound obtained upon 27 h of ball milling the 1:3 mixture was confirmed by temperature programmed synchrotron powder X-ray diffraction (PXRD) measurements. Indeed, the complete set of peaks of the pattern disappeared at once at around 150 °C (Figure S2). Based on the ^{11}B NMR spectrum and the temperature ramping synchrotron PXRD experiment, we deduce that relatively pure Na[B$_3$(MeN)$_2$] with five membered B-MeN-B-MeN-B chains was formed and that the reaction shown in Figure 1A was complete.

The obtained Na[B$_3$(MeN)$_2$] was further characterized by infrared (IR) spectroscopy (Figure 2). On the spectrum, it can be seen that the asymmetry of the N-H stretching band (3162 cm^{-1}) disappeared, compared with the CH$_3$NH$_2$BH$_3$ precursor. This is because one hydrogen on the nitrogen of CH$_3$NH$_2$BH$_3$ is released, combined with a hydride atom from NaH to form H$_2$. In addition, the N-H bending of Na[B$_3$(MeN)$_2$] (1457–1491 cm^{-1}) is redshifted compared to CH$_3$NH$_2$BH$_3$ (1596 cm^{-1}). This can be attributed to the introduction of weak electron donating methyl group on N atom, influencing the electron density of

N and further having an effect on the N-H band. The broad band located in the region of 2000–2500 cm^{-1} belongs to the B-H stretching band. There is no significant difference compared to the CH$_3$NH$_2$BH$_3$ precursor. However, the signal of the B-N stretching is widened and split in Na[B$_3$(MeN)$_2$] (692–716 cm^{-1}), due to the presence of two types of B-N bands in Na[B$_3$(MeN)$_2$], whereas CH$_3$NH$_2$BH$_3$ exhibits only one B-N band.

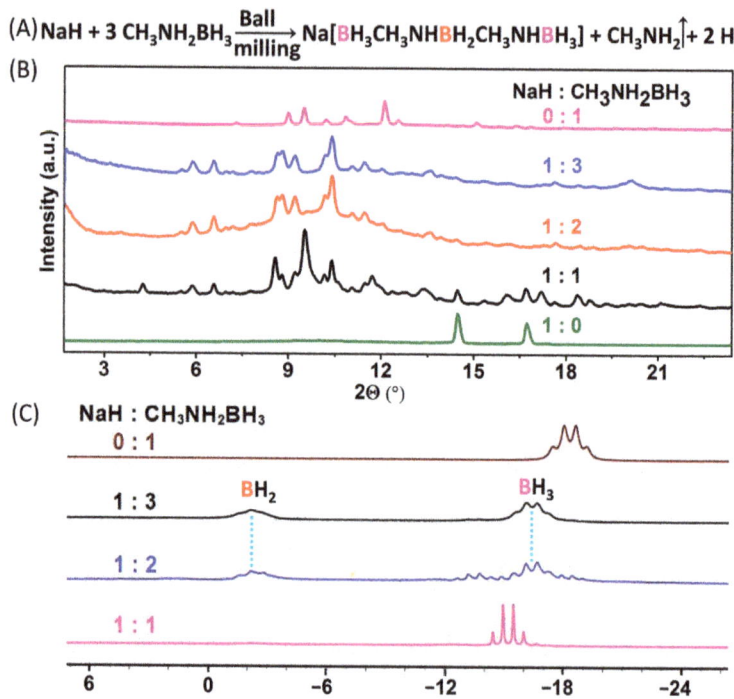

Figure 1. (**A**) Equation of the mechanochemical reaction; (**B**) PXRD patterns of NaH and CH$_3$NH$_2$BH$_3$ ball-milled in different molar ratios, along with patterns of NaH and CH$_3$NH$_2$BH$_3$ (λ = 0.71073 Å); (**C**) ^{11}B NMR spectra of ball-milled NaH and CH$_3$NH$_2$BH$_3$ mixtures in different molar ratios, along with the spectrum of CH$_3$NH$_2$BH$_3$.

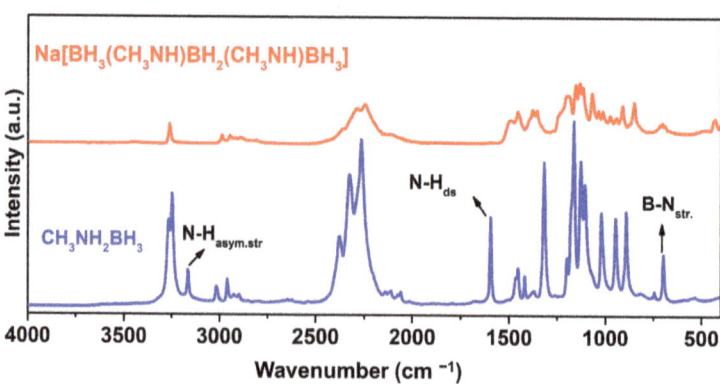

Figure 2. IR spectra of Na[B$_3$(MeN)$_2$] and CH$_3$NH$_2$BH$_3$.

All of the aforementioned differences between Na[B$_3$(MeN)$_2$] and CH$_3$NH$_2$BH$_3$ align with our previous analysis based on XRD and ^{11}B NMR, further confirming the proposed formula of the product as shown in Figure 1A.

The structure of Na[B$_3$(MeN)$_2$] was determined by direct space methods from synchrotron powder X-ray diffraction (PXRD) data, indexed in the monoclinic space group P2$_1$/n; the final Rietveld refinement profile is shown in Figure S3. The [BH$_3$(CH$_3$NH)BH$_2$(CH$_3$NH)BH$_3$]$^-$ anion is a five-membered B-N chain with an alternance of B and N atoms connected in a similar way as in the reported [BH$_3$NH$_2$BH$_2$NH$_2$BH$_3$]$^-$ [33]. Although the N atoms in Na[B$_3$(MeN)$_2$] have four different substituents and are therefore chiral, the crystal structure reveals that the anions are integrated in the solid as a meso compound, as both N atoms possess opposite chirality. This is in agreement with reported DFT calculations, which indicate that the meso isomer is the preferred stereoisomer for this anion [32]. Due to the introduction of the methyl substituents, the skeleton of the B-N-B-N-B chain shows a twisted geometry, which is in contrast with the linear geometry adopted by the [B$_3$N$_2$]$^-$ anion in the Li, Na, and K salts but is similar to the reported Rb and Cs [B$_3$N$_2$]$^-$ salts [34]. This type of geometry enables the formation of intramolecular dihydrogen bonds, of 2.04 and 2.12 Å (Figure 3A). With this type of geometry of the chain anion, an increase in the intramolecular interactions is expected, which should have a positive influence on the hydrogen release properties of Na[B$_3$(MeN)$_2$]. Interanion dihydrogen bonds are also present between H atoms of the NH and terminal BH$_3$ groups, as well as between H atoms of the NH and the ones of the central BH$_2$, as can be seen in Figure 3B and in Table S3. Unlike in the reported K[B$_3$N$_2$] and Rb[B$_3$N$_2$], the N-BH$_2$ distances are not shorter than the N-BH$_3$ ones (Table S4) [34,43]. Na$^+$ cations have a distorted triangular bipyramidal coordination geometry with five B atoms from four distinct [BH$_3$(CH$_3$NH)BH$_2$(CH$_3$NH)BH$_3$]$^-$ anions (Figure 3B,C). The coordination is performed through six hydridic H atoms of the BH$_3$ (green balls in Figure 3C) groups from four different chain anions. Two other hydridic H atoms from the BH$_2$ (red balls in Figure 3C) of one of above four chain anions complete the coordination around Na. This is different from the unsubstituted Na[BH$_3$NH$_2$BH$_2$NH$_2$BH$_3$], where Na atoms are coordinated only to hydrogen atoms of the terminal [BH$_3$] groups. This may be one reason of the release of large undesirable gaseous species during the thermal dehydrogenation of Na[B$_3$N$_2$] [33].

The thermal stability of Na[B$_3$(MeN)$_2$] was investigated by thermogravimetric analysis (TGA) under inert argon atmosphere, from room temperature to 150 °C. A single step decomposition event occurs at about 80 °C, accompanied by a weight oscillation due to the so-called "jet" effect [14,44] (Figure 4A). The solid decomposition products isolated upon heating at 150 °C were identified as being crystalline NaBH$_4$, along with some unknown crystalline and possibly amorphous compounds, based on PXRD and IR analyses (Figure 4B,C). Those by-products are expected to contain B, C, and N atoms, based on 4.6 wt.% the experimental weight loss as compared to 26.2 wt.% B, 19.4 wt.% C, and 22.7 wt.% N in the sample before decomposition. It is interesting to note that thermally decomposing alkali metal salts of the unsubstituted anion (Li – Cs [B$_3$N$_2$]) also leads to the formation of BH$_4^-$ compounds, similarly to the title compound. The observed mass loss during the thermal decomposition of Na[B$_3$(MeN)$_2$], of 4.6 wt.%, is in accordance with the possible release of pure H$_2$, as the compound has a theoretical hydrogen content of 8.09 wt.% (excluding H atoms from the methyl groups). This is interesting, as the parent Na[B$_3$N$_2$] shows a larger mass loss (~20 wt.%) than its theoretical hydrogen content (12.7 wt.%) when heating below 200 °C, resulting in the single-step release of undesirable gaseous decomposition by-products like diborane and ammonia [34].

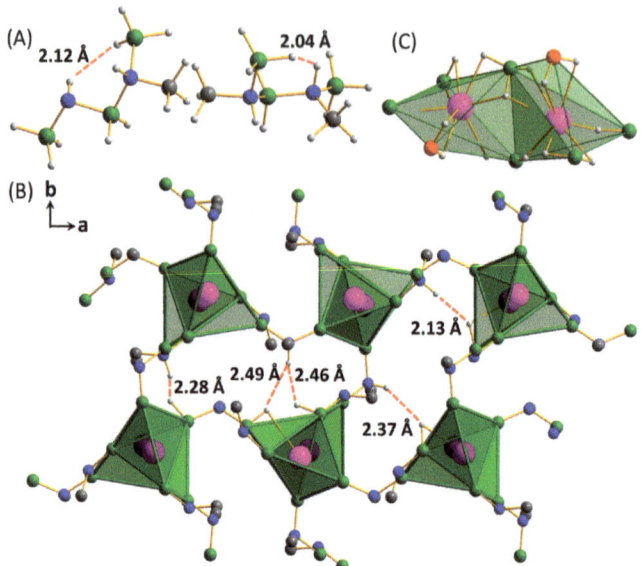

Figure 3. Ball and stick plot of the [BH$_3$(CH$_3$NH)BH$_2$(CH$_3$NH)BH$_3$]$^-$ anion with indication of the intramolecular dihydrogen bond (**A**), crystal packing of Na coordination polyhedra with boron atoms (hydrogen atoms are omitted for clarity) in Na[BH$_3$(CH$_3$NH)BH$_2$(CH$_3$NH)BH$_3$] projected along the c axis, indicating interanion dihydrogen bonds (**B**), and coordination of H atoms around the Na$^+$ cation (central B was highlighted by red color) (**C**). Color code: N = blue, B = green, C = grey, H = white, and Na = pink. Dihydrogen bonds are displayed by red dotted lines.

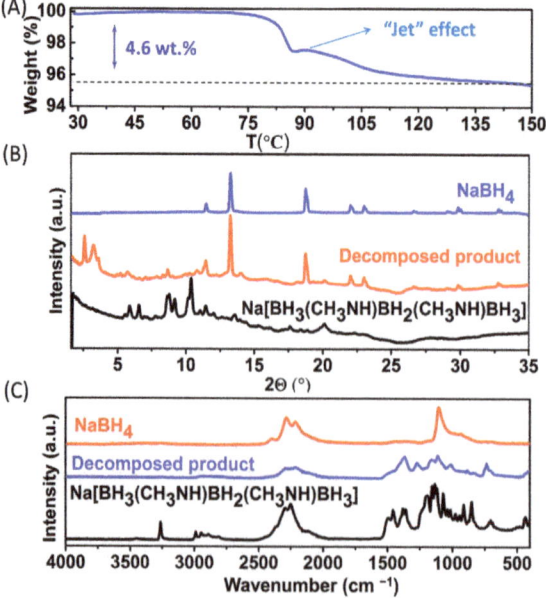

Figure 4. TG analysis of Na[B$_3$(MeN)$_2$] (**A**); PXRD patterns (λ = 0.71073 Å) (**B**) and IR spectra (**C**) of the product upon heating at 150 °C, compared to the starting Na[B$_3$(MeN)$_2$] and NaBH$_4$.

The purity of the gas released during the thermal de-hydrogenation of Na[B$_3$(MeN)$_2$] was analyzed by means of temperature-programmed mass spectrometry between 40 °C and 150 °C. Hydrogen was the only gas detected, and the experiment confirmed that NH$_3$, B$_2$H$_6$, CH$_4$, and CH$_3$NH$_2$ were not released during the decomposition (Figure 5). This confirms that the methyl-substituted Na[B$_3$(MeN)$_2$] indeed releases about 4.6 wt.% of pure hydrogen upon heating to 150 °C. This confirmed that the introduction of a methyl group on the nitrogen atoms efficiently suppresses the release of unwanted by-products during thermal hydrogen desorption.

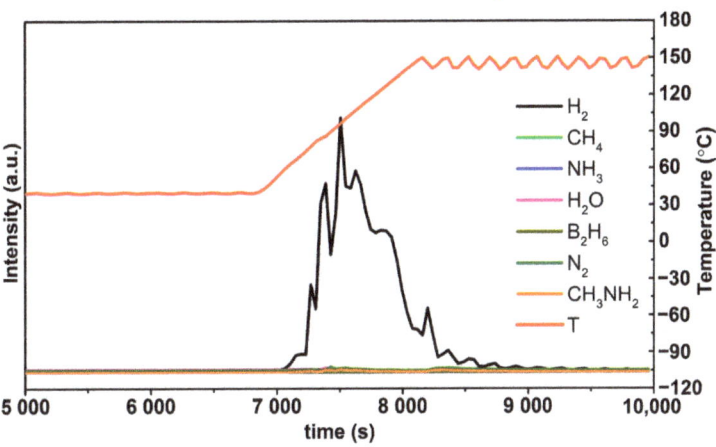

Figure 5. Mass spectrometry analysis of gases released during the thermal decomposition of Na[B$_3$(MeN)$_2$] under argon, between 40 °C and 150 °C.

3. Materials and Methods

All samples were obtained from commercially available NaH (95%), NaBH$_4$ (97%), CH$_3$NH$_2$·HCl (98%), and anhydrous THF ($\geq$99.9%) that were purchased from Sigma Aldrich Co., Ltd. (St. Louis, MI, USA). All operations were performed in gloveboxes with a high purity argon atmosphere.

3.1. Syntheses

Synthesis of CH$_3$NH$_2$BH$_3$: CH$_3$NH$_2$BH$_3$ was synthesized following a procedure adapted from the literature [45]. Initially, powdered NaBH$_4$ (3.79 g, 0.1 mol), CH$_3$NH$_2$·HCl (13.50 g, 0.1 mol), and THF (300 mL) were added to a 500 mL three-neck round-bottom flask. The resulting mixture was then vigorously stirred at ambient temperature under an argon atmosphere for 48 h. Filtration was performed to remove the solid by-product (NaCl) from the reaction mixture, and the collected filtrate was subjected to evaporation under reduced pressure using a rotary evaporator. The resulting white solid of CH$_3$NH$_2$BH$_3$ was then dried under vacuum overnight to eliminate any residual THF. The purity of the product was confirmed through characterization using ^{1}H, ^{11}B, and ^{13}C NMR and PXRD, as depicted in Figures S4–S7.

Synthesis of Na[BH$_3$(CH$_3$NH)BH$_2$(CH$_3$NH)BH$_3$]: Totals of 1 eq. of NaH (30.0 mg) and 3 eqs. of CH$_3$NH$_2$BH$_3$ (168.4 mg) were placed into an 80 mL stainless steel vial with three 10 mm diameter stainless steel balls (ball-to-powder mass ratio of 60:1). The reactants were then milled in a planetary ball mill (Fritsch Pulverisette 7 Premium line), with a rotation speed of 500 rpm for 55 milling cycles of 30 min interrupted by 5 min cooling breaks. The product was obtained as a white powder.

3.2. Instrumental

Samples were carefully filled into 0.7 mm thin-walled glass capillaries (Hilgenberg GmbH, Malsfeld, Germany) within an argon-filled glovebox. To prevent contact with air, the capillaries were sealed with grease before being taken out of the glovebox. The sealed capillaries were then cut and promptly placed into wax on a goniometer head, ensuring that no air entered the capillary. Diffraction data were immediately collected using a MAR345 image-plate detector equipped with an Incoatec Mo (λ = 0.71073 Å) Microfocus (IµS 2.0) X-ray source operating at 50 kV and 1000 µA. The resulting two-dimensional images were azimuthally integrated using the Fit2D software, with LaB_6 serving as a calibrant.

Synchrotron PXRD patterns were collected with a PILATUS@SNBL diffractometer (SNBL, ESRF, Grenoble, France) equipped with a Dectris PILATUS 2M single-photon counting pixel area detector (λ = 0.77509 Å). Powder patterns were obtained by using raw data processed by the SNBL Toolbox software using data for LaB_6 standard. The synchrotron PXRD data for $Na[BH_3(CH_3NH)BH_2(CH_3NH)BH_3]$ were indexed in a monoclinic unit cell, and its structure was solved by global optimization using the FOX software [46]. The anions were modeled by conformationally free z-matrices with restrained bond distances and angles. Since the N-atom of methylamidoborane is chiral, all combinations of these chiral centres were examined. The final structure showed the best fit to the data but also satisfied crystal-chemical expectations, such as the formation of dihydrogen bonds (N-H$\cdots$H-B) and the coordination of Na^+ to H atoms of BH_3 and BH_2 groups. Rietveld refinements were done in Fullprof [47], refining all non-hydrogen atoms of the anions individually using restraints from DFT-refined geometry. Hydrogen atoms were refined using the rigiding model, with Na as free atoms. The symmetry was confirmed with ADDSYM routine in the PLATON software. R_B = 7.9%, R_p = 14.2, R_{wp} = 12.5, χ^2 = 424 (mind that the counting statistics is very high).

Fourier transform infrared spectroscopy (FTIR): Attenuated total reflectance (ATR)-IR spectra were recorded using a Bruker Alpha spectrometer. The spectrometer was equipped with a Platinum ATR sample holder, which featured a diamond crystal for single bounce measurements. The entire experimental setup was located within an argon-filled glovebox to maintain an inert atmosphere during the measurements.

Thermogravimetric analysis (TGA): TGA measurements were conducted using a Netzsch STA 449 F3 TGA/DSC. The TGA/DSC was equipped with a stainless-steel oven and located within an argon-filled glovebox to ensure an inert atmosphere during the measurements. The samples were loaded into Al_2O_3 crucibles and subjected to a heating rate of 5 K/min under an argon flow of 100 mL/min.

Mass spectrometry: Mass spectrometry measurements were conducted using a Hiden Catlab reactor coupled with a Quantitative Gas Analyser (QGA) Hiden quadrupole mass spectrometer. Prior to the experiment, the samples were loaded into a quartz tube with two layers of quartz wool, all within the protective atmosphere of an argon-filled glovebox. The ends of the quartz tube were sealed with Parafilm before being removed from the glovebox. Subsequently, the quartz tube was placed in the sample holder outside the glovebox after quickly removing the Parafilm. The argon flow (40 mL/min) was immediately initiated to prevent any contact of the sample with air. The samples were then heated to 40 °C and held isothermally for approximately 2 h to stabilize the temperature. Heating was then performed at a rate of 5 °C/min until reaching 150 °C, followed by a 1 h isotherm. Gas evolution was monitored by recording the peak with the highest intensity for each gas, specifically the m/z values of 2, 15, 17, 18, 26, 28, and 30, corresponding to H_2, CH_4, NH_3, H_2O, B_2H_6, N_2, and CH_3NH_2, respectively. The absence of H_2O and N_2 signals in the collected data confirmed the absence of leaks, ensuring that the sample remained under a protective argon atmosphere throughout the measurement.

4. Conclusions

We synthesized $Na[BH_3(CH_3NH)BH_2(CH_3NH)BH_3]$ ($Na[B_3(MeN)_2]$, 130.5 g H_2/kg, 126 g H_2/L, Table S5), a methyl-substituted Na salt with five-membered B-N chain anions, by a novel mechanochemical approach from NaH and $CH_3NH_2BH_3$. Its crystal structure was determined for the first time based on synchrotron PXRD, showing that the introduction of -CH_3 groups on the N atoms leads to the introduction of the anion in a kinked geometry into the solid, unlike its unsubstituted parent counterpart ($Na[B_3N_2]$), that possesses straight B-N chains. $Na[B_3(MeN)_2]$ releases up to 4.6 wt.% of pure hydrogen below to 150 °C, contrary to its unsubstituted analogue that releases undesirable gaseous by-products during heating. This indicates that the introduction of methyl (or other) substituents on the nitrogen atoms of similar compounds is a promising approach to suppress the release of unwanted volatile by-products during thermal hydrogen release.

Supplementary Materials: The following supporting information can be downloaded at: https://www.mdpi.com/article/10.3390/inorganics11050202/s1, Figure S1: Chemical formula of VB; Table S1: H-contents, mass losses and by-products formed during thermal treatment of several $M[B_3N_2]$ compounds; Table S2: H-content in $NaNH_2BH_3$, $NaBH_3NH_2BH_3$, $NaBH_3NH_2BH_2NH_2BH_3$, $NaBH_3NH_2BH_2NH_2BH_2NH_2BH_3$, and $NaBH_3NH_2BH(NH_2BH_3)_2$; Table S3. Inter-anion dihydrogen bond lengths and angles in $Na[B_3(MeN)_2]$; Figure S2: temperature ramping synchrotron PXRD patterns of $Na[B_3(MeN)_2]$; Figure S3: Rietveld refinement of the synchrotron PXRD pattern of $Na[B_3(MeN)_2]$; Table S4: B-N bond lengths in $CH_3NH_2BH_3$, $M[B_3N_2]$ (M = Li – Cs), and $Na[B_3(MeN)_2]$; Figures S4–S7: NMR and PXRD of $CH_3NH_2BH_3$; Table S5: The mole mass, density, and gravimetric and volumetric hydrogen density of $Na[B_3(MeN)_2]$.

Author Contributions: Conceptualization, software, T.Z. and Y.F.; methodology, validation, formal analysis, investigation, data curation, writing—original draft preparation, and visualization, T.Z.; writing—review and editing, T.S., M.D. and Y.F.; supervision, M.D. and Y.F.; project administration, funding acquisition, Y.F. All authors have read and agreed to the published version of the manuscript.

Funding: This work was supported by the FNRS (CC J.0073.20, EQP U.N038.13, EQP U.N022.19) and the Communauté Française de Belgique (ARC 18/23-093). Ting Zhang was supported through the China Scholarship Council fellowship (201809370045).

Data Availability Statement: CCDC number: 2254456 contains supplementary crystallographic data for this paper. This data can be obtained free of charge from The Cambridge Crystallographic Data Center.

Acknowledgments: We thank the ESRF for the beamtime allocation at the SNBL. We also thank François Devred for help with the mass spectrometry measurements.

Conflicts of Interest: The authors declare no conflict of interest

References

1. Chen, Z.; Ma, Z.; Zheng, J.; Li, X.; Akiba, E.; Li, H.-W. Perspectives and Challenges of Hydrogen Storage in Solid-State Hydrides. *Chin. J. Chem. Eng.* **2021**, *29*, 1–12. [CrossRef]
2. Milanese, C.; Jensen, T.; Hauback, B.; Pistidda, C.; Dornheim, M.; Yang, H.; Lombardo, L.; Zuettel, A.; Filinchuk, Y.; Ngene, P.; et al. Complex Hydrides for Energy Storage. *Int. J. Hydrogen Energy* **2019**, *44*, 7860–7874. [CrossRef]
3. Dematteis, E.M.; Amdisen, M.B.; Autrey, T.; Barale, J.; E Bowden, M.; E Buckley, C.; Cho, Y.W.; Deledda, S.; Dornheim, M.; de Jongh, P.; et al. Hydrogen Storage in Complex Hydrides: Past Activities and New Trends. *Prog. Energy* **2022**, *4*, 032009. [CrossRef]
4. Huang, Z.; Autrey, T. Boron–Nitrogen–Hydrogen (BNH) Compounds: Recent Developments in Hydrogen Storage, Applications in Hydrogenation and Catalysis, and New Syntheses. *Energy Environ. Sci.* **2012**, *5*, 9257–9268. [CrossRef]
5. Hamilton, C.W.; Baker, R.T.; Staubitz, A.; Manners, I. B–N Compounds for Chemical Hydrogen Storage. *Chem. Soc. Rev.* **2009**, *38*, 279–293. [CrossRef] [PubMed]
6. Wang, K.; Pan, Z.; Yu, X. Metal B-N-H Hydrogen-Storage Compound: Development and Perspectives. *J. Alloys Compd.* **2019**, *794*, 303–324. [CrossRef]
7. Castilla-Martinez, C.A.; Moury, R.; Demirci, U.B. Amidoboranes and Hydrazinidoboranes: State of the Art, Potential for Hydrogen Storage, and Other Prospects. *Int. J. Hydrogen Energy* **2020**, *45*, 30731–30755. [CrossRef]
8. Kumar, R.; Karkamkar, A.; Bowden, M.; Autrey, T. Solid-State Hydrogen Rich Boron–Nitrogen Compounds for Energy Storage. *Chem. Soc. Rev.* **2019**, *48*, 5350–5380. [CrossRef]
9. Dovgaliuk, I.; Filinchuk, Y. Aluminium Complexes of B- and N-Based Hydrides: Synthesis, Structures and Hydrogen Storage Properties. *Int. J. Hydrogen Energy* **2016**, *41*, 15489–15504. [CrossRef]

10. Paskevicius, M.; Jepsen, L.H.; Schouwink, P.; Černý, R.; Ravnsbæk, D.B.; Filinchuk, Y.; Dornheim, M.; Besenbacher, F.; Jensen, T.R. Metal Borohydrides and Derivatives—Synthesis, Structure and Properties. *Chem. Soc. Rev.* **2017**, *46*, 1565–1634. [CrossRef] [PubMed]
11. Stephens, F.H.; Pons, V.; Tom Baker, R. Ammonia–Borane: The Hydrogen Source Par Excellence? *Dalton Trans.* **2007**, 2613–2626. [CrossRef]
12. Demirci, U.B. Ammonia Borane, a Material with Exceptional Properties for Chemical Hydrogen Storage. *Int. J. Hydrogen Energy* **2017**, *42*, 9978–10013. [CrossRef]
13. Akbayrak, S.; Özkar, S. Ammonia Borane as Hydrogen Storage Materials. *Int. J. Hydrogen Energy* **2018**, *43*, 18592–18606. [CrossRef]
14. Staubitz, A.; Robertson, A.P.M.; Manners, I. Ammonia-Borane and Related Compounds as Dihydrogen Sources. *Chem. Rev.* **2010**, *110*, 4079–4124. [CrossRef] [PubMed]
15. Wang, K.; Zhang, J.-G.; Man, T.-T.; Wu, M.; Chen, C.-C. Recent Process and Development of Metal Aminoborane. *Chem. Asian J.* **2013**, *8*, 1076–1089. [CrossRef]
16. Demirci, U.B. Mechanistic Insights into the Thermal Decomposition of Ammonia Borane, a Material Studied for Chemical Hydrogen Storage. *Inorg. Chem. Front.* **2021**, *8*, 1900–1930. [CrossRef]
17. Baitalow, F.; Baumann, J.; Wolf, G.; Jaenicke-Rößler, K.; Leitner, G. Thermal Decomposition of B–N–H Compounds Investigated by Using Combined Thermoanalytical Methods. *Thermochim. Acta* **2002**, *391*, 159–168. [CrossRef]
18. Chua, Y.S.; Chen, P.; Wu, G.; Xiong, Z. Development of Amidoboranes for Hydrogen Storage. *Chem. Commun.* **2011**, *47*, 5116–5129. [CrossRef] [PubMed]
19. Owarzany, R.; Leszczyński, J.P.; Fijalkowski, J.K.; Grochala, W. Mono- and Bimetalic Amidoboranes. *Crystals* **2016**, *6*, 88. [CrossRef]
20. Hügle, T.; Hartl, M.; Lentz, D. The Route to a Feasible Hydrogen-Storage Material: Mofs Versus Ammonia Borane. *Eur. J. Chem.* **2011**, *17*, 10184–10207. [CrossRef]
21. Li, L.; Yao, X.; Sun, C.; Du, A.; Cheng, L.; Zhu, Z.; Yu, C.; Zou, J.; Smith, S.C.; Wang, P.; et al. Lithium-Catalyzed Dehydrogenation of Ammonia Borane within Mesoporous Carbon Framework for Chemical Hydrogen Storage. *Adv. Funct. Mater.* **2009**, *19*, 265–271. [CrossRef]
22. Huang, X.; Liu, Y.; Wen, H.; Shen, R.; Mehdi, S.; Wu, X.; Liang, E.; Guo, X.; Li, B. Ensemble-boosting effect of Ru-Cu alloy on catalytic activity towards hydrogen evolution in ammonia borane hydrolysis. *Appl. Catal. B* **2021**, *287*, 119960. [CrossRef]
23. Kang, N.; Wei, X.; Shen, R.; Li, B.; Cal, E.G.; Moya, S.; Salmon, L.; Wang, C.; Coy, E.; Berlande, M.; et al. Fast Au-Ni@ZIF-8-catalyzed ammonia borane hydrolysis boosted by dramatic volcano-type synergy and plasmonic acceleration. *Appl. Catal. B* **2023**, *320*, 121957. [CrossRef]
24. Mehdi, S.; Liu, Y.; Wei, H.; Zhang, H.; Shen, R.; Guan, S.; Wu, X.; Liu, T.; Wen, H.; Peng, Z.; et al. P-induced Co-based interfacial catalysis on Ni foam for hydrogen generation from ammonia borane. *Appl. Catal. B* **2023**, *325*, 122317. [CrossRef]
25. Himmelberger, D.W.; Yoon, C.W.; Bluhm, M.E.; Carroll, P.J.; Sneddon, L.G. Base-Promoted Ammonia Borane Hydrogen-Release. *J. Am. Chem. Soc.* **2009**, *131*, 14101–14110. [CrossRef]
26. Xiong, Z.; Yong, C.K.; Wu, G.; Chen, P.; Shaw, W.; Karkamkar, A.; Autrey, T.; Jones, M.O.; Johnson, S.R.; Edwards, P.P.; et al. High-Capacity Hydrogen Storage in Lithium and Sodium Amidoboranes. *Nat. Mater.* **2007**, *7*, 138–141. [CrossRef]
27. Diyabalanage, H.V.K.; Nakagawa, T.; Shrestha, R.P.; Semelsberger, T.A.; Davis, B.L.; Scott, B.L.; Burrell, A.K.; David, W.I.F.; Ryan, K.R.; Jones, M.O.; et al. Potassium(I) Amidotrihydroborate: Structure and Hydrogen Release. *J. Am. Chem. Soc.* **2010**, *132*, 11836–11837. [CrossRef]
28. Luo, J.; Kang, X.; Wang, P. Synthesis, Formation Mechanism, and Dehydrogenation Properties of the Long-Sought Mg(NH$_2$BH$_3$)$_2$ Compound. *Energy Environ. Sci.* **2013**, *6*, 1018–1025. [CrossRef]
29. Diyabalanage, H.V.K.; Shrestha, R.P.; Semelsberger, T.A.; Scott, B.L.; Bowden, M.E.; Davis, B.L.; Burrell, A.K. Calcium Amidotrihydroborate: A Hydrogen Storage Material. *Angew. Chem. Int. Ed.* **2007**, *46*, 8995–8997. [CrossRef]
30. Wolf, G.; Baumann, J.; Baitalow, F.; Hoffmann, F.P. Calorimetric Process Monitoring of Thermal Decomposition of B–N–H Compounds. *Thermochim. Acta* **2000**, *343*, 19–25. [CrossRef]
31. Ewing, W.C.; Marchione, A.; Himmelberger, D.W.; Carroll, P.J.; Sneddon, L.G. Syntheses and Structural Characterizations of Anionic Borane-Capped Ammonia Borane Oligomers: Evidence for Ammonia Borane H$_2$ Release Via a Base-Promoted Anionic Dehydropolymerization Mechanism. *J. Am. Chem. Soc.* **2011**, *133*, 17093–17099. [CrossRef]
32. Ewing, W.C.; Carroll, P.J.; Sneddon, L.G. Syntheses and Characterizations of Linear Triborazanes. *Inorg. Chem.* **2013**, *52*, 10690–10697. [CrossRef] [PubMed]
33. Fijalkowski, K.J.; Jaroń, T.; Leszczyński, P.J.; Magos-Palasyuk, E.; Palasyuk, T.; Cyrański, M.K.; Grochala, W. M(BH$_3$NH$_2$BH$_2$NH$_2$BH$_3$) —The Missing Link in the Mechanism of the Thermal Decomposition of Light Alkali Metal Amidoboranes. *Phys. Chem. Chem. Phys.* **2014**, *16*, 23340–23346. [CrossRef] [PubMed]
34. Owarzany, R.; Fijalkowski, K.J.; Jaroń, T.; Leszczyński, P.J.; Dobrzycki, Ł.; Cyrański, M.K.; Grochala, W. Complete Series of Alkali-Metal M(BH$_3$NH$_2$BH$_2$NH$_2$BH$_3$) Hydrogen-Storage Salts Accessed via Metathesis in Organic Solvents. *Inorg. Chem.* **2016**, *55*, 37–45. [CrossRef] [PubMed]
35. Chen, X.-M.; Jiang, X.; Jing, Y.; Chen, X. Synthesis and Dehydrogenation of Organic Salts of a Five-Membered B/N Anionic Chain, a Novel Ionic Liquid. *Chem. Asian J.* **2021**, *16*, 2475–2480. [CrossRef] [PubMed]

36. Ju, M.-Y.; Guo, Y.; Chen, X.-M.; Chen, X. Facile Synthetic Method of Na[BH$_3$(NH$_2$BH$_2$)$_2$H] Based on the Reactions of Sodium Amidoborane (NaNH$_2$BH$_3$) with NiBr$_2$ or CoCl$_2$. *Inorg. Chem.* **2021**, *60*, 7101–7107. [CrossRef]
37. Nawrocka, E.K.; Prus, A.; Owarzany, R.; Koźmiński, W.; Kazimierczuk, K.; Fijalkowski, K.J. The Assignment of ^{11}B and ^{1}H Resonances in the Post-Reaction Mixture from the Dry Synthesis of Li(BH$_3$NH$_2$BH$_2$NH$_2$BH$_3$). *Magn. Reson. Chem.* **2023**, *61*, 49–54. [CrossRef]
38. Owarzany, R.; Jaroń, T.; Kazimierczuk, K.; Malinowski, P.J.; Grochala, W.; Fijałkowski, K.J. In Towards Hydrogen-Rich Ionic (NH$_4$)(BH$_3$NH$_2$BH$_2$NH$_2$BH$_3$) and Related Molecular NH$_3$BH$_2$NH$_2$BH$_2$NH$_2$BH$_3$. *Dalton Trans.* **2023**, *52*, 3586–3595. [CrossRef]
39. James, S.L.; Adams, C.J.; Bolm, C.; Braga, D.; Collier, P.; Friščić, T.; Grepioni, F.; Harris, K.D.M.; Hyett, G.; Jones, W.; et al. Mechanochemistry: Opportunities for New and Cleaner Synthesis. *Chem. Soc. Rev.* **2012**, *41*, 413–447. [CrossRef]
40. Do, J.-L.; Friščić, T. Mechanochemistry: A Force of Synthesis. *ACS Cent. Sci.* **2017**, *3*, 13–19. [CrossRef]
41. Zhang, T.; Steenhaut, T.; Li, X.; Devred, F.; Devillers, M.; Filinchuk, Y. Aluminum Methylamidoborane Complexes: Mechanochemical Synthesis, Structure, Stability, and Reactive Hydride Composites. *Sustain. Energy Fuels* **2023**, *7*, 1119–1126. [CrossRef]
42. Chen, X.-M.; Wang, J.; Liu, S.-C.; Zhang, J.; Wei, D.; Chen, X. Controllable Syntheses of B/N Anionic Aminoborane Chain Complexes by the Reaction of NH$_3$BH$_3$ with NaH and the Mechanistic Study. *Dalton Trans.* **2019**, *48*, 14984–14988. [CrossRef] [PubMed]
43. Bowden, M.E.; Brown, I.W.M.; Gainsford, G.J.; Wong, H. Structure and Thermal Decomposition of Methylamine Borane. *Inorganica Chim. Acta* **2008**, *361*, 2147–2153. [CrossRef]
44. Dovgaliuk, I.; Jepsen, L.H.; Safin, D.A.; Łodziana, Z.; Dyadkin, V.; Jensen, T.R.; Devillers, M.; Filinchuk, Y. A Composite of Complex and Chemical Hydrides Yields the First Al-Based Amidoborane with Improved Hydrogen Storage Properties. *Eur. J. Chem.* **2015**, *21*, 14562–14570. [CrossRef]
45. Dovgaliuk, I.; Møller, K.T.; Robeyns, K.; Louppe, V.; Jensen, T.R.; Filinchuk, Y. Complexation of Ammonia Boranes with Al^{3+}. *Inorg. Chem.* **2019**, *58*, 4753–4760. [CrossRef]
46. Favre-Nicolin, V.; Cerny, R. Fox, 'Free Objects for Crystallography': A Modular Approach to Ab Initio Structure Determination from Powder Diffraction. *J. Appl. Crystallogr.* **2002**, *35*, 734–743. [CrossRef]
47. Rodríguez-Carvajal, J. Recent Advances in Magnetic Structure Determination by Neutron Powder Diffraction. *Phys. B Condens. Matter.* **1993**, *192*, 55–69. [CrossRef]

Disclaimer/Publisher's Note: The statements, opinions and data contained in all publications are solely those of the individual author(s) and contributor(s) and not of MDPI and/or the editor(s). MDPI and/or the editor(s) disclaim responsibility for any injury to people or property resulting from any ideas, methods, instructions or products referred to in the content.

Article

Destabilization of the LiBH$_4$–NaBH$_4$ Eutectic Mixture through Pore Confinement for Hydrogen Storage

Filippo Peru [1,2], Seyedhosein Payandeh [3,4], Torben R. Jensen [3], Georgia Charalambopoulou [1] and Theodore Steriotis [1,*]

1. National Centre for Scientific Research "Demokritos", Ag. Paraskevi Attikis, 15341 Athens, Greece
2. Inorganic Chemistry Laboratory, Department of Chemistry, National and Kapodistrian University of Athens, Panepistimiopolis Zografou, 15771 Athens, Greece
3. Interdisciplinary Nanoscience Center (iNANO) and Department of Chemistry, University of Aarhus, Langelandsgade 140, DK-8000 Aarhus C, Denmark
4. Electronics & Electrification (BEB-S) Department, FEV Group, Neuenhofstraße 181, 52078 Aachen, Germany
* Correspondence: t.steriotis@inn.demokritos.gr

Citation: Peru, F.; Payandeh, S.; Jensen, T.R.; Charalambopoulou, G.; Steriotis, T. Destabilization of the LiBH$_4$–NaBH$_4$ Eutectic Mixture through Pore Confinement for Hydrogen Storage. *Inorganics* **2023**, *11*, 128. https://doi.org/10.3390/inorganics11030128

Academic Editor: Maurizio Peruzzini

Received: 26 January 2023
Revised: 9 March 2023
Accepted: 16 March 2023
Published: 18 March 2023

Copyright: © 2023 by the authors. Licensee MDPI, Basel, Switzerland. This article is an open access article distributed under the terms and conditions of the Creative Commons Attribution (CC BY) license (https://creativecommons.org/licenses/by/4.0/).

Abstract: Both LiBH$_4$ and NaBH$_4$ are well known for having high hydrogen contents, but also high decomposition temperatures and slow hydrogen absorption–desorption kinetics, preventing their use for hydrogen storage applications. The low melting temperature (219 °C) of their eutectic mixture 0.71 LiBH$_4$–0.29 NaBH$_4$ allowed the synthesis of a new composite material through the melt infiltration of the hydrides into the ~5 nm diameter pores of a CMK-3 type carbon. A composite of 0.71 LiBH$_4$–0.29 NaBH$_4$ and non-porous graphitic carbon discs was also prepared by similar methods for comparison. Both composites showed improved kinetics and a partial reversibility of the dehydrogenation/rehydrogenation reactions. However, the best results were observed for the CMK-3 nanoconfined hydrides; a consistent uptake of about 3.5 wt.% H$_2$ was recorded after five hydrogenation/dehydrogenation cycles for an otherwise non-reversible system. The improved hydrogen release kinetics are attributed to carbon–hydride surface interactions rather than nanoconfinement, while enhanced heat transfer due to the carbon support may also play a role. Likewise, the carbon–hydride contact proved beneficial in terms of reversibility, without, however, ruling out the potential positive effect of pore confinement.

Keywords: hydrogen storage; complex hydrides; nanoconfinement; borohydrides; porous carbons

1. Introduction

Climate change and its mitigation is one of the most important challenges that humankind will face in the 21st century. The reduction of greenhouse gas emissions is a fundamental measure in order to reverse the global warming trend, and massive effort is thus directed to moderate the anthropogenic CO$_2$ production [1,2]. The compelling necessity to divert the energy production and consumption from fossil fuel combustion towards more sustainable renewable sources is clear. Nevertheless, the majority of such renewable sources (such as wind, solar and tidal power) have the major drawback of variability, in several cases in an unpredictable fashion, due to their fluctuating nature. In order to have a reliable and sustainable energy supply, it is necessary to adopt an energy carrier able to accumulate the energy produced through renewable energy sources and provide it when and where it is required. 'Green' hydrogen could be such a sustainable energy carrier [3,4] not only for having just water as the byproduct of its combustion reaction, but also due to its high energy density, i.e., 120 MJ kg^{-1} (33.33 kWh kg^{-1}), which is twice as high as that of the common fuels [5]. This property is particularly interesting especially for mobile applications, where energy must be stored onboard. However, the extremely low density of hydrogen gas is a considerable obstacle for its practical use in common applications. For instance, for storing 5 kg of H$_2$ (equivalent to a car autonomy of about 500 km) at 200 bar (typical pressure of commercial hydrogen cylinders), a tank volume of about 340 L would be required [6,7]. H_2 liquefaction at very low temperatures, around 20 K (-253 °C) or pressurization to 700 bar, can reduce the system volume, but then significant amounts of energy are sacrificed (for cooling or pressurization but also due to boil-off), while more advanced and expensive vessels are needed [8]. Cryo-compressed hydrogen storage [9,10] has some significant advantages such as improved gravimetric and volumetric capacities compared with compressed hydrogen, but also reduced boil-off compared with typical liquid H_2 storage systems. In all cases, the required pressure increase or temperature decrease comes along with a significant energy penalty, while serious safety issues emerge [11,12]. A much safer approach, which may avoid the multiple disadvantages of compressed gas storage and operation at very low temperatures, involves the use of solid-state materials with high hydrogen content [13–19]. Among the materials considered as good candidates for low-pressure solid-state hydrogen storage, borohydrides, which are able to release hydrogen simply through their thermal decomposition, have particularly attractive characteristics [16,20–25]. These materials generally have a high gravimetric and volumetric hydrogen density, e.g., $LiBH_4$ and $NaBH_4$, whose gravimetric and volumetric capacities are 18.5 and 10.6 wt.% and 121 and 113 kg m^{-3}, respectively [26–28]. However, these materials are highly stable, a property that leads directly to high decomposition temperatures, slow kinetics and in several cases lack of reversibility of the dehydrogenation/hydrogenation reaction [28–34]. Ion substitution and liquefaction can help reduce the temperatures required for the hydrogen release from borohydrides. Both phenomena have been studied on several combinations of such compounds and the most appreciable effects are observed when using eutectic mixtures [35–40].

This work focuses on the study of a $LiBH_4$–$NaBH_4$ eutectic composite. The pure $LiBH_4$ and $NaBH_4$ are known to melt at 280 °C and 505 °C, respectively, while their main dehydrogenation reactions take place at temperatures higher than 400 °C [35,41,42]. On the other hand, upon mixing $LiBH_4$ (71 mol%) and $NaBH_4$ (29 mol%), a eutectic mixture is formed as reported by Dematteis et al. [43]. This mixture has a gravimetric hydrogen content of 15.25 wt.% and a melting temperature of 219 °C, allowing at the same time a drastic destabilization of both compounds and the possibility to melt the hydrides in relatively mild conditions with minimal risk of decomposition. Other approaches that can further destabilize the material, reduce the decomposition temperature and enhance the pertinent kinetics most commonly involve particle size reduction and contact with other materials' surfaces [44,45]. Particle size reduction, in most cases, is achieved through mechanical milling [46,47]. This approach aims to afford particle sizes in the order of nanometers. For such small particles, the surface to volume ratio increases significantly, providing shorter diffusion paths in the solid state and the hydrogen release kinetics are improved significantly. However, the benefits of nanosizing by high-energy ball milling are rapidly reverted by sintering reactions occurring during the hydrogenation/dehydrogenation processes. One way to address this problem is to protect the hydride nanoparticles from agglomeration by confining them in nanosized pores. This approach is particularly facile for molten phases such as low temperature eutectic mixtures and has multiple advantages: (i) solvent free nano-infiltration and (ii) prevention of particle growth and agglomeration due to the encapsulation. In order to have the minimum mass penalty from the confining material, it is necessary to adopt a porous solid with very high pore volume and low density. Carbonaceous materials are ideal for this purpose, since they are in principle chemically inert, they can be synthesized in a wide variety of forms and their pore size can be tuned in order to obtain a good balance between easy infiltration in the host material and optimal particle size of the active guest material. Upon confinement, the hydride nanoparticles and the carbon matrix are bound to proximity, forcing the different components to interact. Beyond nanosizing, the system may also benefit from the intimate contact of the hydride phase with the carbon surface. For instance, the high thermal conductivity of the carbon matrix may facilitate heat transfer and therefore promote de- and re- hydrogenation reactions. Moreover, the carbon–hydride's contact could also trigger surface-induced catalytic reactions due to the carbon electron affinity on the nanoscale, which can promote a decrease in the hydrogen release energy [44,45,48].

A low melting mixture of $LiBH_4$ and $NaBH_4$ confined in porous carbon materials for hydrogen storage purposes has been studied by Javadian et al. [49]. However, the borohydride mixture investigated in that specific work consisted of 0.62 $LiBH_4$–0.38 $NaBH_4$ according to the phase diagram proposed by Adams [50], while the molten borohydrides were infiltrated in a porous carbon aerogel with large mesopores (diameter of ca. 40 nm) and tested with only four cycles of dehydrogenation/rehydrogenation. In the present work, we adopted a more recently reported eutectic composition, 0.71 $LiBH_4$–0.29 $NaBH_4$, with a higher gravimetric hydrogen content [43] of 15.25 wt.%. Moreover, we used for the first time an ordered mesoporous CMK-3 type carbon as hosting material characterized by much smaller pore sizes (5 nm) with a very narrow distribution, which is ideal for nanoconfinement studies. CMK-3 is a mesoporous carbonaceous material made by nanorods organized in a regular 2D hexagonal pattern and held together by thin carbon strands [51]. The high porosity of the CMK-3 structure results in an overall very high pore volume, a characteristic that allows the confinement of large quantities of hydrides. Here, we compare the properties of (a) $LiBH_4$–$NaBH_4$/CMK-3, (b) the non-confined (bulk) hydrides and (c) a composite of $LiBH_4$–$NaBH_4$ with a non-porous carbon, namely graphitic carbon discs, in order to analyze the effect of nanoconfinement to the destabilization, kinetics and cyclability of hydrogen release and uptake of the $LiBH_4$–$NaBH_4$ eutectic system.

2. Results

The XRPD measurements of bulk, composite and cycled materials are presented in Figure 1. CMK-3 carbon is amorphous, as revealed by the practically featureless powder diffraction pattern. From the analysis of the XRPD pattern of the bulk 0.71 $LiBH_4$–0.29 $NaBH_4$ (denoted LiNa) obtained after the milling process, it was possible to identify the peaks relative to the two pure borohydrides, $LiBH_4$ and $NaBH_4$, while no intermediate compounds were detected; based on Scherrer analysis of the most intense peaks, an average crystallite size of around 20 nm was calculated for both phases. On the other hand, the XRPD pattern of the CMK-3 infiltrated sample (LiNa/CMK-3) showed only the diffraction peaks of $NaBH_4$ with low intensity. This pattern suggests that although a small amount of bulk hydride remained outside the pores, the majority of the hydride phase was infiltrated in the porous network, causing a significant overall decrease in the peak intensity in a way that only the strongest peaks of the eutectic mixture were detected; these are the (111), (200) and (220) reflections of $NaBH_4$. Nevertheless, even these low intensity $NaBH_4$ diffraction peaks disappeared after the five dehydrogenation–rehydrogenation cycles (LiNa/CMK-3 five cycles pattern), implying that upon further thermal treatment all the borohydride material is eventually infiltrated in the CMK-3 pores. The situation was somehow different for the non-porous CD composite. CD revealed a graphitic structure, with a broad (002) reflection at ~26° coupled with another visible feature at ~43°; this pertained to the merged (100) and (101) reflections. The LiNa structure was fully visible in the LiNa/CD composite, while several diffraction peaks were still visible, even after the five dehydrogenation–rehydrogenation cycles (LiNa/CD, five cycles).

The presence of a small amount of unconfined hydrides was confirmed by the scanning electron microscope images (Figure 2), also showing that the CMK-3 carbon retained its typical rod-like elongated structure, fully withstanding multiple heat treatments.

Nevertheless, the confirmation that the majority of the hydride phase was successfully confined in the porous carbon scaffold was given by the comparison of the pore properties before and after melt infiltration. N_2 adsorption–desorption isotherms at 77 K (approx. −196 °C) were measured for the pure CMK-3 type carbon (Figure 3). The isotherm was of type IV(a) according to the IUPAC classification, with an H_2(a) hysteresis loop; both are characteristic for CMK-3 materials with mesopores larger than ~4 nm and a rather sharp size distribution [52]. Based on this isotherm, a Brunauer–Emmett–Teller (BET) surface area (S_{BET}) of 1250 m^2 g^{-1}, a total pore volume (TPV) of 1.2 cm^3 g^{-1} and an average pore size distribution centered at 4.6 nm were deduced for CMK-3 (Table 1).

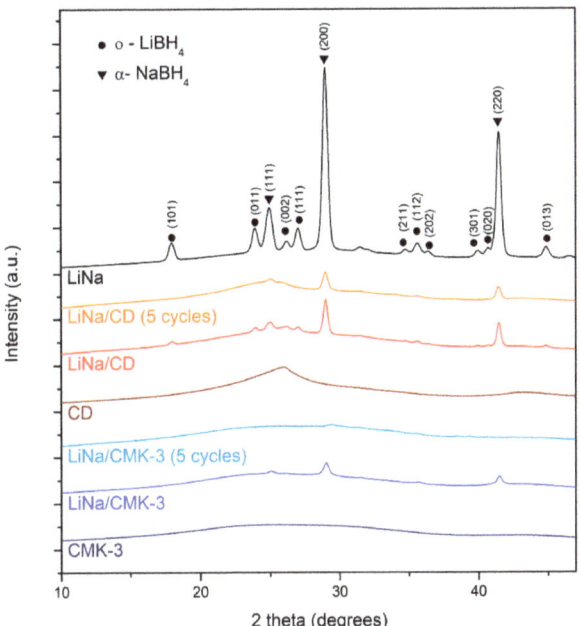

Figure 1. XRPD patterns of pure carbons (CMK-3 and CD), bulk 0.71 LiBH$_4$–0.29 NaBH$_4$ eutectic mixture (LiNa), the carbon/borohydride composites (LiNa/CMK-3 and LiNa/CD) as well as the composites after 5 dehydrogenation and rehydrogenation cycles. The peak labels indicate the following phases: orthorhombic o-LiBH$_4$ (•), α-NaBH$_4$ (▼).

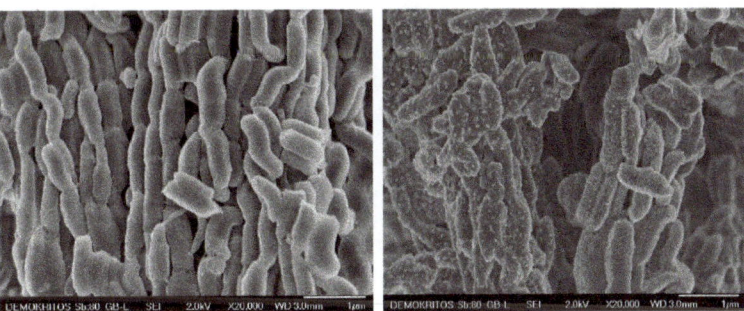

Figure 2. Scanning electron microscope (SEM) pictures showing the typical elongated structure of pure CMK-3 (**left**) and LiNa/CMK-3 after melt infiltration, with traces of borohydrides on the external surface of the carbon particles (**right**).

The general shape of the isotherm of LiNa/CMK-3 was the same; nevertheless, the BET area dropped to 170 m^2 g^{-1} and TPV to 0.21 cm^3 g^{-1}, while the pore size distribution remained centered at 4.6 nm. This implies that after infiltration a large number of pores were lost (i.e., becoming completely inaccessible to the adsorbate), which strongly suggests that the borohydride molten phase indeed fills the pores and does not form a layer on the pore walls (this would decrease the pore size of the sample). It should be noted that the isotherm of the composite material presented in Figure 3 has been scaled to the mass of pure carbon (the same holds for the BET and TPV calculations), thus excluding the weight penalty of the non-porous hydride phase. On the other hand, the pristine carbon discs (CDs) showed a clear type-I isotherm, typical for non-porous materials; a BET area of ~30 m^2 g^{-1} was calculated. Both the isotherm shape and the BET area (~25 m^2 g^{-1}) on a carbon basis remained practically unchanged for LiNa/CD (Table 1).

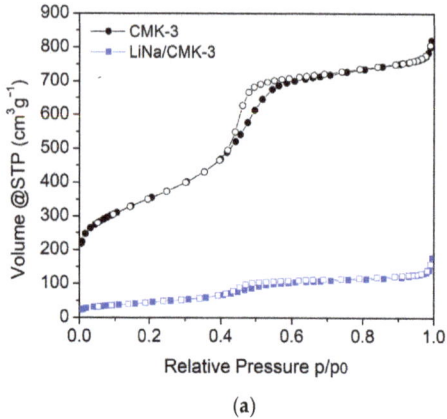
(a)

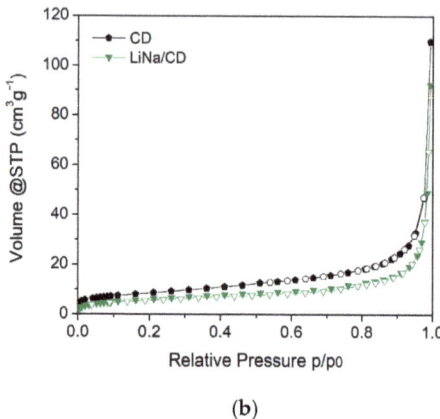
(b)

Figure 3. N$_2$ adsorption/desorption isotherms performed at 77 K (a): on the pure CMK-3 type carbon (black) and on the composite material LiNa/CMK-3 (blue), and (b): on the pure CD (black) and the LiNa/CD samples (green).

Table 1. Surface area (S$_{BET}$), total pore volume (TPV) and average pore size (<d>) of pure CMK-3 and carbon discs (CD), as well as their composites with 0.71 LiBH$_4$–0.29 NaBH$_4$ (LiNa).

	S$_{BET}$ (m^2 g^{-1})	TPV (cm^3 g^{-1})	<d> (nm)
CMK-3	1250	1.2	4.6
LiNa/CMK-3	170	0.21	4.6
CD	~30	-	-
LiNa/CD	~25	-	-

The temperature programmed desorption-mass spectrometry (TPD-MS) measurement of hydrogen release from bulk LiNa showed that dehydrogenation of the eutectic mixture occurred mostly at an intermediate temperature between that of the two pure components, in accordance with previously reported results for the pure LiBH$_4$ and NaBH$_4$ compounds and their mixtures with compositions close to the eutectic one [53]. In more detail, bulk LiNa, after being pre-melted, dehydrogenated in four stages (Figure 4), with a first release at 290 °C (2%), two desorption peaks at about 390 (26%) and 450 °C (39%) and a final decomposition at 490 °C (33%). Based on the shape of the LiNa TPD-MS spectra as compared with the thermal desorption from the pure compounds presented in [54–56], and since the XRPD pattern of LiNa revealed only pure LiBH$_4$ and NaBH$_4$, the first three dehydrogenation steps can be directly associated with the decomposition of LiBH$_4$, while the last one can be attributed to NaBH$_4$. This observation is very similar to that of Liu et al. [53], who studied a similar mixture of LiBH$_4$ and NaBH$_4$ and identified two main decomposition routes: the first (up to ~520 °C) involved the release of H$_2$ followed by the formation of LiH, Li$_2$B$_{12}$H$_{12}$ and B; the second, at higher temperatures, involved the release of H$_2$ together with the formation of Na and B.

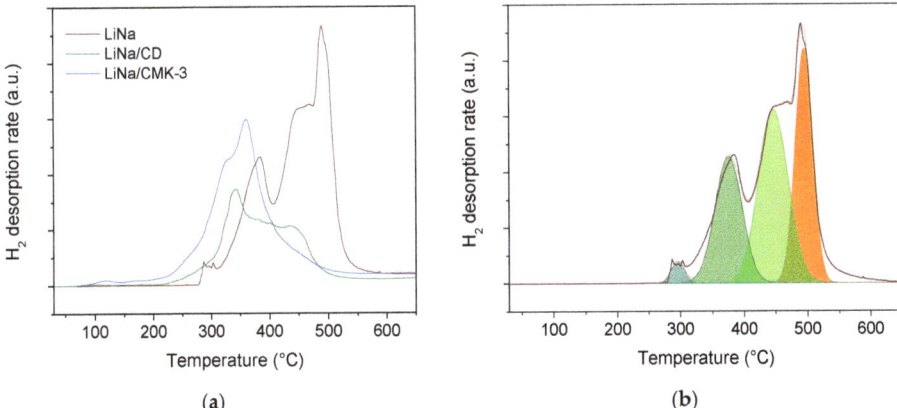

Figure 4. (**a**): Temperature programmed desorption coupled with mass spectrometry (TPD/MS, m/e = 2) results for LiNa (dark red), LiNa/CD (green), LiNa/CMK-3 (blue). (**b**): Deconvolution of the bulk LiNa TPD signal highlighting the different H_2 release reactions. The first three (green) peaks are attributed to the $LiBH_4$ decomposition, covering, respectively, the 2%, 26% and 39% of the total H_2 released, while the orange peak is related to the $NaBH_4$ decomposition (33% of the total H_2 released).

The LiNa/CMK-3 composite released hydrogen in a similar manner to the bulk LiNa and with a similar ratio between the signal intensities. However, the different reaction steps occurred at much lower temperatures and in a shorter temperature range and were somehow merged together, with the main peaks located at 325 and 360 °C, highlighting the contribution from the carbon surface. The substrate's active role in lowering the dehydrogenation temperatures of LiNa was also evident in the behavior of LiNa/CD, which mostly decomposed at 340 and 440 °C. In the latter case, it can be noticed that despite the absence of nanoconfinement, the carbon matrix still played a role in the material destabilization, albeit with less intensity, probably due to the lower surface area of the carbon discs. In all samples, only hydrogen was detected among the gases analyzed.

The hydrogen release and uptake were further investigated for the three samples using a custom-made manometric device (Sieverts apparatus) and a series of dehydrogenation/rehydrogenation cycles (Figure 5). Excessive heating (temperatures higher than 650 °C) led to complete decomposition of $NaBH_4$ with the precipitation of Na, B and the evolution of H_2 ($NaBH_4 \rightarrow Na + B + 2H_2$) [53,57]. Without the aid of catalysts/additives [58–60], this process is highly irreversible and, for this reason, it was decided to limit the cycling experiments to 450 °C and only actually study the decomposition of $LiBH_4$ (in the presence of $NaBH_4$). The decomposition of $LiBH_4$ went through a complex mechanism that can be represented by two main reactions [53,61,62]:

$$LiBH_4 \rightarrow LiH + B + 3/2 H_2 \tag{1}$$

$$LiBH_4 \rightarrow 5/6 LiH + 1/12 Li_2B_{12}H_{12} + 13/12 H_2 \tag{2}$$

This explains the drastic reduction of hydrogen released between the first and second dehydrogenation cycle for all samples. Due to the stability of $Li_2B_{12}H_{12}$ closo-borane, we assumed that only the first reaction was reversible and, based on the amount of hydrogen released during the manometric experiments, it can be deduced that it occurred in a percentage of around 60–70% of the $LiBH_4$ decomposition. It should be noted that attempts to hydrogenate $Li_2B_{12}H_{12}$–10 LiH composites, e.g., at, $p(H_2)$ = 970 bar and T = 400 °C for 48 h, did not reveal any formation of $LiBH_4$ or other compounds [63].

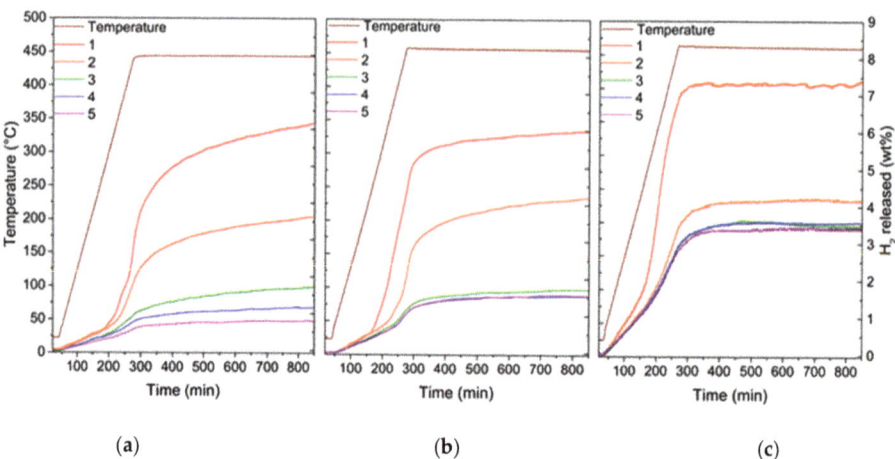

Figure 5. Kinetics and quantitative measurement of hydrogen release and uptake was investigated using a manometric method (Sieverts apparatus). From left to right: (**a**) LiNa, (**b**) LiNa/CD and (**c**) LiNa/CMK-3.

In more detail, the bulk LiNa released about 6.2 wt.% of H_2, an amount that dropped to 3.6 wt.% already at the second dehydrogenation step and continued to decrease at each cycle until it reached about 0.8 wt.% of H_2 at the fifth hydrogenation–dehydrogenation cycle, suggesting strong deactivation of the system (Figure 5a). It is also important to note that the dehydrogenation kinetics were extremely slow, and during the first cycles the system was far from equilibrium even after 800 min. Taking into account only the mass of the active materials in the hydrogen sorption reaction, i.e., that of $LiBH_4$ and $NaBH_4$ and excluding the carbon, the LiNa/CD composite had a similar behavior with the bulk during the first three cycles, i.e., it released about 6 wt.% of H_2 at the first, 4.2 wt.% of H_2 at the second and 1.7 wt.% during the third cycle. Nevertheless, the sample reached a stable value of around 1.5 wt.% of H_2 for the last three cycles (Figure 5b). In this case, the desorption kinetics were significantly faster compared with the bulk, but still quite slow.

The LiNa/CMK-3 composite, on the other hand, released in a very fast manner (equilibrium reached at around 300 min) about 7.3 wt.% of H_2 in the first cycle, which dropped to 4.2 wt.% in the second dehydrogenation step, but stabilized around 3.5 wt.% for cycles three–five, revealing that at least a part of the infiltrated material retained its reversibility (Figure 5c). The increased release during the first cycle can be attributed to partial decomposition of $NaBH_4$ below 450 °C. This kinetic enhancement was probably due to nanoconfinement, and was also detected in our TPD measurements [64]. However, the decomposed $NaBH_4$, could not be re-hydrogenated and the second cycle finally showed a similar release with the bulk and the LiNa/CD samples. On the other hand, contrary to the CDs, the nanoconfinement in CMK-3 provided more protection, and therefore a higher degree of reversibility, probably by minimizing the formation of closo-boranes.

The derivatives of the kinetic desorption curves of Figure 5 measured with a manometric (Sieverts) approach also reveal some interesting details and are presented in Figure 6 for the first and third cycle. The ramp rates between TPD and cycling experiments were quite different; however, the dehydrogenation curve of bulk LiNa also showed different dehydrogenation events during the first decomposition. Probably due to the much lower amount of hydrogen released, for the third dehydrogenation the signals merged in a broad one in an average position. Additionally, again in accordance with the TPD-MS results, the derivative curve of the first dehydrogenation curve of the composite LiNa/CMK-3 appeared much earlier than the bulk, showing the contribution of the carbon. This, however, seemed to disappear upon cycling and the derivative of the third cycle showed that the main hydrogen desorption occurred later, similarly to that of the bulk. As expected, the LiNa/CD composite showed an intermediate behavior.

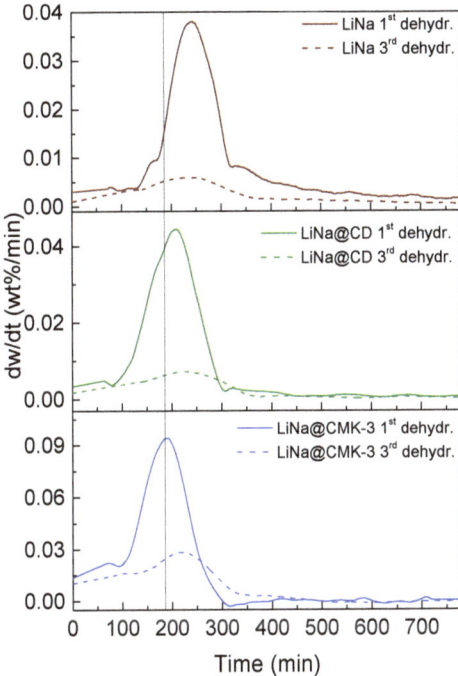

Figure 6. Derivative curves of the H$_2$ release versus time measured with a manometric device (Sieverts method, data shown in Figure 5) for LiNa (dark red), LiNa/CD (green) and LiNa/CMK-3 (blue).

3. Discussion

The XRPD pattern, as well as the SEM study of LiNa/CMK-3, suggest that although most of the borohydride particles were successfully introduced into the porous scaffold, a small part of the hydride material remained outside the pores. Nevertheless, the proof of an overall successful infiltration was provided by the pore network properties pre- and post infiltration as derived from the N$_2$ adsorption/desorption isotherms at 77 K (approx. −196 °C). The TPV of the CMK-3 was reduced from 1.2 cm^3 g^{-1} to 0.21 cm^3 g^{-1} with a loss of about the 82.5% of the pore volume. It should be noted that the volume of the composite borohydride (LiNa) used for the infiltration was 60% of the CMK-3 pore volume, and thus in principle a corresponding reduction of the pore volume was expected. This difference can be attributed to pore blocking caused by the molten phase in a way that some extra closed porosity (empty yet inaccessible to N$_2$ molecules) was created. A similar picture was obtained by examining the dramatic decrease in the BET area of the carbon, from 1250 m^2 g^{-1} to only 170 m^2 g^{-1} after the melt infiltration. This reduction was even more pronounced (86.5%) in accordance with the pore blocking mechanism. In brief, the blocking of smaller pores reduced the BET area more than the pore volume as the surface to volume ratio increased with decreasing pore size.

The TPD-MS experiments showed the H$_2$ release during the thermal decomposition, without detectable traces of undesired gases such as B$_2$H$_6$ (all the recorded m/z signals analyzed are reported in Section 4). The experimental results allowed us to resolve the multiple decomposition stages particularly evident in the bulk LiNa, from about 280 °C until the last main dehydrogenation occurring at about 490 °C. Additionally, the role of the carbons on the borohydrides' decomposition is highlighted by the TPD-MS results.

The presence of a high surface mesoporous carbon such as CMK-3 allowed the decomposition to take place at temperatures even 130 °C lower than the bulk hydride mixture, in accordance with previous observations for carbon aerogels [49]. The relative intensity of the two main dehydrogenation signals was maintained, suggesting that all the reactive components were affected by the carbon presence. On the other hand, the composite with the non-porous carbon (LiNa/CD) also showed lower decomposition temperatures, with the dehydrogenation events occurring in a range of temperatures between the bulk and the LiNa/CMK composite, suggesting an intermediate behavior. Contrary to what is generally believed, the decrease in the decomposition temperature suggests that even without nanoconfinement the carbon provided a kinetic improvement to the decomposition reaction. The inferior performance of LiNa/CD compared with LiNa/CMK may perhaps be explained by the significantly lower surface area of the CD substrate. Similar results were obtained by nanoconfinement of sodium aluminum hydride, $NaAlH_4$, in carbons with different pore sizes in the range 7 to 100 nm. A significant contribution to the observed improvements of kinetics and reversibility was assigned to the catalytic effects of the carbon scaffolds along with a minor contribution from the nanosizing, i.e., confinement in pores in the range of 7 to 39 nm [65].

A similar behavior was observed by the manometric study of the dehydrogenation/rehydrogenation of the samples. While the pure eutectic LiNa mixture decomposed at high temperatures and showed a non-reversible behavior, the carbon composites allowed lower decomposition temperatures and provided the system with some degree of reversibility. Especially for the case of LiNa/CMK-3, it can be seen that after the first two cycles the amount of hydrogen exchanged was constant, with a H_2 uptake of about 3.5 wt.%. The carbon discs had a comparable effect on LiNa; however, as expected due to the smaller surface area, the reversible hydrogen uptake was limited and did not exceed 1.5 wt.% of H_2. Unlike the kinetic enhancement that was attributed to the catalytic effect of the carbon surface, the stability of the system under cycling may be attributed to confinement since the pore network provided a space where extended aggregation was difficult. The reversible part of the LiNa/CD can, in this respect, be associated with the part of the molten phase that was directly "bound" to the surface of the CDs, thus hampering agglomeration.

From the study of the derivatives of the dehydrogenation curves it is possible to conclude that the decomposition of the carbon composites starts at lower temperatures than the bulk eutectic mixture. There is an obvious kinetic improvement for LiNa/CMK-3, which is less pronounced for LiNa/CD. However, this behavior is somehow limited to the first dehydrogenation cycles, and the kinetic enhancement is practically lost for the third cycle. This behavior may be related to the loss of the destabilizing interaction between the hydrides and the carbon surface after cycling. It is quite obvious from the integral curves that in all cases there is a major loss of activity between the first and second cycle, implying that at least a part of the material is dehydrogenated irreversibly in the bulk but also under confinement.

4. Materials and Methods

Sample preparation. The CMK-3 carbon scaffold was synthesized through nanocasting using SBA-15 porous silica (Claytec Inc., East Lansing, MI, USA) as hard template. In order to achieve an adequately narrow pore size distribution and a smaller mean pore size, a procedure slightly different from the one originally proposed by Ryoo et al. [51] was adopted; this approach involved three impregnation steps with the carbon precursor instead of just two [66], and allowed a very good control of the pore size of CMK-3, at the cost of a small reduction of the surface area and pore volume. More specifically, 1 g of dried SBA-15 powder was impregnated with 5 mL of an aqueous solution containing 1.25 g of sucrose (Merck, Darmstadt, Germany) as carbon precursor and 0.07 mL of sulfuric acid (Merck, Darmstadt, Germany) in order to promote the polymerization-caramelization reactions that were carried out in an oven. The sample was initially heated from room temperature to 100 °C with a heating rate of 10 °C min^{-1}. After 6 h at a constant temperature, the sample was further heated up to 160 °C (10 °C min^{-1}) and left overnight. The whole procedure of impregnation with the sucrose solution and thermal treatment was then repeated two more times, while gradually reducing (by 65% each time) the solution concentration. In order to complete the carbonization process, the carbon–silica composite material was pyrolyzed in a tubular furnace (GSL 1100 X, MTI, CA, USA) at 900 °C under nitrogen flow. During this process, which lasted approx. 6 h, the residual hydrogen and oxygen were eliminated and the carbon structure was consolidated. The silica template was then dissolved by washing the composite material with a 40% solution of hydrofluoric acid. In order to completely remove the acidic solution from the porous carbon structure, the material was subjected to a series of centrifugations at 9000 RPM (Universal 320, Hettich, Tuttlingen, Germany) and water additions. The resulting CMK-3 type carbon was then washed and filtered under vacuum with water and ethanol and dried overnight in an oven at 100 °C. The mesoporous carbon obtained with this procedure may still have residual oxygen functional groups on the surface that may react and oxidize the borohydrides. For this reason, the as-produced CMK-3 was further treated in a tubular furnace at 700 °C under a stream of argon. This process removes the surface functional groups and may increase the amount of accessible micropores, thus increasing the material's total surface area [67]. The sample was stored at ambient conditions; however, before using it as a host for infiltrating borohydrides it was thoroughly outgassed overnight under high vacuum at 250 °C and consequently stored in an argon-filled glove box (LABstar, MBraun, Garching, Germany).

The non-porous carbon used (n-TEC, Norway) consisted of carbon discs (CD) made by multiple graphitic layers having an average thickness of 20–50 nm and diameter of about 0.8–3 µm. This type of carbon was synthesized through the pyrolysis of hydrocarbons using a plasma torch process. Similarly to the CMK-3 material, the CDs were also heat treated before use in a tubular furnace at 700 °C under inert atmosphere for about 3 h.

The eutectic mixture of the two borohydrides 0.71 LiBH$_4$–0.29 NaBH$_4$ (referred to as LiNa) was prepared according to the previously described procedures [49] using the eutectic molar ratio studied by Dematteis et al. [43]. The two pure salts, LiBH$_4$ (95%, Sigma-Aldrich, St. Louis, MO, USA) and NaBH$_4$ (97%, Sigma-Aldrich, St. Louis, MO, USA), were stored in an argon-filled glove box and mixed at a molar ratio 0.71:0.29, first manually with a pestle and mortar, and then mechano-chemically using a planetary ball mill (Pulverisette 4, Fritch GmbH, Idar-Oberstein, Germany). The powders were milled in a gas-tight tungsten carbide jar with internal volume of 80 mL, using tungsten carbide spheres with 10 mm diameter and a 1:30 powder to balls ratio. The total mixing time necessary to obtain a homogeneous mixture was about 1 h at 350 RPM. However, due to the low melting temperature of the LiNa mixture and in order to avoid unwanted reactions during the high energy process, processing was divided into 30 cycles of 2 min of milling followed by 2 min of pause, which was necessary to let the powders cool down to room temperature. In addition, the sample treatment as well as the ball milling reactor loading and unloading was performed under inert argon atmosphere, with H$_2$O and O$_2$ concentrations <0.5 ppm.

The low melting temperature of the obtained eutectic borohydride mixture allowed the filling of the CMK-3 carbon pores via melt infiltration. A calculated amount of LiNa with a volume equal to 60% of the total pore volume (TPV) of the carbon scaffold was used for the melt infiltration. Taking into account the total pore volume of the carbon (TPV = 1.2 cm^3 g^{-1}) and the density of the eutectic mixture (0.78 g cm^{-3}), the carbon and hydrides were mixed together with a mass ratio of CMK-3: LiNa equal to 1.76:1. All the handling processes necessary to obtain a homogeneous mixture of carbon and hydrides were performed in an argon-filled glove box. The manually mixed LiNa/CMK-3 composite was transferred to a high pressure stainless steel gas tight reactor under argon (1 atm). The reactor was then sealed, removed from the glove box and attached to a custom-made manometric apparatus, where the actual melt infiltration took place. After being meticulously outgassed, the system was loaded with an overpressure of ~100 bar of H$_2$ in order to avoid any unwanted dehydrogenation reactions during melt infiltration.

The sample in the reactor was heated up to 250 °C with a heating ramp of 3 °C min^{-1} and the final temperature was kept constant for 30 min, allowing the infiltration of the molten hydride inside the carbon pores. The approach for CMK-3, LiNa and LiNa/CMK-3 preparation is schematically presented in Figure 7. The same procedure was also used for the synthesis of an "equivalent" composite material combining the LiNa mixture with the non-porous CDs (namely LiNa/CD), where nanoconfinement does not take place. In this case, since the carbon disks are non-porous, the carbon material and the hydrides were mixed with the same mass ratio used for the LiNa/CMK-3 composite, i.e., 1.76:1.

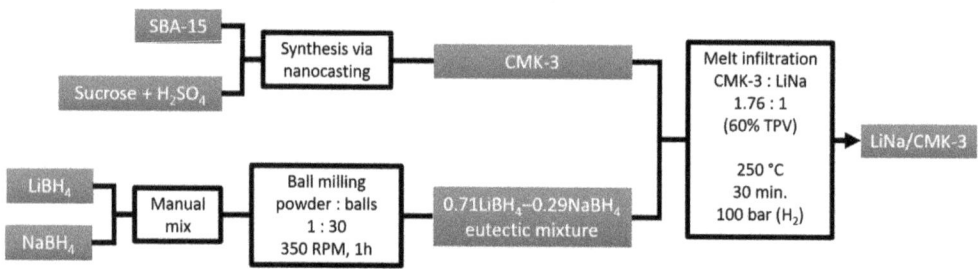

Figure 7. Schematic representation of the experimental procedure leading to the synthesis of the LiNa/CMK-3 composite.

Finally, for reference, the bulk LiNa was also heat treated at 250 °C for 30 min under 100 bar of H$_2$, simulating the infiltration process.

The efficiency of the infiltration of the eutectic mixture in the pores of the CMK-3 carbon was evaluated through a series of powder X-ray diffraction (XRPD) measurements, using a Rigaku RAXIS IV Imaging Plate Detector mounted on a Rigaku RU-H3R Rotating Copper Anode Xray Generator (λ = 1.54 Å) and N$_2$ sorption measurements at 77 K, using a volumetric gas adsorption analyzer (Autosorb-1-MP, Quantachrome Instruments, Boynton Beach, FL, USA). High vacuum (10^{-6} mbar) was applied to the materials in order to outgas the samples before analysis The specific surface area of the materials was calculated according to the Brunauer–Emmett–Teller (BET) method, employing the BET consistency criteria (ISO 9277:2010). The pore size distribution was calculated through the QSDFT (quenched solid density functional theory) method assuming slit/cylindrical pore models. The total pore volume was calculated at p/p_0 = 0.98. The morphology of the infiltrated scaffold was also examined on a JSM 7401F (JEOL Ltd., Akishima, Tokyo, Japan) field emission scanning electron microscope (SEM) using the Gentle Beam mode.

The thermal decomposition of both the bulk LiNa as well as the carbon-based composites was studied with temperature programmed desorption coupled with mass spectrometry (TPD/MS). The sample was contained in a quartz cell, part of a custom-made set-up, and was heated up using a furnace under a constant carrier gas stream (argon) regulated by a mass flow controller (80 mL min^{-1}). In a typical run, the sample was heated up to 700 °C with a heating ramp of 5 °C min^{-1}. The decomposition products were transferred with the aid of the carrier gas to an OmniStar GSD 301 O1 (Pfeiffer Vacuum Technology AGm Aßlar, Germany) quadrupole mass spectrometer and then vented. The temporal change of intensity at m/z: 2, 4, 15–18, 23–28, 32, 44, 79 and 80 was continuously recorded.

The absorption–desorption capacity of the materials under five dehydrogenation/hydrogenation cycles were investigated with a custom-made high pressure manometric device (Sieverts apparatus). The respective experiments involved a dehydrogenation step under an initial back-pressure of ~1.5 bar, during which the sample was heated up with a heating rate of 2 °C min^{-1} from room temperature up to 450 °C, then the system was kept at this temperature for 10 h and consequently cooled down freely back to room temperature. The second part of the experiment involved the rehydrogenation of the sample, performed under an initial pressure of about 100 bar of hydrogen. The sample was heated up with a rate of 5 °C min^{-1} from room temperature up to 450 °C. The high temperature was kept constant for 12 h in order to allow the material to absorb as much hydrogen as possible before cooling down to room temperature. The dehydrogenation and hydrogenation procedure (Figure 8) were repeated 5 times for each sample.

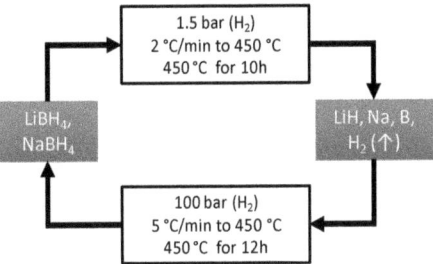

Figure 8. Schematic representation of the LiNa dehydrogenation/rehydrogenation process carried out with a custom-made Sievert apparatus. The amount of H_2 released by LiNa and LiNa composites was calculated after heating the material to 450 °C under 1.5 bar of H_2 pressure. The products were then re-hydrogenated at 450 °C under 100 bar of H_2 pressure. The cycle was repeated five times.

5. Conclusions

The low melting temperature of the 0.71 $LiBH_4$–0.29 $NaBH_4$ (LiNa) eutectic mixture allowed the nanoconfinement of the hydrides in the ~5 nm diameter pores of a CMK-3 type carbon, as well as the synthesis of a borohydride/non-porous carbon composite. For both the porous and non-porous carbon hosts an improvement on the hydrogen absorption-release properties over the bulk hydrides (LiNa) was observed. The carbon-LiNa composites revealed faster decomposition kinetics compared with bulk LiNa, while the effect is more obvious for the nanoconfined material. However, the kinetic improvements decrease with cycling. Moreover, the presence of carbon increased the reaction reversibility compared with the bulk material. The nanoconfined LiNa shows a consistent uptake of H_2 of about 3.5 wt.% after five hydrogenation/dehydrogenation cycles. A similar behavior but with minor intensity was revealed by the composite material with non-porous carbon, which can exchange about 1.5 wt.% of H_2 after five cycles. Both the kinetic improvement observed for the carbon composite materials and the reversibility are probably associated with surface catalytic interactions; confinement in porous materials with high surface area seems to amplify such effects, leading to overall better performance.

Author Contributions: Conceptualization, T.R.J., G.C. and T.S.; Methodology, F.P.; Investigation, F.P. and S.P.; Data curation, F.P.; Writing—original draft, F.P.; Writing—review & editing, T.R.J., G.C. and T.S.; Visualization, S.P.; Supervision, T.R.J., G.C. and T.S. All authors have read and agreed to the published version of the manuscript.

Funding: This research received funding from the European Marie Curie Actions under ECOSTORE grant agreement No. 607040. The work of F.P. was also partially supported by the H2020 project ENDURUNS (Grant agreement: 824348).

Data Availability Statement: The data presented in this study are available on request from the corresponding author.

Acknowledgments: F.P. would like to acknowledge P. Kyritsis and Ch.A. Mitsopoulou (Inorganic Chemistry Laboratory, Department of Chemistry of the National and Kapodistrian University of Athens, Greece) for the fruitful discussions and their support during the realization of this work.

Conflicts of Interest: The authors declare no conflict of interest.

References

1. Asghar, U.; Rafiq, S.; Anwar, A.; Iqbal, T.; Ahmed, A.; Jamil, F.; Khurram, M.S.; Akbar, M.M.; Farooq, A.; Shah, N.S.; et al. Review on the Progress in Emission Control Technologies for the Abatement of CO_2, SO_x and NO_x from Fuel Combustion. *J. Environ. Chem. Eng.* **2021**, *9*, 106064. [CrossRef]
2. Beringer, S.L. Energy, Climate Change and EU Development Policy. In *EU Development Policies*; Palgrave Macmillan: Cham, Switzerland, 2019; pp. 17–34. ISBN 9783030013073.
3. Zhou, Y.; Li, R.; Lv, Z.; Liu, J.; Zhou, H.; Xu, C. Green Hydrogen: A Promising Way to the Carbon-Free Society. *Chin. J. Chem. Eng.* **2022**, *43*, 2–13. [CrossRef]
4. Bongartz, D.; Dore', L.; Eichler, K.; Grube, T.; Heuser, B.; Hombach, L.E.; Robinius, M.; Pischinger, S.; Stolten, D.; Walther, G. Comparison of Light-Duty Transportation Fuels Produced from Renewable Hydrogen and Green Carbon Dioxide. *Appl. Energy* **2018**, *231*, 757–767. [CrossRef]
5. Abe, J.O.; Popoola, A.P.I.; Ajenifuja, E.; Popoola, O.M. Hydrogen Energy, Economy and Storage: Review and Recommendation. *Int. J. Hydrogen Energy* **2019**, *44*, 15072–15086. [CrossRef]
6. Ball, M.; Weeda, M. The Hydrogen Economy - Vision or Reality? *Int. J. Hydrogen Energy* **2015**, *40*, 7903–7919. [CrossRef]
7. Lai, Q.; Paskevicius, M.; Sheppard, D.A.; Buckley, C.E.; Thornton, A.W.; Hill, M.R.; Gu, Q.; Mao, J.; Huang, Z.; Liu, H.K.; et al. Hydrogen Storage Materials for Mobile and Stationary Applications: Current State of the Art. *ChemSusChem* **2015**, *8*, 2789–2825. [CrossRef] [PubMed]
8. Hua, T.Q.; Roh, H.S.; Ahluwalia, R.K. Performance Assessment of 700-Bar Compressed Hydrogen Storage for Light Duty Fuel Cell Vehicles. *Int. J. Hydrogen Energy* **2017**, *42*, 25121–25129. [CrossRef]
9. Zhang, J.; Yan, Y.; Zhang, C.; Xu, Z.; Li, X.; Zhao, G.; Ni, Z. Properties Improvement of Composite Layer of Cryo-Compressed Hydrogen Storage Vessel by Polyethylene Glycol Modified Epoxy Resin. *Int. J. Hydrogen Energy* **2022**, *48*, 5576–5594. [CrossRef]
10. Zhao, X.; Yan, Y.; Zhang, J.; Zhang, F.; Wang, Z.; Ni, Z. Analysis of Multilayered Carbon Fiber Winding of Cryo-Compressed Hydrogen Storage Vessel. *Int. J. Hydrogen Energy* **2022**, *47*, 10934–10946. [CrossRef]
11. Rivard, E.; Trudeau, M.; Zaghib, K. Hydrogen Storage for Mobility: A Review. *Materials* **2019**, *12*, 1973. [CrossRef]
12. Balasooriya, W.; Clute, C.; Schrittesser, B.; Pinter, G. A Review on Applicability, Limitations, and Improvements of Polymeric Materials in High-Pressure Hydrogen Gas Atmospheres. *Polym. Rev.* **2022**, *62*, 175–209. [CrossRef]
13. Grinderslev, J.B.; Amdisen, M.B.; Skov, L.N.; Møller, K.T.; Kristensen, L.G.; Polanski, M.; Heere, M.; Jensen, T.R. New Perspectives of Functional Metal Borohydrides. *J. Alloys Compd.* **2022**, *896*, 163014. [CrossRef]
14. Schneemann, A.; White, J.L.; Kang, S.; Jeong, S.; Wan, L.F.; Cho, E.S.; Heo, T.W.; Prendergast, D.; Urban, J.; Wood, B.C.; et al. Nanostructured Metal Hydrides for Hydrogen Storage. *Chem. Rev.* **2018**, *118*, 10775–10839. [CrossRef]
15. Hadjixenophontos, E.; Dematteis, E.M.; Berti, N.; Wołczyk, A.R.; Huen, P.; Brighi, M.; Le, T.T.; Santoru, A.; Payandeh, S.H.; Peru, F.; et al. A Review of the MSCA ITN ECOSTORE-Novel Complex Metal Hydrides for Efficient and Compact Storage of Renewable Energy as Hydrogen and Electricity. *Inorganics* **2020**, *8*, 17. [CrossRef]
16. Hirscher, M.; Yartys, V.A.; Baricco, M.; Bellosta von Colbe, J.; Blanchard, D.; Bowman, R.C.J.; Broom, D.P.; Buckley, C.E.; Chang, F.; Chen, P.; et al. Materials for Hydrogen-Based Energy Storage—Past, Recent Progress and Future Outlook. *J. Alloys Compd.* **2020**, *827*, 153548. [CrossRef]
17. Dematteis, E.M.; Amdisen, M.B.; Autrey, T.; Barale, J.; Bowden, M.E.; Buckley, C.E.; Cho, Y.W.; Deledda, S.; Dornheim, M.; De Jongh, P.; et al. Hydrogen Storage in Complex Hydrides: Past Activities and New Trends. *Prog. Energy* **2022**, *4*, 032009. [CrossRef]
18. Eghbali, P.; Nişancı, B.; Metin, Ö. Graphene Hydrogel Supported Palladium Nanoparticles as an Efficient and Reusable Heterogeneous Catalysts in the Transfer Hydrogenation of Nitroarenes Using Ammonia Borane as a Hydrogen Source. *Pure Appl. Chem.* **2018**, *90*, 327–335. [CrossRef]
19. Eghbali, P.; Gürbüz, M.U.; Ertürk, A.S.; Metin, Ö. In Situ Synthesis of Dendrimer-Encapsulated palladium(0) Nanoparticles as Catalysts for Hydrogen Production from the Methanolysis of Ammonia Borane. *Int. J. Hydrogen Energy* **2020**, *45*, 26274–26285. [CrossRef]
20. Bellosta von Colbe, J.; Ares, J.R.; Barale, J.; Baricco, M.; Buckley, C.; Capurso, G.; Gallandat, N.; Grant, D.M.; Guzik, M.N.; Jacob, I.; et al. Application of Hydrides in Hydrogen Storage and Compression: Achievements, Outlook and Perspectives. *Int. J. Hydrogen Energy* **2019**, *44*, 7780–7808. [CrossRef]
21. Manickam, K.; Mistry, P.; Walker, G.; Grant, D.; Buckley, C.E.; Humphries, T.D.; Paskevicius, M.; Jensen, T.; Albert, R.; Peinecke, K.; et al. Future Perspectives of Thermal Energy Storage with Metal Hydrides. *Int. J. Hydrogen Energy* **2019**, *44*, 7738–7745. [CrossRef]
22. Mazzucco, A.; Dornheim, M.; Sloth, M.; Jensen, T.R.; Oluf, J.; Rokni, M. Bed Geometries, Fueling Strategies and Optimization of Heat Exchanger Designs in Metal Hydride Storage Systems for Automotive Applications: A Review. *Int. J. Hydrogen Energy* **2014**, *39*, 17054–17074. [CrossRef]
23. Yartys, V.A.; Lototskyy, M.V.; Akiba, E.; Albert, R.; Antonov, V.E.; Ares, J.R.; Baricco, M.; Bourgeois, N.; Buckley, C.E.; Bellaosta von Colbe, J.M.; et al. Magnesium Based Materials for Hydrogen Based Energy Storage: Past, Present and Future. *Int. J. Hydrogen Energy* **2019**, *44*, 7809–7859. [CrossRef]
24. Callini, E.; Aguey-Zinsou, K.F.; Ahuja, R.; Ares, J.R.; Bals, S.; Biliškov, N.; Chakraborty, S.; Charalambopoulou, G.; Chaudhary, A.L.; Cuevas, F.; et al. Nanostructured Materials for Solid-State Hydrogen Storage: A Review of the Achievement of COST Action MP1103. *Int. J. Hydrogen Energy* **2016**, *41*, 14404–14428. [CrossRef]

25. Møller, K.T.; Jensen, T.R.; Akiba, E.; Li, H. wen Hydrogen - A Sustainable Energy Carrier. *Prog. Nat. Sci. Mater. Int.* **2017**, *27*, 34–40. [CrossRef]
26. Puszkiel, J.; Gasnier, A.; Amica, G.; Gennari, F. Tuning LiBH$_4$ for Hydrogen Storage: Destabilization, Additive, and Nanoconfinement Approaches. *Molecules* **2020**, *25*, 163. [CrossRef]
27. Demirci, U.B.; Miele, P. Sodium Borohydride versus Ammonia Borane, in Hydrogen Storage and Direct Fuel Cell Applications. *Energy Environ. Sci.* **2009**, *2*, 627–637. [CrossRef]
28. Møller, K.T.; Sheppard, D.; Ravnsbæk, D.B.; Buckley, C.E.; Akiba, E.; Li, H.; Jensen, T.R. Complex Metal Hydrides for Hydrogen, Thermal and Electrochemical Energy Storage. *Energies* **2017**, *10*, 1645. [CrossRef]
29. Lai, Q.; Aguey-Zinsou, K.-F.; Lai, Q.; Aguey-Zinsou, K.-F. Borohydrides as Solid-State Hydrogen Storage Materials: Past, Current Approaches and Future Perspectives. *Gen. Chem.* **2018**, *4*, 180017. [CrossRef]
30. Milanese, C.; Jensen, T.R.; Hauback, B.C.; Pistidda, C.; Dornheim, M.; Yang, H.; Lombardo, L.; Zuettel, A.; Filinchuk, Y.; Ngene, P.; et al. Complex Hydrides for Energy Storage. *Int. J. Hydrogen Energy* **2019**, *44*, 7860–7874. [CrossRef]
31. Paskevicius, M.; Jepsen, L.H.; Schouwink, P.; Černý, R.; Ravnsbæk, D.B.; Filinchuk, Y.; Dornheim, M.; Besenbacher, F.; Jensen, T.R. Metal Borohydrides and Derivatives-Synthesis, Structure and Properties. *Chem. Soc. Rev.* **2017**, *46*, 1565–1634. [CrossRef]
32. Callini, E.; Özlem Kocabas Atakli, Z.; Hauback, B.C.; Orimo, S.-I.; Jensen, C.; Dornheim, M.; Grant, D.; Cho, Y.W.; Chen, P.; Hjörvarsson, B.; et al. Complex and Liquid Hydrides for Energy Storage. *Appl. Phys. A* **2016**, *122*, 353. [CrossRef]
33. Javadian, P.; Payandeh, S.G.; Sheppard, D.A.; Buckley, C.E.; Jensen, T.R. Reversibility of LiBH$_4$ Facilitated by the LiBH$_4$–Ca(BH$_4$)$_2$ Eutectic. *J. Phys. Chem. C* **2017**, *121*, 18439–18449. [CrossRef]
34. Černý, R.; Murgia, F.; Brighi, M. Metal Hydroborates: From Hydrogen Stores to Solid Electrolytes. *J. Alloys Compd.* **2022**, *895*, 162659. [CrossRef]
35. Paskevicius, M.; Ley, M.B.; Sheppard, D.A.; Jensen, T.R.; Buckley, C.E. Eutectic Melting in Metal Borohydrides. *Phys. Chem. Chem. Phys.* **2013**, *15*, 19774–19789. [CrossRef] [PubMed]
36. Nakamori, Y.; Orimo, S.I. Destabilization of Li-Based Complex Hydrides. *J. Alloys Compd.* **2004**, *370*, 271–275. [CrossRef]
37. Nickels, E.A.; Jones, M.O.; David, W.I.F.; Johnson, S.R.; Lowton, R.L.; Sommariva, M.; Edwards, P.P. Tuning the Decomposition Temperature in Complex Hydrides: Synthesis of a Mixed Alkali Metal Borohydride. *Angew. Chemie Int. Ed.* **2008**, *47*, 2817–2819. [CrossRef] [PubMed]
38. Shao, J.; Xiao, X.; Fan, X.; Huang, X.; Zhai, B.; Li, S.; Ge, H.; Wang, Q.; Chen, L. Enhanced Hydrogen Storage Capacity and Reversibility of LiBH$_4$ Nanoconfined in the Densified Zeolite-Templated Carbon with High Mechanical Stability. *Nano Energy* **2015**, *15*, 244–255. [CrossRef]
39. Zhang, L.; Zheng, J.; Xiao, X.; Fan, X.; Huang, X.; Yang, X.; Chen, L. Enhanced Hydrogen Storage Properties of a Dual-Cation (Li$^+$, Mg^{2+}) Borohydride and Its Dehydrogenation Mechanism. *RSC Adv.* **2017**, *7*, 36852–36859. [CrossRef]
40. Zhang, L.; Zheng, J.; Chen, L.; Xiao, X.; Qin, T.; Jiang, Y.; Li, S.; Ge, H.; Wang, Q. Remarkable Enhancement in Dehydrogenation Properties of Mg(BH$_4$)$_2$ Modified by the Synergetic Effect of Fluorographite and LiBH$_4$. *Int. J. Hydrogen Energy* **2015**, *40*, 14163–14172. [CrossRef]
41. El Kharbachi, A.; Pinatel, E.; Nuta, I.; Baricco, M. A Thermodynamic Assessment of LiBH$_4$. *Calphad Comput. Coupling Phase Diagrams Thermochem.* **2012**, *39*, 80–90. [CrossRef]
42. Urgnani, J.; Torres, F.J.; Palumbo, M.; Baricco, M. Hydrogen Release from Solid State NaBH$_4$. *Int. J. Hydrogen Energy* **2008**, *33*, 3111–3115. [CrossRef]
43. Dematteis, E.M.; Roedern, E.; Pinatel, E.R.; Corno, M.; Jensen, T.R.; Baricco, M. A Thermodynamic Investigation of the LiBH$_4$-NaBH$_4$ System. *RSC Adv.* **2016**, *6*, 60101–60108. [CrossRef]
44. Plerdsranoy, P.; Kaewsuwan, D. Effects of Specific Surface Area and Pore Volume of Activated Carbon Nanofibers on Nanoconfinement and Dehydrogenation of LiBH$_4$. *Int. J. Hydrogen Energy* **2017**, *42*, 6189–6201. [CrossRef]
45. Sofianos, M.V.; Chaudhary, A.; Paskevicius, M.; Sheppard, D.A.; Humphries, T.D.; Dornheim, M.; Buckley, C.E. Hydrogen Storage Properties of Eutectic Metal Borohydrides Melt- Infiltrated into Porous Al Scaffolds. *J. Alloys Compd.* **2019**, *775*, 474–480. [CrossRef]
46. Suryanarayana, C. Mechanical Alloying and Milling. *Prog. Mater. Sci.* **2001**, *46*, 1–184. [CrossRef]
47. Le Caër, G.; Delcroix, P.; Bégin-Colin, S.; Ziller, T. High-Energy Ball-Milling of Alloys and Compounds. *Hyperfine Interact.* **2002**, *141–142*, 63–72. [CrossRef]
48. Wu, C.; Cheng, H.M. Effects of Carbon on Hydrogen Storage Performances of Hydrides. *J. Mater. Chem.* **2010**, *20*, 5390–5400. [CrossRef]
49. Javadian, P.; Sheppard, D.A.; Buckley, C.E.; Jensen, T.R. Hydrogen Storage Properties of Nanoconfined LiBH$_4$ -NaBH$_4$. *Int. J. Hydrogen Energy* **2015**, *40*, 14916–14924. [CrossRef]
50. Adams, R.M. Preparation of Diborane. *Adv. Chem.* **1961**, *32*, 60–68. [CrossRef]
51. Jun, S.; Joo, S.H.; Ryoo, R.; Kruk, M. Synthesis of New, Nanoporous Carbon with Hexagonally Ordered Mesostructure. *J. Am. Chem. Soc.* **2000**, *122*, 10712–10713. [CrossRef]
52. Thommes, M.; Kaneko, K.; Neimark, A.V.; Olivier, J.P.; Rodriguez-Reinoso, F.; Rouquerol, J.; Sing, K.S.W. Physisorption of Gases, with Special Reference to the Evaluation of Surface Area and Pore Size Distribution (IUPAC Technical Report). *Pure Appl. Chem.* **2015**, *87*, 1051–1069. [CrossRef]

53. Liu, Y.; Reed, D.; Paterakis, C.; Contreras, L.; Baricco, M.; Book, D. Study of the Decomposition of a 0.62LiBH$_4$–0.38NaBH$_4$ Mixture. *Int. J. Hydrogen Energy* **2017**, *42*, 22480–22488. [CrossRef]
54. Orimo, S.; Nakamori, Y.; Eliseo, J.R.; Züttel, A.; Jensen, C.M. Complex Hydrides for Hydrogen Storage. *Chem. Rev.* **2007**, *107*, 4111–4132. [CrossRef] [PubMed]
55. Zuttel, A.; Wenger, P.; Rentsch, S.; Sudan, P.; Mauron, P.; Emmenegger, C. LiBH$_4$ a New Hydrogen Storage Material. *J. Power Sources* **2003**, *118*, 1–7. [CrossRef]
56. Zuttel, A.; Rentsch, S.; Fischer, P.; Wenger, P.; Sudan, P.; Mauron, P.; Emmenegger, C. Hydrogen Storage Properties of LiBH$_4$. *J. Alloys Compd.* **2003**, *356-357*, 515–520. [CrossRef]
57. Humphries, T.D.; Kalantzopoulos, G.N.; Llamas-Jansa, I.; Olsen, J.E.; Hauback, B.C. Reversible Hydrogenation Studies of NaBH$_4$ Milled with Ni-Containing Additives. *J. Phys. Chem. C* **2013**, *117*, 6060–6065. [CrossRef]
58. Mao, J.; Gregory, D.H. Recent Advances in the Use of Sodium Borohydride as a Solid State Hydrogen Store. *Energies* **2015**, *8*, 430–453. [CrossRef]
59. Orłowski, P.A.; Grochala, W. Effect of Vanadium Catalysts on Hydrogen Evolution from NaBH$_4$. *Solids* **2022**, *3*, 21. [CrossRef]
60. Mao, J.; Guo, Z.; Nevirkovets, I.P.; Liu, H.K.; Dou, S.X. Hydrogen De-/absorption Improvement of NaBH$_4$ Catalyzed by Titanium-Based Additives. *J. Phys. Chem. C* **2012**, *116*, 1596–1604. [CrossRef]
61. Orimo, S.I.; Nakamori, Y.; Ohba, N.; Miwa, K.; Aoki, M.; Towata, S.I.; Züttel, A. Experimental Studies on Intermediate Compound of LiBH$_4$. *Appl. Phys. Lett.* **2006**, *89*, 87–90. [CrossRef]
62. Reed, D.; Book, D. In-Situ Raman Study of the Thermal Decomposition of LiBH$_4$. *Mater. Res. Soc. Symp. Proc.* **2010**, *1216*, 2–8. [CrossRef]
63. Jensen, S.R.H.; Paskevicius, M.; Hansen, B.R.S.; Jakobsen, A.S.; Møller, K.T.; White, J.L.; Allendorf, M.D.; Stavila, V.; Skibsted, J.; Jensen, T.R. Hydrogenation Properties of Lithium and Sodium Hydride - Closo-Borate, [B$_{10}$H$_{10}$]$^{2-}$ and [B$_{12}$H$_{12}$]$^{2-}$, Composites. *Phys. Chem. Chem. Phys.* **2018**, *20*, 16266–16275. [CrossRef]
64. Ampoumogli, A.; Steriotis, T.; Trikalitis, P.; Giasafaki, D.; Bardaji, E.G.; Fichtner, M.; Charalambopoulou, G. Nanostructured Composites of Mesoporous Carbons and Boranates as Hydrogen Storage Materials. *J. Alloys Compd.* **2011**, *509*, S705–S708. [CrossRef]
65. Nielsen, T.K.; Javadian, P.; Polanski, M.; Besenbacher, F.; Bystrzycki, J.; Jensen, T.R. Nanoconfined NaAlH$_4$: Determination of Distinct Prolific Effects from Pore Size, Crystallite Size, and Surface Interactions. *J. Phys. Chem. C* **2012**, *116*, 21046–21051. [CrossRef]
66. Peru, F.; Payandeh, S.; Charalambopoulou, G.; Jensen, T.R.; Steriotis, T. Hydrogen Sorption and Reversibility of the LiBH$_4$—KBH$_4$ Eutectic System Confined in a CMK-3 Type Carbon via Melt Infiltration. *J. Carbon Res.* **2020**, *2*, 19. [CrossRef]
67. Ampoumogli, A.; Charalambopoulou, G.; Javadian, P.; Richter, B.; Jensen, T.R.; Steriotis, T. Hydrogen Desorption and Cycling Properties of Composites Based on Mesoporous Carbons and a LiBH$_4$–Ca(BH$_4$)$_2$ Eutectic Mixture. *J. Alloys Compd.* **2015**, *645*, S480–S484. [CrossRef]

Disclaimer/Publisher's Note: The statements, opinions and data contained in all publications are solely those of the individual author(s) and contributor(s) and not of MDPI and/or the editor(s). MDPI and/or the editor(s) disclaim responsibility for any injury to people or property resulting from any ideas, methods, instructions or products referred to in the content.

Article

Hydrogen Release and Uptake of MgH$_2$ Modified by Ti$_3$CN MXene

Xiantun Huang [1,†], Chenglin Lu [2,†], Yun Li [3,*], Haimei Tang [2], Xingqing Duan [2], Kuikui Wang [4] and Haizhen Liu [2,5,*]

1. Department of Materials Science and Engineering, Baise University, Baise 533000, China
2. Guangxi Novel Battery Materials Research Center of Engineering Technology, Guangxi Colleges and Universities Key Laboratory of Blue Energy and Systems Integration, School of Physical Science and Technology, Guangxi University, Nanning 530004, China
3. School of Mechanical and Electrical Engineering, Quzhou College of Technology, Quzhou 324000, China
4. College of Materials Science and Engineering, Qingdao University, Qingdao 266071, China; kkwang@qdu.edu.cn
5. Key Laboratory of Advanced Energy Materials Chemistry (Ministry of Education), Nankai University, Tianjin 300071, China
* Correspondence: shanxiliyun2006@163.com (Y.L.); liuhz@gxu.edu.cn (H.L.)
† These authors contribute equally to this work.

Abstract: MgH$_2$ has a high hydrogen content of 7.6 wt% and possesses good reversibility under normal conditions. However, pristine MgH$_2$ requires a high temperature above 300 °C to release hydrogen, with very slow kinetics. In this work, we utilized Ti$_3$CN MXene to reduce the operating temperature and enhance the kinetics of MgH$_2$. The initial temperature of MgH$_2$ decomposition can be lowered from 322 °C for pristine MgH$_2$ to 214 °C through the employment of Ti$_3$CN. The desorbed MgH$_2$ + 7.5 wt% Ti$_3$CN can start absorption at room temperature, while the desorbed pristine MgH$_2$ can only start absorption at 120 °C. The employment of Ti$_3$CN can significantly improve the hydrogen release kinetics of MgH$_2$, with the desorption activation energy decreasing from 121 to 80 kJ mol^{-1}. Regarding thermodynamics, the desorption enthalpy changes of MgH$_2$ and MgH$_2$ + 7.5 wt% Ti$_3$CN were 79.3 and 78.8 kJ mol^{-1}, respectively. This indicates that the employment of Ti$_3$CN does not alter the thermal stability of MgH$_2$. Phase evolution studies through the use of X-ray diffraction and electron diffraction both confirm that Ti$_3$CN remains stable during the hydrogen release and uptake process of the composite. This work will help understand the impact of a transition metal carbonitride on the hydrogen storage of MgH$_2$.

Keywords: hydrogen storage materials; magnesium hydride; transition metal carbonitride; 2D materials; layered materials

1. Introduction

Hydrogen energy is acknowledged as an ideal strategy to solve energy shortages and environmental pollution issues. However, hydrogen under ambient conditions is a gas of low density (0.089 kg m^{-3}) [1]. In addition, it is flammable and combustible with a wide explosion limit of 4–75 vol%. Therefore, the safe and compact storage of hydrogen is an important issue when utilizing hydrogen energy on a large scale [2–4].

Solid-state hydrogen storage, with the hydrogen bonded in a hydrogen storage material, is a good method to store hydrogen since it has a very large capacity (>50 kg m^{-3}). In addition, the method is safe since it can be operated under low hydrogen pressure (generally <5 MPa). Construction of high-performance materials for hydrogen storage is the key issue in developing a solid-state hydrogen storage system [5–11].

MgH$_2$ has attracted extensive attention as a material for hydrogen storage due to its large capacity of 7.6 wt% and the ability to reversibly store hydrogen [12–14]. In addition, there is an abundant resource of Mg on Earth, which makes large-scale application possible. However, MgH$_2$ with high thermal stability requires a high temperature to desorb hydrogen. Moreover, the hydrogen sorption process is very slow for MgH$_2$ when the temperature is not high enough. These two drawbacks have severely limited the practical application of MgH$_2$. Constructing nanoscale Mg-based materials [12,15–20], alloying Mg with other metals [8,21–24], or introducing additives [25–38] are the commonly utilized strategies to modify the hydrogen sorption properties of MgH$_2$.

In the past decade, MXenes (transition metal carbides/nitrides with layered structures) have received much attention in catalysis, energy storage, and conversion. MXene has also been demonstrated to show the positive impact on MgH$_2$ [35,39–47]. In 2016, Liu et al. [47] first reported the enhancing impact of Ti$_3$C$_2$ MXene on MgH$_2$. It was shown that the employment of 7 wt% Ti$_3$C$_2$ can reduce the starting hydrogen desorption temperature of MgH$_2$ to 180 °C. Li et al. [44] used Ti$_2$C MXene to reduce the temperature of MgH$_2$ by 37 °C. It was suggested that the Ti elements with multivalences will enhance the electron transfer during hydrogen sorption. Lu et al. [31] showed that V$_2$C MXene can tailor both the kinetics and thermodynamics of MgH$_2$. Liu et al. [40] demonstrated that the hybrid of Ti$_3$C$_2$ and V$_2$C MXenes exhibits a synergistic impact on MgH$_2$. The starting temperature of the hydrogen release of MgH$_2$–Ti$_3$C$_2$/V$_2$C can be reduced by 140 °C. Bimetallic MXene which contains two transition metals also has a good enhancing impact on MgH$_2$. For example, Shen et al. [46] reported that MgH$_2$ + 10 wt% (Ti$_{0.5}$V$_{0.5}$)$_3$C$_2$ can start desorption at 196 °C. Wang et al. [42] displayed that NbTiC MXene reduces the starting hydrogen desorption temperature of MgH$_2$ to 195 °C. It has been supposed by many researchers that the unique layered structures and the active transition metals contained within both contribute to the enhanced hydrogen storage properties of MgH$_2$ [40,42–44,46,47].

Based on the above introduction, MXene materials have shown excellent enhancing influence on MgH$_2$. However, the studies mainly focus on carbides. The impact of transition metal nitrides or carbonitrides on MgH$_2$ is not clear currently. In this work, we first synthesized a layered transition metal carbonitride (Ti$_3$CN MXene) and then used it to modify the hydrogen sorption properties of MgH$_2$. The hydrogen release and uptake kinetics and thermodynamics of MgH$_2$ modified by Ti$_3$CN MXene will be investigated. Microstructures will be studied to reveal the role of Ti$_3$CN MXene in modifying MgH$_2$.

2. Results

Ti$_3$CN MXene was synthesized by the exfoliation of Ti$_3$AlCN MAX (hexagonal layered transition metal carbides and nitrides). A hydrofluoric acid solution was used to remove the Al layers from Ti$_3$AlCN to synthesize the layered Ti$_3$CN MXene. Figure 1a shows the XRD spectrum of Ti$_3$AlCN MAX and Ti$_3$CN MXene. The diffraction peak of the (002) crystalline plane shifting to a lower angle indicates the exfoliation of Ti$_3$AlCN MAX to form the layered Ti$_3$CN MXene. The SEM picture in Figure 1b indicates that Ti$_3$CN MXene has a layered structure. In Figure 1c, the elemental mappings show that the Ti, C, and N elements are all distributed uniformly in the material. Some traces of the Al element were also observed in the material. The above characterizations indicate the successful synthesis of the layered Ti$_3$CN MXene.

The Ti$_3$CN MXene was mixed with MgH$_2$ by ball milling to obtain MgH$_2$ + m wt% Ti$_3$CN (m = 0, 5, 7.5, 10) composites. Figure 2a shows the hydrogen release curves of the MgH$_2$ + m wt% Ti$_3$CN (m = 0, 5, 7.5, 10) composites when the temperature was increased from room temperature (RT) to about 400 °C at 2 °C min^{-1}. The as-milled MgH$_2$ without additive starts desorbing hydrogen at 322 °C and could offer a capacity of 7.0 wt% when the temperature reached 400 °C. Excitingly, the addition of Ti$_3$CN can significantly lower the starting temperature of MgH$_2$ desorption to 214 °C. This means a reduction of 108 °C in the starting temperature. The 7.5 wt% Ti$_3$CN-doped MgH$_2$ has a slightly lower hydrogen desorption temperature than the 5 wt% Ti$_3$CN-doped MgH$_2$. However, further increasing the Ti$_3$CN content to 10 wt% does not further reduce the temperature of MgH$_2$ but will slightly reduce the capacity of the composite. Considering achieving both low temperature and high capacity, the MgH$_2$ with the addition of 7.5 wt% of Ti$_3$CN was selected for further absorption studies. Figure 2b shows the hydrogen absorption curves of the desorbed MgH$_2$ + 7.5 wt% Ti$_3$CN composite and the pristine MgH$_2$ at 4 MPa H$_2$. During absorption, the temperature was increased from RT to 400 °C at 2 °C min^{-1}. The desorbed MgH$_2$ starts to absorb hydrogen at about 120 °C and could absorb 7.4 wt% H$_2$ after the temperature was ramped to 400 °C. It is exciting that the desorbed MgH$_2$ + 7.5 wt% Ti$_3$CN sample can start to absorb hydrogen at RT and absorb 7.0 wt% H$_2$ at 400 °C. Therefore, Ti$_3$CN MXene can significantly improve the non-isothermal hydrogen desorption and absorption performance of MgH$_2$.

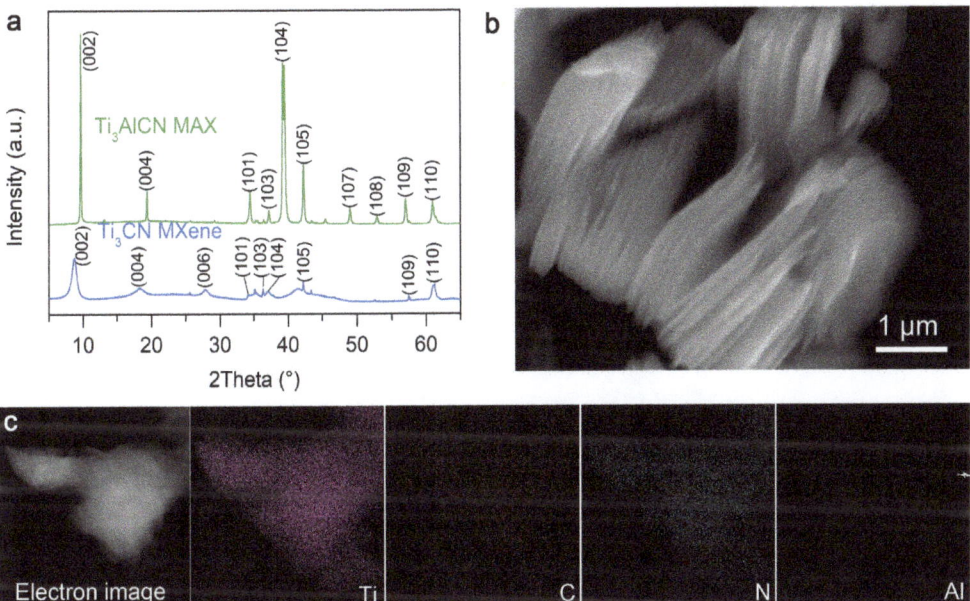

Figure 1. (a) XRD spectrum of Ti$_3$AlCN and Ti$_3$CN. (b) SEM image of Ti$_3$CN. (c) EDS elemental distributions of Ti$_3$CN.

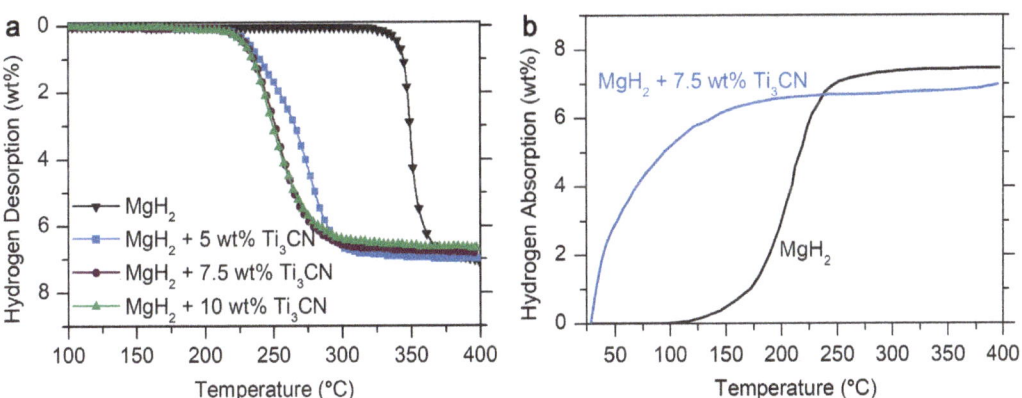

Figure 2. (a) Hydrogen release curves of MgH$_2$ + m wt% Ti$_3$CN (m = 0, 5, 7.5, 10) with the temperature rising from RT to 400 °C at 2 °C min^{-1}. (b) Hydrogen uptake curves of MgH$_2$ and MgH$_2$ + 7.5 wt% Ti$_3$CN at 6 MPa H$_2$ with the same temperature program as (a).

The hydrogen release kinetics of MgH$_2$ and MgH$_2$ + 7.5 wt% Ti$_3$CN were studied by testing the isothermal hydrogen desorption curves, as shown in Figure 3a,d, respectively. The MgH$_2$ without an additive can achieve fast kinetics only at a temperature higher than 350 °C. However, the MgH$_2$ + 7.5 wt% Ti$_3$CN composite has fast hydrogen desorption kinetics even at a lower temperature below 300 °C. At a constant temperature of 300 °C, MgH$_2$ + 7.5 wt% Ti$_3$CN can desorb 6.6 wt% H$_2$ within 10 min and 6.9 wt% within 60 min. Therefore, the hydrogen release kinetics were greatly improved by Ti$_3$CN addition. The curves in Figure 3a,d were further studied by the Johnson–Mehl–Avrami (JMA) equation and the Arrhenius equation. The JMA equation is:

$$\ln[-\ln(1-\alpha)] = n\ln k + n\ln t, \tag{1}$$

where α refers to the extent of the reaction; n represents the Avrami index; t is the time; k stands for the reaction rate constant. The isothermal hydrogen desorption curves were converted to JMA plots ($\ln[-\ln(1-\alpha)]$ vs. $\ln t$) as shown in Figure 3b,e. Then, linear fitting was performed to obtain the n and $n\ln k$ from the slopes and the intercepts. The $\ln k$ values were then plotted vs. $1000/T$ based on the Arrhenius equation, which is:

$$\ln k = -E_a/RT + \ln A, \tag{2}$$

where E_a refers to the activation energy; R represents the universal gas constant; and A stands for a constant. The Arrhenius plots ($\ln k$ vs. $1000/T$) are shown in Figure 3c,f. Then, linear fitting was performed to obtain the values of E_a from the slope. The desorption activation energy for MgH$_2$ + 7.5 wt% Ti$_3$CN was estimated to be 80 kJ mol^{-1}, which is much lower compared to MgH$_2$ without an additive (121 kJ mol^{-1}). This indicates that Ti$_3$CN improved the hydrogen release kinetics of MgH$_2$.

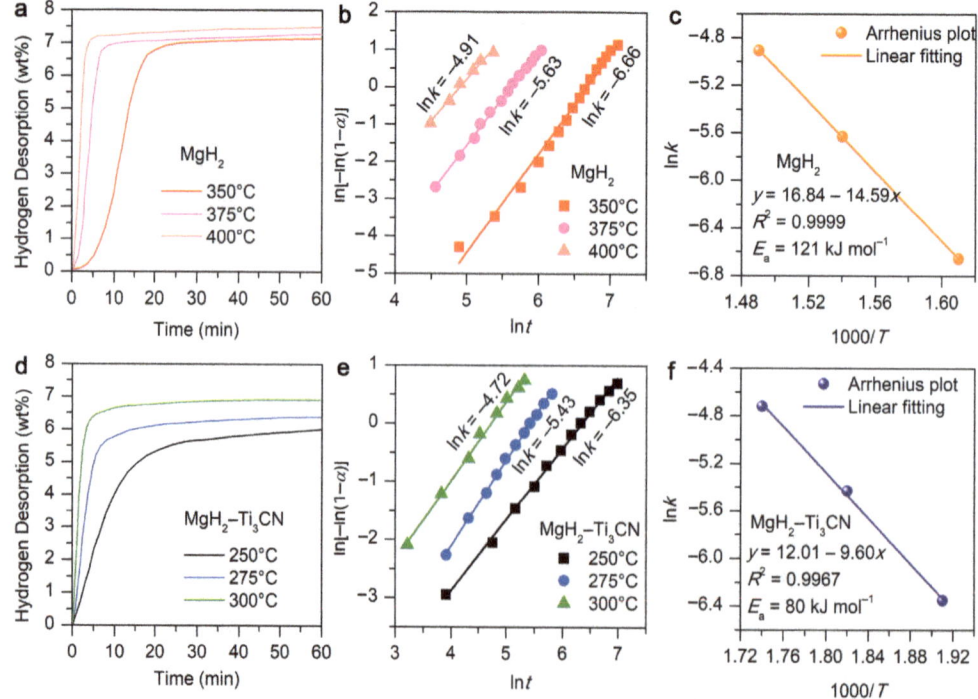

Figure 3. Hydrogen release curves at various temperatures (**a,d**), JMA plots (**b,e**), and Arrhenius plots (**c,f**) of MgH$_2$ without addition (upper) and MgH$_2$ + 7.5 wt% Ti$_3$CN (down).

The thermodynamics of MgH$_2$ were further studied by testing the pressure–concentration isotherms (PCT) and using the van't Hoff equation written as:

$$\ln(p/p_0) = -\Delta H/RT + \Delta S/R, \quad (3)$$

where p refers to the plateau hydrogen pressure; p_0 stands for the standard atmosphere pressure; ΔH represents the enthalpy change of the reaction; and ΔS represents the entropy changes of the reaction. Figure 4a,c shows the hydrogen desorption PCT curves of the two samples at various temperatures. From the PCT curves, the plateau hydrogen pressures (p) can be obtained. Then, the van't Hoff plots ($\ln(p/p_0)$ vs. $1000/RT$) can be made (Figure 4b,d). The slopes of the linear fitting lines give the values of ΔH. The enthalpy change for the hydrogen release reaction of MgH$_2$ + 7.5 wt% Ti$_3$CN was estimated to be 78.8 kJ mol^{-1}, which is very equal to that of MgH$_2$ without an additive (79.3 kJ mol^{-1}). Therefore, Ti$_3$CN addition does not alter the thermodynamics of MgH$_2$.

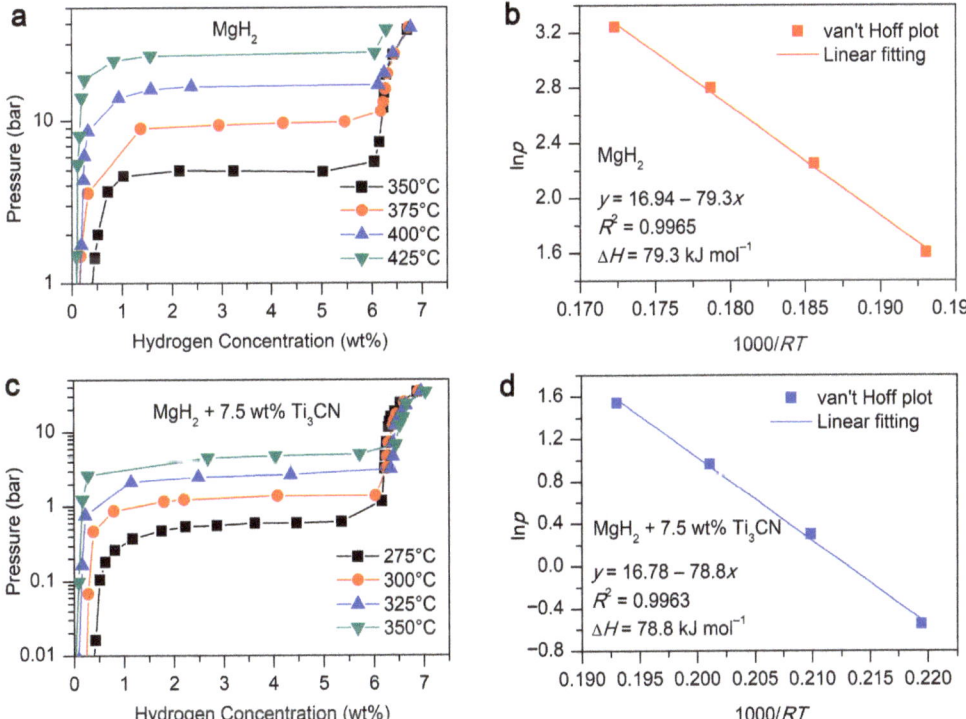

Figure 4. Hydrogen desorption PCT curves (a,c) and van't Hoff plots (b,d) of MgH$_2$ without addition (upper) and MgH$_2$ + 7.5 wt% Ti$_3$CN (down).

To reveal the role of Ti$_3$CN MXene in tailoring the hydrogen storage of MgH$_2$, the structures of MgH$_2$ + 7.5 wt% Ti$_3$CN in different states were studied by X-ray diffraction (XRD). Figure 5b–d shows the XRD profiles of MgH$_2$ + 7.5 wt% Ti$_3$CN at different stages, with the as-synthesized Ti$_3$CN MXene for reference (Figure 5a). After ball milling (Figure 5b), MgH$_2$ and Ti$_3$CN were observed in the sample, suggesting that it is a physical mixture of the starting materials. After hydrogen desorption (Figure 5c), MgH$_2$ decomposes and Mg forms. Ti$_3$CN is still observed in the desorbed sample, which indicates that Ti$_3$CN does not react with other components and stays stable in the sample. It should be noted that MgO is observed in the sample, which may be due to the partial oxidation of MgH$_2$/Mg during sample transfer or testing. After hydrogen absorption (Figure 5d), MgH$_2$ is fully recovered and Ti$_3$CN is still observed in the sample. From the above structure evolution studies, it can be seen that Ti$_3$CN stays stable during the hydrogen release and uptake process. Therefore, Ti$_3$CN mainly plays the role of an efficient catalyst for the hydrogen release and uptake of MgH$_2$. This is consistent with the results in Figure 4 in that the thermodynamics of MgH$_2$ is not altered by the addition of Ti$_3$CN.

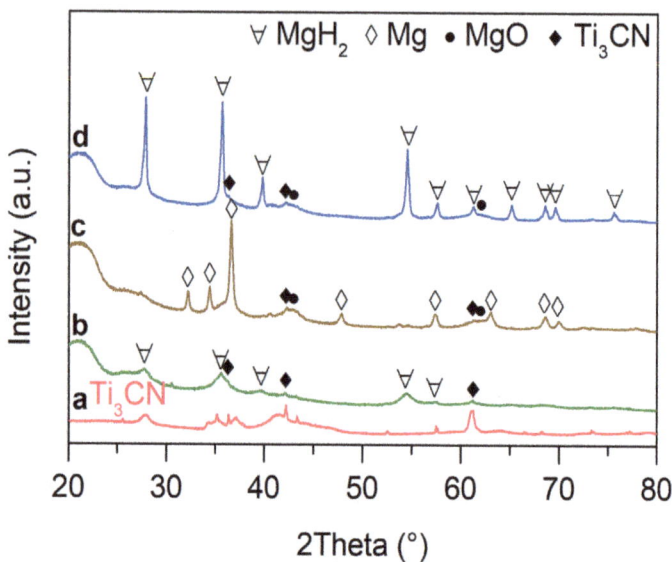

Figure 5. XRD profiles of Ti$_3$CN MXene (**a**) and MgH$_2$ + 7.5 wt% Ti$_3$CN after ball milling (**b**), after hydrogen desorption (**c**), and after hydrogen absorption (**d**).

The microstructures of the MgH$_2$ + 7.5 wt% Ti$_3$CN composite after rehydrogenation were further studied by SEM, TEM, EDS, and SAED methods. Figure 6a shows the SEM image of the composite, which displays that the particles of the composite are of several microns. Figure 6b shows the EDS elemental mappings of the composite. The Mg, Ti, C, and N elements are all distributed very uniformly in the composite. Figure 6c shows the TEM image of the composite with its SAED pattern shown in Figure 6d. In the SAED pattern, MgH$_2$, Mg, and Ti$_3$CN are observed. These three components are also observed in the HRTEM images in Figure 6e−i. The presence of Ti$_3$CN is consistent with the XRD results in Figure 5d, which again suggests that Ti$_3$CN mainly plays the role of an efficient catalyst for MgH$_2$. It is interesting that Mg is detected in the rehydrogenated composite, which is different from Figure 5d. In Figure 5d, Mg is not observed in the XRD pattern. This indicates that the high-energy electron beam may have stimulated the partial decomposition of the Ti$_3$CN-modified MgH$_2$. It should be also noted that only those MgH$_2$ particles that are contacting with Ti$_3$CN can be stimulated to decompose by the high-energy electron beam, as shown in regions 1 and 3 of Figure 6e. In region 4 of Figure 6e, MgH$_2$ without contacting with Ti$_3$CN is not decomposed. Therefore, Ti$_3$CN indeed is an excellent catalyst for MgH$_2$.

Figure 6. SEM image (**a**), EDS elemental mappings (**b**), TEM image (**c**), SAED pattern (**d**), and HRTEM image (**e–i**) of the MgH$_2$ + 7.5 wt% Ti$_3$CN composite after rehydrogenation.

3. Discussion

From the above results, it can be said that Ti$_3$CN MXene can greatly enhance the hydrogen sorption kinetics of MgH$_2$. The addition of Ti$_3$CN can lower the initial hydrogen release temperature of MgH$_2$ from 322 °C to 214 °C, with a reduction of 108 °C. Moreover, the desorbed MgH$_2$ starts to absorb hydrogen at about 120 °C, while the desorbed MgH$_2$ + 7.5 wt% Ti$_3$CN sample can start to absorb hydrogen at RT. The MgH$_2$ + 7.5 wt% Ti$_3$CN has a desorption activation energy of 80 kJ mol^{-1}, which is significantly lower than that of pristine MgH$_2$ (121 kJ mol^{-1}).

However, it seems that Ti$_3$CN does not alter the thermodynamics of MgH$_2$. Many published papers have demonstrated that MXene materials such as Ti$_3$C$_2$ [41,45,47], Ti$_2$C [44], NbTiC [42], (Ti$_{0.5}$V$_{0.5}$)$_3$C$_2$ [46], etc., can enhance the hydrogen sorption kinetics of MgH$_2$. However, there is barely any work that has reported that MXene materials can reduce the thermal stability of MgH$_2$ except for V$_2$C MXene [31]. Therefore, it can be deduced that most MXene materials do not change the thermodynamics of MgH$_2$ but mainly alter the kinetics of MgH$_2$.

4. Materials and Methods

Ti$_3$AlCN MAX (500 mesh, 98% purity) was purchased from Laizhou Kaixi Ceramic Co., Ltd., Laizhou, China. MgH$_2$ (98% purity) was purchased from Langfang Beide Commerce and Trade Co., Ltd., Langfang, China. HF (analytical purity, 40%) was purchased from Aladdin, Shanghai, China. These reagents were used as received without any further treatment.

HF-etching was used to synthesize the layered Ti$_3$CN MXene. In the experiment, 3 g of Ti$_3$AlCN MAX was added into a 40 mL HF solution with a concentration of 40%. The solution was then stirred at 30 °C for 18 h followed by centrifugation three times. The rotation speed used for centrifugation was 3500 rpm. After that, the sediment was washed until the pH value of the deionized water used was higher than 6. Then, the sediment was dried in a freeze-dryer for 24 h. After that, Ti$_3$CN MXene can finally be obtained.

Ti$_3$CN was then mixed with MgH$_2$ by ball milling under an argon atmosphere to prepare MgH$_2$ + m wt% Ti$_3$CN (m = 0, 5, 7.5, 10) samples at a planetary ball mill (Pulverisette 7, Fritsch, Germany). The as-received MgH$_2$ and the as-synthesized Ti$_3$CN were first weighted based on the compositions in a glove box filled with high-purity argon and then placed in a milling jar. Some milling balls were also placed in the milling jar with a ball-to-powder ratio of 40:1. After sealing, the milling jar was transferred to the planetary ball mill. All samples were milled at 400 rpm for 10 h.

An X-ray diffraction (XRD) instrument (Miniflex 600, Rigaku, Japan) was utilized to determine the phase structures. The incident ray was Cu Kα radiation and the scanning speed was 2 °C min^{-1}. A working current of 200 mA and a working voltage of 40 kV were used during the tests. The samples for the XRD test were sealed with transparent tape to prevent the samples from oxidizing during the sample transfer and test. Scanning electron microscopy (SEM, JSM-6510A, JEOL, Japan) was employed to analyze the morphologies. The samples were adhered to conductive tape. The transfer of the samples was carried out carefully to protect the samples from contacting the air. An attached X-ray energy dispersive detector (EDS) was employed to collect the elemental distributions. A transition electronic microscope (TEM, Tecnai G2 F20, FEI, The Netherlands) with a voltage of 200 kV was used to study the microstructures of the samples. Anhydrous acetone was used to disperse the sample on Cu grids.

A Sievert-type apparatus built by the Institute of Metallic Materials, Zhejiang University, Hangzhou, China, was utilized to study the hydrogen release and uptake behavior of the samples. During the non-isothermal hydrogen release tests, the samples were heated gradually from RT to 400 °C at 2 °C min^{-1} from an initial pressure of 10^{-4} MPa. During the non-isothermal hydrogen uptake tests, the temperature program was the same as the isothermal hydrogen release test. At the starting point of the heating program, hydrogen of 6 MPa was charged into the sample holder. During the isothermal hydrogen release tests, the samples were first heated to the target temperature with a hydrogen back pressure of 6 MPa. When the temperature was stabilized, hydrogen gas was rapidly vented to start hydrogen desorption. An automatic Sievert-type apparatus (IMI-Flow, Hiden, UK) was used to collect the PCT curves of the samples.

5. Conclusions

Layered Ti$_3$CN MXene was successfully synthesized by exfoliation of Ti$_3$AlCN MAX with HF as the etching solution. The layered Ti$_3$CN can significantly improve the kinetics of MgH$_2$. In particular, MgH$_2$ + 7.5 wt% Ti$_3$CN shows good hydrogen desorption performance, with an initial hydrogen release temperature of 214 °C and a low hydrogen release reaction activation energy of 80 kJ mol^{-1}. Moreover, the desorbed MgH$_2$ + 7.5 wt% Ti$_3$CN can absorb hydrogen at RT, while the desorbed pristine MgH$_2$ can only start absorption at 120 °C. The layered Ti$_3$CN barely changes the thermodynamics of MgH$_2$ since the enthalpy changes of the hydrogen release reactions of MgH$_2$ and MgH$_2$ + 7.5 wt% Ti$_3$CN are very close (79.3 and 78.8 kJ mol^{-1}, respectively). Ti$_3$CN stays stable during the hydrogen release and uptake process of the MgH$_2$–Ti$_3$CN composite, which means that Ti$_3$CN mainly plays the role of an efficient catalyst for MgH$_2$. This work confirms that transition metal carbonitrides also have a good catalytic impact on the hydrogen release and uptake properties of MgH$_2$.

Author Contributions: Conceptualization, C.L. and X.H.; methodology, X.D.; validation, H.T.; formal analysis, C.L.; investigation, X.H.; writing—original draft preparation, H.L.; writing—review and editing, Y.L. and K.W.; supervision, H.L.; project administration, H.L.; funding acquisition, Y.L., K.W. and H.L. All authors have read and agreed to the published version of the manuscript.

Funding: This research was funded by the Science and Technology Department of Guangxi Zhuang Autonomous Region, grant number GuiKeAD21238022, the National Natural Science Foundation of China, grant number 52001079, the Quzhou Science and Technology Project, grant number 2022K103, and the open foundation of State Key Laboratory of Featured Metal Materials and Life-cycle Safety for Composite Structures, Guangxi University, grant number 2022GXYSOF16.

Data Availability Statement: The data presented in this study are available from the corresponding author upon request.

Conflicts of Interest: The authors declare no conflict of interest.

References

1. Yang, Z.-X.; Li, X.-G.; Yao, Q.-L.; Lu, Z.-H.; Zhang, N.; Xia, J.; Yang, K.; Wang, Y.-Q.; Zhang, K.; Liu, H.-Z.; et al. 2022 roadmap on hydrogen energy from production to utilizations. *Rare Met.* **2022**, *41*, 3251–3267. [CrossRef]
2. Schlapbach, L.; Zuttel, A. Hydrogen-storage materials for mobile applications. *Nature* **2001**, *414*, 353–358. [CrossRef] [PubMed]
3. Berstad, D.; Gardarsdottir, S.; Roussanaly, S.; Voldsund, M.; Ishimoto, Y.; Nekså, P. Liquid hydrogen as prospective energy carrier: A brief review and discussion of underlying assumptions applied in value chain analysis. *Renew. Sust. Energy Rev.* **2022**, *154*, 111772. [CrossRef]
4. Hassan, I.A.; Ramadan, H.S.; Saleh, M.A.; Hissel, D. Hydrogen storage technologies for stationary and mobile applications: Review, analysis and perspectives. *Renew. Sust. Energy Rev.* **2021**, *149*, 111311. [CrossRef]
5. Liu, L.; Ilyushechkin, A.; Liang, D.; Cousins, A.; Tian, W.; Chen, C.; Yin, J.; Schoeman, L. Metal Hydride Composite Structures for Improved Heat Transfer and Stability for Hydrogen Storage and Compression Applications. *Inorganics* **2023**, *11*, 181. [CrossRef]
6. Yao, J.; Wu, Z.; Wang, H.; Yang, F.; Ren, J.; Zhang, Z. Application-oriented hydrolysis reaction system of solid-state hydrogen storage materials for high energy density target: A review. *J. Energy Chem.* **2022**, *74*, 218–238. [CrossRef]
7. Simanullang, M.; Prost, L. Nanomaterials for on-board solid-state hydrogen storage applications. *Int. J. Hydrogen Energy* **2022**, *47*, 29808–29846. [CrossRef]
8. Lin, H.J.; Lu, Y.S.; Zhang, L.T.; Liu, H.Z.; Edalati, K.; Révész, Á. Recent advances in metastable alloys for hydrogen storage: A review. *Rare Met.* **2022**, *41*, 1797–1817. [CrossRef]
9. Zhao, L.; Xu, F.; Zhang, C.; Wang, Z.; Ju, H.; Gao, X.; Zhang, X.; Sun, L.; Liu, Z. Enhanced hydrogen storage of alanates: Recent progress and future perspectives. *Prog. Nat. Sci. Mater. Int.* **2021**, *31*, 165–179. [CrossRef]
10. Liu, H.; Zhang, L.; Ma, H.; Lu, C.; Luo, H.; Wang, X.; Huang, X.; Lan, Z.; Guo, J. Aluminum hydride for solid-state hydrogen storage: Structure, synthesis, thermodynamics, kinetics, and regeneration. *J. Energy Chem.* **2021**, *52*, 428–440. [CrossRef]
11. Jiang, W.; Wang, H.; Zhu, M. AlH$_3$ as a hydrogen storage material: Recent advances, prospects and challenges. *Rare Met.* **2021**, *40*, 3337–3356. [CrossRef]
12. Sui, Y.; Yuan, Z.; Zhou, D.; Zhai, T.; Li, X.; Feng, D.; Li, Y.; Zhang, Y. Recent progress of nanotechnology in enhancing hydrogen storage performance of magnesium-based materials: A review. *Int. J. Hydrogen Energy* **2022**, *47*, 30546–30566. [CrossRef]
13. Shang, Y.; Pistidda, C.; Gizer, G.; Klassen, T.; Dornheim, M. Mg-based materials for hydrogen storage. *J. Magnes. Alloy.* **2021**, *9*, 1837–1860. [CrossRef]
14. Grigorova, E.; Nihtianova, D.; Tsyntsarski, B.; Stoycheva, I. Investigation of Hydrogen Storage Characteristics of MgH$_2$ Based Materials with Addition of Ni and Activated Carbon. *Inorganics* **2020**, *8*, 12. [CrossRef]
15. Ren, L.; Zhu, W.; Zhang, Q.; Lu, C.; Sun, F.; Lin, X.; Zou, J. MgH$_2$ confinement in MOF-derived N-doped porous carbon nanofibers for enhanced hydrogen storage. *Chem. Eng. J.* **2022**, *434*, 134701. [CrossRef]
16. Zhang, X.; Liu, Y.; Ren, Z.; Zhang, X.; Hu, J.; Huang, Z.; Lu, Y.; Gao, M.; Pan, H. Realizing 6.7 wt% reversible storage of hydrogen at ambient temperature with non-confined ultrafine magnesium hydrides. *Energy Environ. Sci.* **2021**, *14*, 2302–2313. [CrossRef]
17. Yan, N.; Lu, X.; Lu, Z.; Yu, H.; Wu, F.; Zheng, J.; Wang, X.; Zhang, L. Enhanced hydrogen storage properties of Mg by the synergistic effect of grain refinement and NiTiO$_3$ nanoparticles. *J. Magnes. Alloy.* **2022**, *10*, 3542–3552. [CrossRef]
18. Si, T.-Z.; Zhang, X.-Y.; Feng, J.-J.; Ding, X.-L.; Li, Y.-T. Enhancing hydrogen sorption in MgH$_2$ by controlling particle size and contact of Ni catalysts. *Rare Met.* **2021**, *40*, 995–1002. [CrossRef]

19. Le, T.T.; Pistidda, C.; Nguyen, V.H.; Singh, P.; Raizada, P.; Klassen, T.; Dornheim, M. Nanoconfinement effects on hydrogen storage properties of MgH_2 and $LiBH_4$. *Int. J. Hydrogen Energy* **2021**, *46*, 23723–23736. [CrossRef]
20. Zhang, X.L.; Liu, Y.F.; Zhang, X.; Hu, J.J.; Gao, M.X.; Pan, H.G. Empowering hydrogen storage performance of MgH_2 by nanoengineering and nanocatalysis. *Mater. Today Nano* **2020**, *9*, 100064. [CrossRef]
21. Pang, X.; Ran, L.; Chen, Y.A.; Luo, Y.; Pan, F. Enhancing hydrogen storage performance via optimizing Y and Ni element in magnesium alloy. *J. Magnes. Alloy.* **2022**, *10*, 821–835. [CrossRef]
22. Liu, P.; Lian, J.; Chen, H.; Liu, B.; Zhou, S. In situ formation of Mg_2Ni on magnesium surface via hydrogen activation for improving hydrogen sorption performance. *ACS Appl. Energy Mater.* **2022**, *5*, 6043–6049. [CrossRef]
23. Ali, N.A.; Ismail, M. Advanced hydrogen storage of the Mg-Na-Al system: A review. *J. Magnes. Alloy.* **2021**, *9*, 1111–1122. [CrossRef]
24. Yong, H.; Guo, S.; Yuan, Z.; Qi, Y.; Zhao, D.; Zhang, Y. Phase transformation, thermodynamics and kinetics property of $Mg_{90}Ce_5RE_5$ (RE = La, Ce, Nd) hydrogen storage alloys. *J. Mater. Sci. Technol.* **2020**, *51*, 84–93. [CrossRef]
25. Tian, G.; Wu, F.; Zhang, H.; Wei, J.; Zhao, H.; Zhang, L. Boosting the hydrogen storage performance of MgH_2 by Vanadium based complex oxides. *J. Phys. Chem. Solids* **2023**, *174*, 111187. [CrossRef]
26. Lu, Z.Y.; He, J.H.; Song, M.C.; Zhang, Y.; Wu, F.Y.; Zheng, J.G.; Zhang, L.T.; Chen, L.X. Bullet-like vanadium-based MOFs as a highly active catalyst for promoting the hydrogen storage property in MgH_2. *Int. J. Min. Met. Mater.* **2023**, *30*, 44–53. [CrossRef]
27. Duan, X.-Q.; Li, G.-X.; Zhang, W.-H.; Luo, H.; Tang, H.-M.; Xu, L.; Sheng, P.; Wang, X.-H.; Huang, X.-T.; Huang, C.-K.; et al. Ti_3AlCN MAX for tailoring MgH_2 hydrogen storage material: From performance to mechanism. *Rare Met.* **2023**, in press.
28. Yuan, Z.; Li, S.; Wang, K.; Xu, N.; Sun, W.; Sun, L.; Cao, H.; Lin, H.; Zhu, Y.; Zhang, Y. In-situ formed Pt nano-clusters serving as destabilization-catalysis bi-functional additive for MgH_2. *Chem. Eng. J.* **2022**, *435*, 135050. [CrossRef]
29. Shao, Y.; Gao, H.; Tang, Q.; Liu, Y.; Liu, J.; Zhu, Y.; Zhang, J.; Li, L.; Hu, X.; Ba, Z. Ultra-fine TiO_2 nanoparticles supported on three-dimensionally ordered macroporous structure for improving the hydrogen storage performance of MgH_2. *Appl. Surf. Sci.* **2022**, *585*, 152561. [CrossRef]
30. Pukazhselvan, D.; Sandhya, K.S.; Ramasamy, D.; Shaula, A.; Bdikin, I.; Fagg, D.P. Active catalytic species generated in situ in zirconia incorporated hydrogen storage material magnesium hydride. *J. Magnes. Alloy.* **2022**, *10*, 786–796. [CrossRef]
31. Lu, C.; Liu, H.; Xu, L.; Luo, H.; He, S.; Duan, X.; Huang, X.; Wang, X.; Lan, Z.; Guo, J. Two-dimensional vanadium carbide for simultaneously tailoring the hydrogen sorption thermodynamics and kinetics of magnesium hydride. *J. Magnes. Alloy.* **2022**, *10*, 1051–1065. [CrossRef]
32. Lan, Z.; Fu, H.; Zhao, R.; Liu, H.; Zhou, W.; Ning, H.; Guo, J. Roles of in situ-formed NbN and Nb_2O_5 from N-doped Nb_2C MXene in regulating the re/hydrogenation and cycling performance of magnesium hydride. *Chem. Eng. J.* **2022**, *431*, 133985. [CrossRef]
33. Dan, L.; Wang, H.; Liu, J.; Ouyang, L.; Zhu, M. H_2 plasma reducing Ni nanoparticles for superior catalysis on hydrogen sorption of MgH_2. *ACS Appl. Energy Mater.* **2022**, *5*, 4976–4984. [CrossRef]
34. Dai, M.; Lei, G.T.; Zhang, Z.; Li, Z.; Cao, H.J.; Chen, P. Room temperature hydrogen absorption of V_2O_5 catalyzed MgH_2/Mg. *Acta Chim. Sin.* **2022**, *80*, 303–309. [CrossRef]
35. Bolarin, J.A.; Zou, R.; Li, Z.; Zhang, Z.; Cao, H. MXenes for magnesium-based hydrides: A review. *Appl. Mater. Today* **2022**, *29*, 101570. [CrossRef]
36. Lu, Z.-Y.; Yu, H.-J.; Lu, X.; Song, M.-C.; Wu, F.-Y.; Zheng, J.-G.; Yuan, Z.-F.; Zhang, L.-T. Two-dimensional vanadium nanosheets as a remarkably effective catalyst for hydrogen storage in MgH_2. *Rare Met.* **2021**, *40*, 3195–3204. [CrossRef]
37. Liu, X.-S.; Liu, H.-Z.; Qiu, N.; Zhang, Y.-B.; Zhao, G.-Y.; Xu, L.; Lan, Z.-Q.; Guo, J. Cycling hydrogen desorption properties and microstructures of MgH_2-AlH_3-NbF_5 hydrogen storage materials. *Rare Met.* **2021**, *40*, 1003–1007. [CrossRef]
38. Zhou, C.; Zhang, J.; Bowman, R.C.; Fang, Z.Z. Roles of Ti-Based Catalysts on Magnesium Hydride and Its Hydrogen Storage Properties. *Inorganics* **2021**, *9*, 36. [CrossRef]
39. Gao, H.; Shi, R.; Liu, Y.; Zhu, Y.; Zhang, J.; Hu, X.; Li, L. Enhanced hydrogen storage performance of magnesium hydride with incompletely etched $Ti_3C_2T_x$: The nonnegligible role of Al. *Appl. Surf. Sci.* **2022**, *600*, 154140. [CrossRef]
40. Liu, H.; Lu, C.; Wang, X.; Xu, L.; Huang, X.; Wang, X.; Ning, H.; Lan, Z.; Guo, J. Combinations of V_2C and Ti_3C_2 MXenes for boosting the hydrogen storage performances of MgH_2. *ACS Appl. Mater. Interfaces* **2021**, *13*, 13235–13247. [CrossRef]
41. Kong, Q.; Zhang, H.; Yuan, Z.; Liu, J.; Li, L.; Fan, Y.; Fan, G.; Liu, B. Hamamelis-like $K_2Ti_6O_{13}$ Synthesized by Alkali Treatment of Ti_3C_2 MXene: Catalysis for Hydrogen Storage in MgH_2. *ACS Sust. Chem. Eng.* **2020**, *8*, 4755–4763. [CrossRef]
42. Wang, Z.Y.; Zhang, X.L.; Ren, Z.H.; Liu, Y.; Hu, J.J.; Li, H.W.; Gao, M.X.; Pan, H.G.; Liu, Y.F. In situ formed ultrafine NbTi nanocrystals from a NbTiC solid-solution MXene for hydrogen storage in MgH_2. *J. Mater. Chem. A* **2019**, *7*, 14244–14252. [CrossRef]
43. Liu, Y.; Gao, H.; Zhu, Y.; Li, S.; Zhang, J.; Li, L. Excellent catalytic activity of a two-dimensional $Nb_4C_3T_x$ (MXene) on hydrogen storage of MgH_2. *Appl. Surf. Sci.* **2019**, *493*, 431–440. [CrossRef]
44. Li, J.X.; Wang, S.; Du, Y.L.; Liao, W.H. Catalytic effect of Ti_2C MXene on the dehydrogenation of MgH_2. *Int. J. Hydrogen Energy* **2019**, *44*, 6787–6794. [CrossRef]
45. Gao, H.; Liu, Y.; Zhu, Y.; Zhang, J.; Li, L. Catalytic Effect of Sandwich-Like Ti_3C_2/TiO_2(A)-C on Hydrogen Storage Performance of MgH_2. *Nanotechnology* **2019**, *31*, 115404. [CrossRef]

46. Shen, Z.; Wang, Z.; Zhang, M.; Gao, M.; Hu, J.; Du, F.; Liu, Y.; Pan, H. A novel solid-solution MXene ($Ti_{0.5}V_{0.5})_3C_2$ with high catalytic activity for hydrogen storage in MgH_2. *Materialia* **2018**, *1*, 114–120. [CrossRef]
47. Liu, Y.; Du, H.; Zhang, X.; Yang, Y.; Gao, M.; Pan, H. Superior catalytic activity derived from a two-dimensional Ti_3C_2 precursor towards the hydrogen storage reaction of magnesium hydride. *Chem. Commun.* **2016**, *52*, 705–708. [CrossRef]

Disclaimer/Publisher's Note: The statements, opinions and data contained in all publications are solely those of the individual author(s) and contributor(s) and not of MDPI and/or the editor(s). MDPI and/or the editor(s) disclaim responsibility for any injury to people or property resulting from any ideas, methods, instructions or products referred to in the content.

Article

Liquid Channels Built-In Solid Magnesium Hydrides for Boosting Hydrogen Sorption

Zhi-Kang Qin [1], Li-Qing He [2], Xiao-Li Ding [1,*], Ting-Zhi Si [1], Ping Cui [1,*], Hai-Wen Li [2] and Yong-Tao Li [1,3]

[1] School of Materials Science and Engineering & Low-Carbon New Materials Research Center, Anhui University of Technology, Maanshan 243002, China
[2] Hefei General Machinery Research Institute, Hefei 230031, China
[3] Key Laboratory of Green Fabrication and Surface Technology of Advanced Metal Materials of Ministry of Education, Anhui University of Technology, Maanshan 243002, China
* Correspondence: dingxiaoli@ahut.edu.cn (X.-L.D.); cokecp@ahut.edu.cn (P.C.)

Abstract: Realizing rapid and stable hydrogen sorption at low temperature is critical for magnesium-based hydrogen storage materials. Herein, liquid channels are built in magnesium hydride by introducing lithium borohydride ion conductors as an efficient route for improving its hydrogen sorption. For instance, the 5 wt% LiBH$_4$-doped MgH$_2$ can release about 7.1 wt.% H$_2$ within 40 min at 300 °C but pure MgH$_2$ only desorbs less than 0.7 wt.% H$_2$, and more importantly it delivers faster desorption kinetics with more than 10 times enhancement to pure MgH$_2$. The hydrogen absorption capacity of LiBH$_4$-doped MgH$_2$ can still be well kept at approximately 7.2 wt.% without obvious capacity degradation even after six absorption and desorption cycles. This approach is not only through building ion transfer channels as a hydrogen carrier for kinetic enhancement but also by inhibiting the agglomeration of MgH$_2$ particles to obtain stable cyclic performance, which brings further insights to promoting the hydrogen ab-/desorption of other metal hydrides.

Keywords: hydrogen storage materials; magnesium hydrides; borohydrides; liquid channels; kinetics

Citation: Qin, Z.-K.; He, L.-Q.; Ding, X.-L.; Si, T.-Z.; Cui, P.; Li, H.-W.; Li, Y.-T. Liquid Channels Built-In Solid Magnesium Hydrides for Boosting Hydrogen Sorption. *Inorganics* 2023, 11, 216. https://doi.org/10.3390/inorganics11050216

Academic Editor: Maurizio Peruzzini

Received: 20 March 2023
Revised: 25 April 2023
Accepted: 2 May 2023
Published: 17 May 2023

Copyright: © 2023 by the authors. Licensee MDPI, Basel, Switzerland. This article is an open access article distributed under the terms and conditions of the Creative Commons Attribution (CC BY) license (https:// creativecommons.org/licenses/by/ 4.0/).

1. Introduction

Green and renewable energy development is the key to reducing carbon dioxide emissions and fossil fuel overuse [1–5]. Of these, hydrogen energy has garnered the most interest because of its abundant sources, high combustion heat value, and pollution-free combustion products [6]. However, achieving safe and efficient hydrogen storage is a key challenge. Solid-state hydrogen storage has relatively high storage volume density and transport safety, which have become the focus of hydrogen storage research in recent years [7–11]. Metal hydrides are widely used as solid hydrogen storage materials because they can store large quantities of hydrogen under milder conditions in a reversible manner. MgH$_2$ is a candidate with sufficient reserves, a broad application, and high efficiency and safety. The Department of Energy considers its high reversible hydrogen storage capacity (7.6 wt.%) and volume hydrogen storage density (106 kg·m^{-3}) to be among the most promising of the solid-state materials for meeting the technical requirements for onboard hydrogen storage [12–15]. However, high thermodynamic stability, high oxidation reactivity, and slow hydrogen sorption kinetics have become the primary obstacles to practical application of hydrogen storage in vehicles. In the last two decades, numerous techniques for modifying the kinetic and thermodynamic properties of MgH$_2$ have been developed to circumvent these issues: (i) alloying Mg with single transition metals and other metallic elements such as Ni [16,17], V [18], and Ti [19]; and (ii) nanoscale adjustment by confinement into single-walled carbon nanotubes [20] or graphene nanosheets (GNS) [21]. Unfortunately, these solutions typically have several drawbacks, including (i) low hydrogen storage capacity due to the addition of metals without hydrogen affinity, and (ii) irrepressible nanostructure agglomeration and instability [22–25]. Consequently, further investigation of a novel strategy to improve the hydrogen sorption performance of Mg-based materials at lower temperatures with faster kinetics is necessary.

In recent years, complex hydrides have been introduced into magnesium-based systems, proving to be a promising strategy for enhancing the kinetics of hydrogen storage [26–30]. For example, Liu et al. [31] reported a favorable desorption capacity of 4.5 wt.% at a relatively low temperature of 250 °C for the $MgH_2 + Li_3AlH_6$ mixture. Li et al. [32] further discovered that the $LiNH_2-MgH_2$ system began desorbing hydrogen at 150 °C and exhibited improved reversibility, but this combined system released quantities of undesirable ammonia (NH_3) gas. In light of these findings, introducing borohydrides into the MgH_2 system will also improve the kinetic properties, albeit at the expense of inflexible thermodynamic properties and the formation of byproducts. Kato et al. [33] discovered that the altered hydrogen desorption in the $NaBH_4$ and MgH_2 systems could be attributed to the migration of metallic Mg into the surface of $NaBH_4$. The intrinsic mechanisms of these combined systems are still unknown, but these results suggest that the enhanced properties are primarily a result of the rapid migration of ionic hydrogen in MgH_2 [34].

In light of these findings, we present a novel method for improving the hydrogen sorption of MgH_2 by introducing lithium borohydrides, namely the construction of an ion transfer channel in MgH_2. Complex borohydrides such as $LiBH_4$ and $Li_2B_{12}H_{12}$ are known to be fast ion conductors composed of Li^+, $[BH_4]^-$, and $[B_{12}H_{12}]^{2-}$ [35], which can serve as intermediates for high ionic conduction and activity in the diffusion of H- from MgH_2. This new method involves the construction of ion transfer channels as hydrogen carriers for kinetic enhancement and inhibiting the aggregation of MgH_2 particles from achieving stable cyclic performance. The desorption process of the pseudo-eutectic $LiBH_4-MgH_2$ system involves liquid borohydride phases in particular. In addition, the science underlying the remarkable kinetic enhancements of MgH_2 brought about by the introduction of liquid borohydride channels was elucidated.

2. Results and Discussion

2.1. Hydrogen Storage Properties of LiBH$_4$-Doped MgH$_2$

We first introduce $LiBH_4$ as an ionic conductor to enhance the milling kinetic performance of MgH_2 and then compare it with pure and $Li_2B_{12}H_{12}$-doped MgH_2 systems. Figure 1 compares the hydrogen absorption and desorption kinetics of pure, $Li_2B_{12}H_{12}$-doped, and $LiBH_4$-doped MgH_2 at 300 °C, which are detected by their sixth hydrogen cycle. As depicted in Figure 1a, the desorption hydrogen kinetic properties of MgH_2 are sluggish, and less than 0.7 wt.% hydrogen is desorbed within 40 min. In contrast, the addition of complex borohydrides improves the desorption kinetics of MgH_2. The sixth dehydrogenation can be completed rapidly and releases ~7.1 wt.% hydrogen within 40 min, representing enhancement by ten-fold compared to pure MgH_2. Similar desorption enhancements were observed in the $L_2B_{12}H_{12}$-doped MgH_2 system. Figure 1b further indicates that $LiBH_4$-doped MgH_2 has superior hydrogen absorption kinetic properties. As compared to the capacity of ~1.6 wt.% for pure MgH_2 within 10 min, the absorption capacities of $L_2B_{12}H_{12}$-doped and $LiBH_4$-doped MgH_2 are significantly increased by three and four times, respectively.

In order to explore and compare the initial desorption behaviors and cyclic performances of the pure MgH_2 and $LiBH_4$-doped MgH_2, they were tested for six cycles at 300 °C and the curve of hydrogen ab-/desorption at a constant temperature was shown in Figure 2. It can be seen that the desorbed capacity of pure MgH_2 is low, with only ~1.5 wt.% H_2, and the absorbed capacity was ~1.8 wt.% H_2 after the sixth cycle. More excitingly, that of the $LiBH_4$-doped MgH_2 can directly increase by releasing ~7.1 wt.%, and the hydrogen absorption capacity can still be kept at ~7.2 wt.% even after six cycles. These results indicate that the kinetics and cyclic performance of MgH_2 can be significantly enhanced by introducing ionic borohydrides.

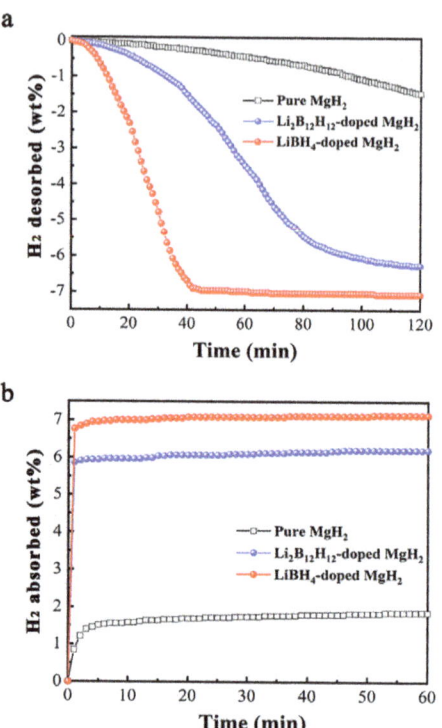

Figure 1. (**a**) Hydrogen desorption and (**b**) absorption kinetic curves of pure, $Li_2B_{12}H_{12}$–doped and $LiBH_4$–doped MgH_2 at 300 °C; all sample data are detected by the sixth cycle.

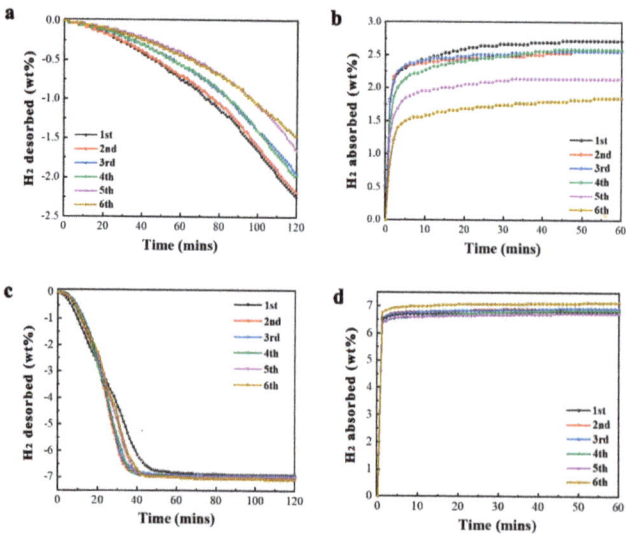

Figure 2. Kinetics and cycling performance of hydrogen desorption and absorption of (**a**,**b**) MgH_2 and (**c**,**d**) $LiBH_4$—doped MgH_2 upon six cycles at 300 °C.

2.2. Structural Features of LiBH$_4$-Doped MgH$_2$

The high-resolution transmission electron microscopy (HRTEM) technique is used to reveal the microstructure characteristics of the as-prepared LiBH$_4$-doped MgH$_2$ to clarify the reason for these improvements. As shown in Figure 3a, the light gray massive structures are embedded homogeneously in the dark gray zonal distribution, forming a transmission channel corresponding to the selected regions in Figure 3b,d. From the selected regions, we obtain lattice stripes by transposing the selection using the Fourier transform, in which all the d-spacings of approximately 0.3801, 0.322 and 0.252 nm can be easily indexed to the (011), (111), and (112) planes of the LiBH$_4$ phase, respectively. In addition, an amorphous layer is visible at the edge of the associated structures for LiBH$_4$-doped MgH$_2$ (yellow dotted line in Figure 3a); LiBH$_4$ and MgH$_2$ have a relatively well-balanced distribution within this structure, as depicted in Figure 3f–i. This novel structure strongly indicated that the LiBH$_4$-doped MgH$_2$ system had successfully constructed the zonal channel for hydrogen transfer. We also found the presence of the element O from the EDS spectrum in Figure 3g, indicating that the passivation layer on the surface of MgH$_2$ is unavoidable even for commercial MgH$_2$. Moreover, the existence of associated structures of LiBH$_4$-doped MgH$_2$ and the formation of an amorphous layer collectively inhibit the Mg grain particle agglomeration during kinetic cycling, thereby facilitating the diffusion of hydrogen for improved kinetic and cyclic properties.

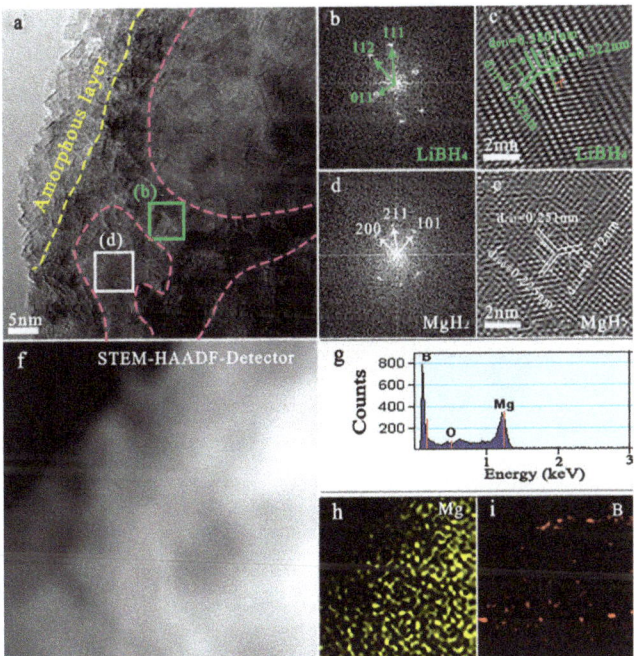

Figure 3. Microstructural features of as-milled LiBH$_4$-doped MgH$_2$: (**a**) TEM images; (**b,d**) fast Fourier transform (FFT) and (**c,e**) their lattice images; (**f**) STEM-HAADF image; (**g**) EDS spectrum and its corresponding (**h**) Mg and (**i**) B elemental mapping.

In order to determine the current state and distribution of the LiBH$_4$-doped MgH$_2$ system upon desorption or absorption, we analyzed the morphological and structural characteristics of the as-prepared LiBH$_4$-doped MgH$_2$ materials before and after six cycles. Figure 4a shows that, except for the MgH$_2$/Mg phases, no LiBH$_4$ peaks were detected in the XRD patterns. Intriguingly, Figure 4b of the FTIR results reveals that the characteristic bands of B-H vibration and stretching bonding of LiBH$_4$ at ~2359, 2293, 2225 and 1128 cm^{-1}

can always be detected before and after cycling, although their intensities diminish slightly. Apparently, LiBH$_4$ indeed exists in amorphous and/or nanocrystal states during both ball milling and cycling, rather than decomposing or reacting to form a new phase, as further demonstrated by the XPS analysis in Figure 4c, where the electronic binding energy of B^{1s} at approximately 188 eV demonstrates the stable existence of LiBH$_4$ in the cyclic LiBH$_4$-doped MgH$_2$. Moreover, Figure 4d–i shows the morphological evolution and elemental distribution of LiBH$_4$-doped MgH$_2$ during cycling. Before and after cycling, a particle morphology with an average size of 1~2 μm was observed, along with sintering-induced connection phenomena upon heating for sorption. Furthermore, Mg and B elemental distributions were well dispersed. These results indicate that the good dispersion of LiBH$_4$ significantly inhibits the growth of Mg grains, which explains why superior kinetic and cyclic properties were obtained in Figure 1.

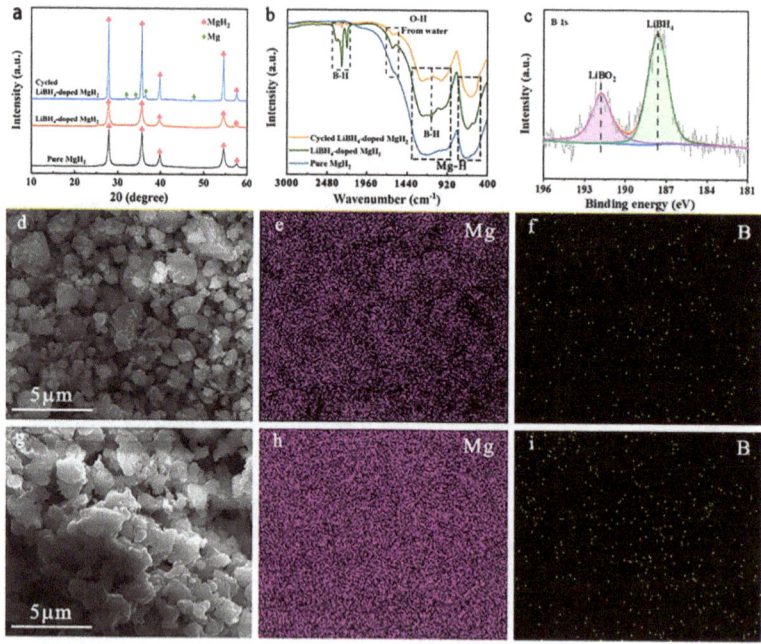

Figure 4. (a) XRD patterns, (b) FTIR and (c) XPS spectra of the pure MgH$_2$, as-milled and cyclic LiBH$_4$-doped MgH$_2$ samples, as well as the FESEM images of LiBH$_4$-doped MgH$_2$ sample (d) before and (g) after cycling and their corresponding (e,h) Mg and (f,i) B elemental mapping.

2.3. Electrochemical Analysis of LiBH$_4$-Doped MgH$_2$

Electrochemical impedance spectroscopy and in situ morphological analysis were utilized to elucidate the intrinsic role of LiBH$_4$ as an ionic conductor in enhancing hydrogen sorption on MgH$_2$. The semicircle Nyquist plots of LiBH$_4$-doped MgH$_2$ are significantly smaller than those of pure MgH$_2$ (see Figure 5a), indicating lower electrolyte resistance and exceptionally fast electron conductivity in a simulation of an all-solid-state battery. Figure 5b depicts the Nyquist plots of LiBH$_4$-doped MgH$_2$ at various temperatures, where the characteristic impedance semicircle gradually decreases with increasing temperatures, corresponding to the resistance value decreasing from approximately 4 × 10^6 Ω at 55 °C to approximately 1.05 × 10^5 Ω at 125 °C. Figure 5c compares the Arrhenius curves of pure MgH$_2$, LiBH$_4$, and LiBH$_4$-doped MgH$_2$. The enhanced ionic conductivity apparently indicates that MgH$_2$ shifts from an insulator to a conductor with ionic conductivity of approximately 3.2 × 10^{-7} S cm^{-1} at 125 °C by introducing LiBH$_4$. In addition, the in situ optical images of the LiBH$_4$-doped MgH$_2$ sample during heating shown in Figure 5d reveal that above 275 °C (i.e., the melting point of LiBH$_4$ [36]), liquid droplets form on the surface of the matrix. Intriguingly, hydrogen desorption from MgH$_2$ was captured by continuous and rapid bubbling from the liquid LiBH$_4$ phase (Figures S6 and S7, ESI).

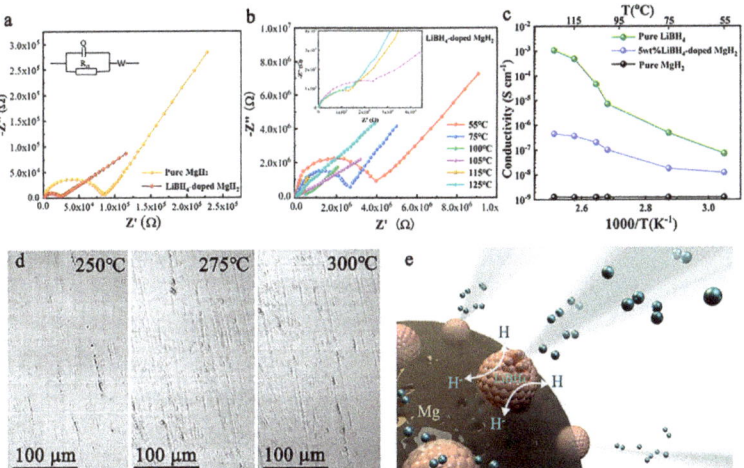

Figure 5. (a) Nyquist plots of pure and LiBH$_4$–doped MgH$_2$ as well as their (b) Nyquist plots at different temperatures and (c) corresponding Arrhenius plots; (d) In situ optical images of LiBH$_4$–doped MgH$_2$ sample detected by the high temperature laser confocal microscope and (e) its proposed model for formation of fast ions migration channels in MgH$_2$ induced by LiBH$_4$ melting.

Figure 5e proposes a model of hydrogen desorption enhanced by liquid ion channels incorporated into solid magnesium hydrides based on previous results. The liquid LiBH$_4$ droplets uniformly embedded in the MgH$_2$ matrix serve as channels for the rapid carrier of hydrogen, which can be understood from three perspectives: (i) the H$_2$ bubbles can move more rapidly in the liquid phase compared to the solid phase due to lower migration energy; (ii) as we know, the LiBH$_4$ would transfer from a low-temperature phase into high-temperature phase at ~120 °C, corresponding to the significant increase in ion conduction as shown in Figure 5a–c, and its enhanced ion conduction of composites would also be kept even at higher temperature. This further forces us to deduce that the enhanced desorption of MgH$_2$ is related to the faster ion conduction of LiBH$_4$; and (iii) the introduction of fast Li$^+$ ions from LiBH$_4$ can easily carry H$^-$ from ionic MgH$_2$ via the ionic interactions and can also alter the ionization degree and ionic density of the Mg-H bond to improve the migration of H atoms, as further confirmed by in situ spectroscopy. Along with enhancing the Li-ion migration, the H$_2$ sorption is also significantly promoted above phase-transition temperature. Thus, we deduce that introducing LiBH$_4$ into MgH$_2$ matrix would enhance the migration of H$^-$ by interacting with fast Li$^+$ transfer based on positive and negative ionic attraction to each other. In this regard, more experiments with advanced technology are needed to directly detect the important correlation that will provide insight into property improvement.

3. Experimental Procedure

3.1. Material Preparation

MgH$_2$ (purity > 95% from anabai Medicine Co., Ltd., Wuhan, China) and LiBH$_4$ (purity > 95% from China Aladdin) were combined in a weight ratio of 19:1 and loaded into a 300 mL stainless-steel ball milling tank (SUS304). The milling tank filled with 1 g of mixed material, and different masses and diameters of stainless steel (DECO-304-B) balls (5 mm—18.6 g; 6 mm—9.2 g; 8 mm—8 g; 10 mm—4.2 g), with a ball-to-sample mass ratio of 40:1 was mechanically milled for 10 h at 400 rpm with a planetary ball mill (QM-3SP2). All prepared samples were milled for twenty periods of 30 min, with a 2 min interval between each period. The same experimental procedures and parameters were used on the control group of pure MgH_2 and that with 5 wt% $Li_2B_{12}H_{12}$ doping (purity > 95% from Aladdin). All mechano-chemical treatments were performed in an Argon-filled glovebox ($\rho(O_2)$ < 0.1 ppm, ρ (H_2O) < 0.1 ppm).

3.2. Material Preparation

The XRD analysis was performed on a MiniFlex 600 XRD unit (Rigaku, Japan), Cu Kα radiation (λ = 0.154056 nm) utilized at 40 kV and 15 mA. The 2θ angle ranged from 10° to 90° with increments of 0.02°. The powder samples were placed in custom-made molds and sealed with polyimide thin-film tape, ensuring that they were under an argon atmosphere during the measurement process. The morphologies of the samples were observed by scanning electron microscopy (SEM, Zeiss Sigma 300) and transmission electron microscopy coupled with an EDS (TEM, FEI Talos F200X). The FTIR analyses were conducted on a TENSOR27 with a wavelength range from 400 to 3000 cm^{-1}. A Thermo Fisher Scientific Spectrometer K-Alpha was used to conduct XPS analyses. The powder sample was contained in an argon-filled glove box before being mounted on a sample holder and transferred to the XPS facility using a special container to prevent air exposure.

Electrochemical impedance spectroscopy (EIS) was used to measure the ionic conductivities of pure MgH_2 and 5 wt.% $LiBH_4$-doped MgH_2 by using a Solartron impedance analyzer. The material was weighed as 120 mg by electronic balance in a high-purity argon glove box, and then poured into a metal sheet mold with a press at 7 MPa for 5 min. The pressing tablet was the positive material, while the negative was a lithium tablet that was wrapped in a plastic bag and then rolled with a metal rod into a uniform sheet. Before electrochemical testing, all experimental and control groups were mixed with ~12 mg of acetylene black to increase their electrical conductivity.

The de-/absorption kinetics of samples were measured by using an automated Sieverts-type apparatus that allowed for accurate determination of the evolved hydrogen amount. Approximately 2 g of sample was loaded into an evacuated stainless-steel autoclave that connected with automatic PCT measurements. After rapid heating of the sample to the desired temperatures, the autoclave was immersed into the heating furnace. After activation, the hydrogen de-/absorption curves at 300 °C were successively performed by a back pressure of 0.01 and 4 MPa, respectively.

4. Conclusions

In conclusion, we successfully incorporated 5 wt.% $LiBH_4$ into the MgH_2 hydrogen storage system to significantly improve the hydrogen kinetic and stable cyclic properties. Our experimental findings support the design of a hydrogen-optimized ion transfer channel, which is facilitated by forming associated structures. For superior cyclic performance, an amorphous layer and the uniform dispersion of $LiBH_4$ are the two most important factors inhibiting the growth of Mg grains. During the heating process, the $LiBH_4$ droplets uniformly embedded in the MgH_2 matrix serve as ion migration channels for rapid transport of hydrogen. These findings suggest a new strategy for enhancing the hydrogen sorption of ionic hydrides and other hydride systems, as well as for accelerating the search for candidate materials that are suitable for hydrogen storage.

Supplementary Materials: The following supporting information can be downloaded at: https://www.mdpi.com/article/10.3390/inorganics11050216/s1, Figure S1: Temperature-programmed kinetics of desorption; Figure S2: Isothermal absorption of pure MgH_2; Figure S3: Isothermal absorption of $LiBH_4$-doped MgH_2; Figure S4: SEM images of $LiBH_4$-doped MgH_2 before kinetic cycles; Figure S5: SEM images of LiBH4-doped MgH2 after six kinetic cycles; Figure S6: High temperature laser confocal image for liquid phase transition during hydrogen desorption at 276 °C; Figure S7: High temperature laser confocal image for liquid phase transition during hydrogen desorption at 281 °C.

Author Contributions: Conceptualization, Z.-K.Q. and X.-L.D.; methodology, L.-Q.H.; validation, P.C. and T.-Z.S.; formal analysis, Z.-K.Q.; investigation, H.-W.L.; writing—original draft preparation, Z.-K.Q.; writing—review and editing, X.-L.D.; visualization, P.C.; supervision, H.-W.L.; project administration, Y.-T.L. All authors have read and agreed to the published version of the manuscript.

Funding: This work was financially supported by the Key Program for International S&T Cooperation Projects of China (No. 2017YFE0124300), National Natural Science Foundation of China (Nos. 51971002, 52171205, 52101249 and 52171197), Scientific Research Foundation of Anhui Provincial Education Department (Nos. KJ2020ZD26, KJ2021A0360), Anhui Provincial Natural Science Foundation for Excellent Youth Scholars (No. 2108085Y16), and the Provincial University Outstanding Youth Research Project (No. 2022AH020033).

Data Availability Statement: Data is contained within the article.

Conflicts of Interest: The authors declare no conflict of interest.

References

1. Nielsen, M.; Alberico, E.; Baumann, W.; Drexler, H.J.; Junge, H.; Gladiali, S.; Beller, M. Low-temperature aqueous-phase methanol dehydrogenation to hydrogen and carbon dioxide. *Nature* **2013**, *495*, 85–89. [CrossRef] [PubMed]
2. Zhang, Q.; Fang, M.; Si, T.; Fang, F.; Sun, D.; Ouyang, L.; Zhu, M. Phase stability, structural transition, and hydrogen absorption-desorption features of the polymorphic La_4MgNi_{19} compound. *J. Phys. Chem. C* **2010**, *114*, 11686–11692. [CrossRef]
3. Ding, X.; Li, Y.; Fang, F.; Sun, D.L.; Zhang, Q. Hydrogen-induced magnesium–zirconium interfacial coupling: Enabling fast hydrogen sorption at lower temperatures. *J. Mater. Chem. A* **2017**, *5*, 5067–5076. [CrossRef]
4. Zhang, Q.; Tang, C.; Fang, C.; Fang, F.; Sun, D.L.; Ouyang, L.; Zhu, M. Synthesis, Crystal Structure, and Thermal Decomposition of Strontium Amidoborane. *J. Phys. Chem. C* **2010**, *114*, 1709–1714. [CrossRef]
5. Si, T.; Yin, F.; Zhang, X.; Zhang, Q.; Liu, D.; Li, Y. In-situ formation of medium-entropy alloy nanopump to boost hydrogen storage in Mg-based alloy. *Scr. Mater.* **2023**, *222*, 115052. [CrossRef]
6. Zhang, Q.; Liu, D.; Wang, Q.; Fang, F.; Sun, D.; Ouyang, L.; Zhu, M. Superior hydrogen storage kinetics of $Mg_{12}YNi$ alloy with a long-period stacking ordered phase. *Scr. Mater.* **2011**, *65*, 233–236. [CrossRef]
7. Xia, G.; Tan, Y.; Chen, X.; Fang, F.; Sun, D.; Li, X.; Guo, Z.; Yu, X. Oxygen-free Layer-by-Layer Assembly of Lithiated Composites on Graphene for Advanced Hydrogen Storage. *Adv. Sci.* **2017**, *4*, 1600257. [CrossRef]
8. Li, Y.; Ding, X.; Zhang, Q. Self-Printing on Graphitic Nanosheets with Metal Borohydride Nanodots for Hydrogen Storage. *Sci. Rep.* **2016**, *6*, 31144. [CrossRef]
9. Zhang, Q.; Zhao, B.; Fang, M.; Liu, C.; Hu, Q.; Fang, F.; Sun, D.; Ouyang, L.; Zhu, M. $(Nd_{1.5}Mg_{0.5})Ni_7$-based compounds: Structural and hydrogen storage properties. *Inorg. Chem.* **2012**, *51*, 2976–2983. [CrossRef]
10. Li, Z.; Qiu, H.; Wang, S.; Jiang, L.; Du, J.; Zhang, J.; Latroche, M.; Cuevas, F. Mechanochemistry and hydrogen storage properties of $2Li_3N+Mg$ mixture. *Rare. Met.* **2022**, *41*, 4223–4229. [CrossRef]
11. Zhang, M.; Xiao, X.; Luo, B.; Liu, M.; Chen, M.; Chen, L. Superior de/hydrogenation performances of MgH_2 catalyzed by 3D flower-like TiO_2@C nanostructures. *J. Energy Chem.* **2020**, *46*, 191–198. [CrossRef]
12. Lin, H.; Lu, Y.; Zhang, L.; Liu, H.; Edalati, K. Recent advances in metastable alloys for hydrogen storage: A review. *Rare. Met.* **2022**, *41*, 1797–1817. [CrossRef]
13. Lu, Z.; Yu, H.; Lu, X.; Song, M.; Wu, F.; Zheng, J.; Yuan, Z.; Zhang, L. Two-dimensional vanadium nanosheets as a remarkably effective catalyst for hydrogen storage in MgH_2. *Rare. Met.* **2021**, *40*, 3195–3204. [CrossRef]
14. Yang, Z.; Li, X.; Yao, Q.; Lu, Z.; Zhang, N.; Xia, J.; Yang, K.; Wang, Y.; Zhang, K.; Liu, H.; et al. 2022 roadmap on hydrogen energy from production to utilizations. *Rare. Met.* **2022**, *41*, 3251–3267. [CrossRef]
15. Lu, C.; Ma, Y.; Li, F.; Zhu, H.; Zeng, X.; Ding, W.; Deng, T.; Wu, J.; Zou, J. Visualization of fast "hydrogen pump" in core–shell nanostructured Mg@Pt through hydrogen-stabilized Mg_3PtC. *J. Mater. Chem. A* **2019**, *7*, 14629–14637. [CrossRef]
16. Zhang, J.; Liu, H.; Sun, P.; Guo, X.; Zhou, C.; Fang, F. The effects of crystalline defects on hydrogen absorption kinetics of catalyzed MgH_2 at ambient conditions. *J. Alloys Compd.* **2022**, *927*, 167090. [CrossRef]
17. Ouyang, L.; Cao, Z.; Wang, H.; Liu, J.; Sun, D.; Zhang, Q. Dual-tuning effect of In on the thermodynamic and kinetic properties of Mg_2Ni dehydrogenation. *Int. J. Hydrogen Energy* **2013**, *38*, 8881–8887. [CrossRef]
18. Conceição, M.; Brum, M.; Santos, D. The effect of V, VCl_3 and VC catalysts on the MgH_2 hydrogen sorption properties. *J. Alloys Compd.* **2014**, *586*, S101–S104. [CrossRef]
19. Rousselot, S.; Guay, D.; Roue, L. Synthesis of fcc Mg–Ti–H alloys by high energy ball milling: Structure and electrochemical hydrogen storage properties. *J. Power Sources* **2010**, *195*, 4370–4373. [CrossRef]
20. Zhang, Q.; Huang, Y.; Ma, T.; Li, K.; Ye, F.; Wang, X.; Jiao, L.; Yuan, H.; Wang, Y. Facile synthesis of small MgH_2 nanoparticles confined in different carbon materials for hydrogen storage. *J. Alloys Compd.* **2020**, *825*, 153953. [CrossRef]
21. Bhatnagar, A.; Pandey, S.; Dixit, V.; Shukla, V.; Shahi, R.; Shaz, M. Srivastava, O.N. Catalytic effect of carbon nanostructures on the hydrogen storage properties of MgH_2–$NaAlH_4$ composite. *Int. J. Hydrogen Energy* **2014**, *39*, 14240–14246. [CrossRef]

22. Inoishi, A.; Sato, H.; Chen, Y.; Saito, H.; Sakamoto, R.; Sakaebe, H.; Okada, S. High capacity all-solid-state lithium battery enabled by in situ formation of an ionic conduction path by lithiation of MgH$_2$. *RSC Adv.* **2022**, *12*, 10749–10754. [CrossRef] [PubMed]
23. Zhao, B.; Fang, F.; Sun, D.; Zhang, Q.; Wei, S.; Cao, F.; Sun, H.; Ouyang, L.; Zhu, M. Formation of Mg$_2$Ni with enhanced kinetics: Using MgH$_2$ instead of Mg as a starting material. *J. Solid State Chem.* **2012**, *192*, 210–214. [CrossRef]
24. Wang, Y.; Fan, G.; Zhang, D.; Fan, Y.; Liu, B. Striking enhanced effect of PrF$_3$ particles on Ti$_3$C$_2$ MXene for hydrogen storage properties of MgH$_2$. *J. Alloys Compd.* **2022**, *914*, 165291. [CrossRef]
25. Kammerer, J.; Duan, X.; Neubrech, F.; Schroder, R.; Liu, N.; Pfannmoller, M. Stabilizing γ-MgH$_2$ at Nanotwins in Mechanically Constrained Nanoparticles. *Adv. Mater.* **2021**, *33*, 2008259. [CrossRef]
26. Liu, X.; Liu, H.; Qiu, N.; Zhang, Y.; Zhao, G.; Xu, L.; Lan, Z.; Guo, J. Cycling hydrogen desorption properties and microstructures of MgH$_2$–AlH$_3$–NbF$_5$ hydrogen storage materials. *Rare. Mat.* **2021**, *40*, 1003–1007. [CrossRef]
27. Si, T.; Zhang, X.; Feng, J.; Ding, X.; Li, Y. Enhancing hydrogen sorption in MgH$_2$ by controlling particle size and contact of Ni catalysts. *Rare. Met.* **2021**, *40*, 995–1002. [CrossRef]
28. Hanada, N.; Ichikawa, T.; Fujii, H. Catalytic Effect of Nanoparticle 3d-Transition Metals on Hydrogen Storage Properties in Magnesium Hydride MgH$_2$ Prepared by Mechanical Milling. *J. Phys. Chem. B* **2005**, *109*, 7188–7194. [CrossRef]
29. Lu, J.; Choi, Y.J.; Fang, Z.; Sohn, H.; Ronnebro, E. Hydrogen Storage Properties of Nanosized MgH$_2$−0.1TiH$_2$ Prepared by Ultrahigh-Energy−High-Pressure Milling. *J. Am. Chem. Soc.* **2009**, *131*, 15843–15852. [CrossRef]
30. Cuan, J.; Zhou, Y.; Zhou, T.; Ling, S.; Rui, K.; Guo, Z.; Liu, H.; Yu, X. Borohydride-Scaffolded Li/Na/Mg Fast Ionic Conductors for Promising Solid-State Electrolytes. *Adv. Mater.* **2019**, *31*, 1803533. [CrossRef]
31. Liu, S.; Sun, L.; Zhang, J.; Zhang, Y.; Xu, F.; Xing, Y.; Li, F.; Zhao, J.; Du, Y.; Hu, W.; et al. Hydrogen storage properties of destabilized MgH$_2$–Li$_3$AlH$_6$ systemInt. *J. Hydrog. Energy* **2010**, *35*, 8122–8129. [CrossRef]
32. Li, B.; Liu, Y.; Li, C.; Gao, M.; Pan, H. In situ formation of lithium fast-ion conductors and improved hydrogen desorption properties of the LiNH$_2$–MgH$_2$ system with the addition of lithium halides. *J. Mater. Chem. A* **2014**, *2*, 3155–3162. [CrossRef]
33. Kato, S.; Borgschulte, A.; Bielmann, M.; Zuttel, A. Interface reactions and stability of a hydride composite (NaBH$_4$ + MgH$_2$). *Phys. Chem. Chem. Phys.* **2012**, *14*, 8360–8368. [CrossRef] [PubMed]
34. Paik, B.; Li, H.; Wang, J.; Akiba, E. A Li–Mg–N–H composite as H$_2$ storage material: A case study with Mg(NH$_2$)$_2$–4LiH–LiNH$_2$. *Chem. Commun.* **2015**, *51*, 10018–10021. [CrossRef]
35. Yan, Y.G.; Grinderslev, J.B.; Lee, Y.; Jorgensen, M.; Cho, Y.; Cerny, R.; Jensen, T. Ammonia-assisted fast Li-ion conductivity in a new hemiammine lithium borohydride, LiBH$_4$·1/2NH$_3$. *Chem. Commun.* **2020**, *28*, 3971–3974. [CrossRef]
36. Zhou, H.; Wang, X.; Liu, H.; Gao, S.; Yan, M. Improved hydrogen storage properties of LiBH$_4$ confined with activated charcoal by ball milling. *Rare Met.* **2019**, *38*, 321–326. [CrossRef]

Disclaimer/Publisher's Note: The statements, opinions and data contained in all publications are solely those of the individual author(s) and contributor(s) and not of MDPI and/or the editor(s). MDPI and/or the editor(s) disclaim responsibility for any injury to people or property resulting from any ideas, methods, instructions or products referred to in the content.

Article

Study of Phase Composition in TiFe + 4 wt.% Zr Alloys by Scanning Photoemission Microscopy

Sabrina Sartori [1,*], Matteo Amati [2], Luca Gregoratti [2], Emil Høj Jensen [1], Natalia Kudriashova [3] and Jacques Huot [4,*]

[1] Department of Technology Systems, University of Oslo, 2027 Kjeller, Norway
[2] Elettra-Sincrotrone Trieste ScPA, Area Science Park, SS14-Km163.5, 34149 Trieste, Italy
[3] School of Physical Sciences, University of Cambridge, Trinity Lane, Cambridge CB2 1TN, UK
[4] Institut de Recherche sur L'hydrogène, Université du Québec à Trois-Rivières, 3351 Boul. des Forges, Trois-Rivières, QC G9A 5H7, Canada
* Correspondence: sabrina.sartori@its.uio.no (S.S.); jacques.huot@uqtr.ca (J.H.)

Abstract: The alloy TiFe is widely used as hydrogen storage material. However, the first hydrogenation is difficult. It was found that the addition of zirconium greatly improves the kinetic of first hydrogenation, but the mechanism is not well understood. In this paper, we report the use of scanning photoemission microscopy to investigate the composition and chemical state of the various phases present in this alloy and how they change upon hydrogenation/dehydrogenation. We found the presence of different oxide phases that were not seen by conventional SEM investigation. The nature of these oxides phases seems to change upon hydrogenation/dehydrogenation cycle. This indicates that oxide phases may play a more significant role in the hydrogen absorption as what was previously believed.

Keywords: scanning photoemission microscope; metal hydride; phase composition; first hydrogenation; TiFe alloy

Citation: Sartori, S.; Amati, M.; Gregoratti, L.; Jensen, E.H.; Kudriashova, N.; Huot, J. Study of Phase Composition in TiFe + 4 wt.% Zr Alloys by Scanning Photoemission Microscopy. *Inorganics* **2023**, *11*, 26. https://doi.org/10.3390/inorganics11010026

Academic Editor: Maurizio Peruzzini

Received: 24 November 2022
Revised: 25 December 2022
Accepted: 28 December 2022
Published: 3 January 2023

Copyright: © 2023 by the authors. Licensee MDPI, Basel, Switzerland. This article is an open access article distributed under the terms and conditions of the Creative Commons Attribution (CC BY) license (https://creativecommons.org/licenses/by/4.0/).

1. Introduction

Intermetallic TiFe-compounds are attractive hydrogen storage materials due to their low cost, elemental abundance, and near ambient temperature and mild hydrogen pressure absorption. In recent years, their use has been demonstrated to be successful in a number of large-scale stationary applications [1–4]. The main disadvantage is the slow activation (first hydrogenation) properties of TiFe, requiring high temperatures (about 400 °C) and pressure of ca. 50 bars [5]. Usually, the first hydrogenation has a long incubation period that may vary with alloy composition and pressure/temperature conditions. This behaviour is associated with an oxide surface passivation layer. Improvement of activation could be achieved in many ways. One is mechanical activation through ball milling, high-pressure torsion and other severe plastic deformation techniques [6–10]. Presently, the most common solution for better first hydrogenation and cycle life is to partly substitute Fe by Mn. This result in the formation of secondary phases reactive to hydrogen that enhances the first hydrogenation properties [11].

Recently it was shown that TiFe with 4 wt.% Zr greatly improved the first hydrogenation kinetics, after processing the sample via ball milling [12]. The faster hydrogenation kinetics of the processed samples compared to the as-cast sample has been attributed to the reduction of crystallite sizes and formation of new grain boundaries (increase of their relative volume) with longer ball milling times. However, the ball milled samples also showed a decreased hydrogen storage capacity compared to the as-cast sample, likely due to the formation of grain boundaries that enhanced hydrogen diffusion but did not store hydrogen in their structures. Analysis concluded that the improved kinetics was not related to the rate-limiting step of the first hydrogenation, since all samples fitted well with a 3D growth, with the growth interface velocity of diffusion decreasing with time.

Another effective approach to improve the activation of TiFe is via element substitution using transition metals (TMs), such as Mn, Cr, and Zr [13,14]. Doing this also affects kinetics, thermodynamics and cyclability of the TiFe-based materials. Substituents can influence the structural properties of the alloys, for instance via formation of secondary phases [15–17], or altering hydride phase stability. Studies have shown that the first hydrogenation of TiFe was achieved at room temperature by adding Zr [18–20].

Previous investigation has shown that the TiFe + 4 wt.% Zr alloy is composed of a TiFe main phase with a Zr-rich secondary phase [12,18,20–22]. This mechanism is somewhat similar to the one seen for the Ti (Fe, Mn) alloys [11]. In this paper, we investigate the role of the composition of the phases in the hydrogenation behaviour. Our hypothesis is that the improvement in kinetics is caused by the specific microstructure [21] and chemical composition of the secondary phase. The zirconium-rich phase acts as a gateway for hydrogen to reach the TiFe phase and thus makes the hydrogenation much faster than in pure TiFe. Additionally, the variation of composition at the interface matrix/secondary phase may play an important role in the transfer of hydrogen from the secondary phase to the matrix.

2. Results and Discussion

2.1. As-Cast Sample

Figure 1 shows a survey spectrum of the cast TiFe + 4 wt.% Zr. Beside the constituent elements oxygen indicates oxidation at the surface and there is a small amount of carbon contamination coming from handling the sample. As the fine chemical state, e.g., oxide vs. metallic, cannot be distinguished in such a survey, in the following, we will focus on the detailed core level spectra.

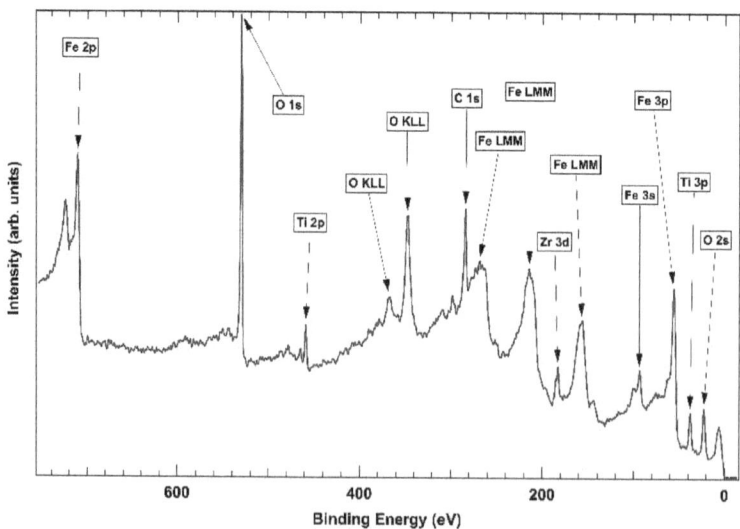

Figure 1. Survey spectrum of as-cast TiFe + 4 wt.% Zr.

The elemental distribution of as-cast TiFe + 4 wt.% Zr is shown in Figure 2. From previous investigations it is known that this alloy is made of two phases: a main TiFe phase along with a Zr-rich secondary phase [20,22]. This is confirmed by the present measurement. It should be pointed out that the region of interest was selected to mainly investigate the secondary phase. Therefore, these mappings do not reflect the true relative abundance of the main (TiFe) and secondary phase. The figure shows that iron is relatively uniformly distributed through the alloy while titanium abundance is slightly reduced in the TiFe phase. The most interesting feature is the higher abundance of zirconium at the edge of the secondary phase.

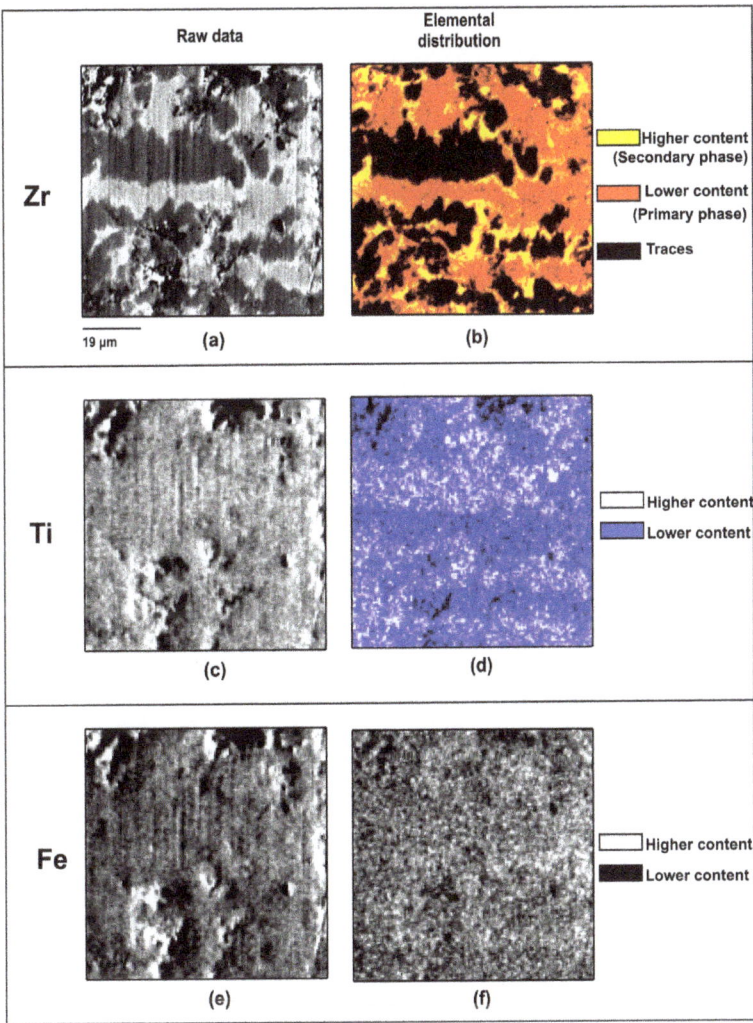

Figure 2. Element distribution in as-cast TiFe + 4 wt.% Zr. Zr (**a,b**), Ti (**c,d**), Fe (**e,f**). Raw data: (**a,c,e**), Elemental distribution (**b,d,f**).

The abundance of zirconium at the edge of the secondary phase was quantified in a previous investigation [19]. Figure 3 is a typical micrograph and table of element concentration in the different regions. The matrix is TiFe phase (Space group Pm-3 m, structure type CsCl) with small amount of Zr substituting for Fe. The grey secondary phase is also close to the TiFe phase with more zirconium replacing iron. At the edge of the secondary phase, the bright area contains zirconium and also more titanium. This is in agreement with Figure 2.

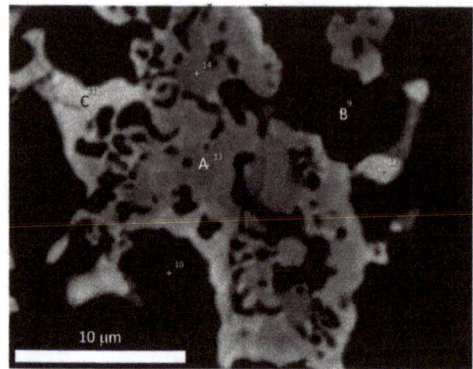

Figure 3. Element distribution in as-cast TiFe + 4 wt.% Zr. (adapted with permission from ref. [19], 2016, Elsevier).

This concentration of zirconium at the edge of the secondary phase is most likely due to the metastable state of the alloy. As shown by Patel et al., after heat treatment at 1173 K for 24 h, the secondary phase is more concentrated in zirconium while the amount of zirconium in the TiFe phase is almost zero [23]. The secondary phase was also found to be uniform with no sign of zirconium concentration at the edges. They also show that the first hydrogenation was practically impossible for the heat-treated alloy. Therefore, the metastable state and the variation of composition at the interface matrix/secondary phase may play an important role in the transfer of hydrogen from the secondary phase to the matrix. The exact crystal structure of the secondary phase is under investigation. Preliminary results from neutron diffraction points toward a hexagonal $MgZn_2$-type structure.

As the bulk composition of the matrix and secondary phases are relatively close, it is important to probe the electronic structure of each element in these phases to see if there is a difference (for instance if the chemical state of iron is different when it is situated in the matrix or in the secondary phase).

The ESCA (Electron Spectroscopy for Chemical Analysis) microscopy measurements have been performed to accurately measure the element's concentration at the boundary between these two phases and also to probe the chemical state of each element. This will possibly help us understand if zirconium-rich phase can act as a gateway for hydrogen to reach the TiFe phase and thus making the hydrogenation much faster than in pure TiFe.

The measurements showed that a good spatial resolution was achieved and that the chemical state of the elements could be verified, with the chemical state of iron being different when situated in the matrix or in the zirconium-rich secondary phase. Zirconium content changes within the zirconium-rich phase, with more zirconium at the edge between the bright phase and TiFe phase.

2.2. As Cast Powder

Figure 4 shows the SPEM (scanning photoemission microscope) analysis of the as-cast powder; images in the first column (panels (a), (e) and (i)) show the photoemission maps recorded by collecting photoelectrons at the Zr 3d, Ti 2p and Fe 2p core levels, respectively. The three maps appear very similar because the contrast is dominated by the topography of the powder sample. In the second column (panels (b), (f) and (j)) the corresponding background subtracted maps are shown (see experimental section). They represent the elemental concentration of each atomic species in the probed area. The brighter the region, the higher the concentration of elements. The areas at the left and right side appearing very dark in the contrast are regions completely shadowed where the applied algorithm is not able to properly remove the topographic contribution. They should not be considered in the analysis.

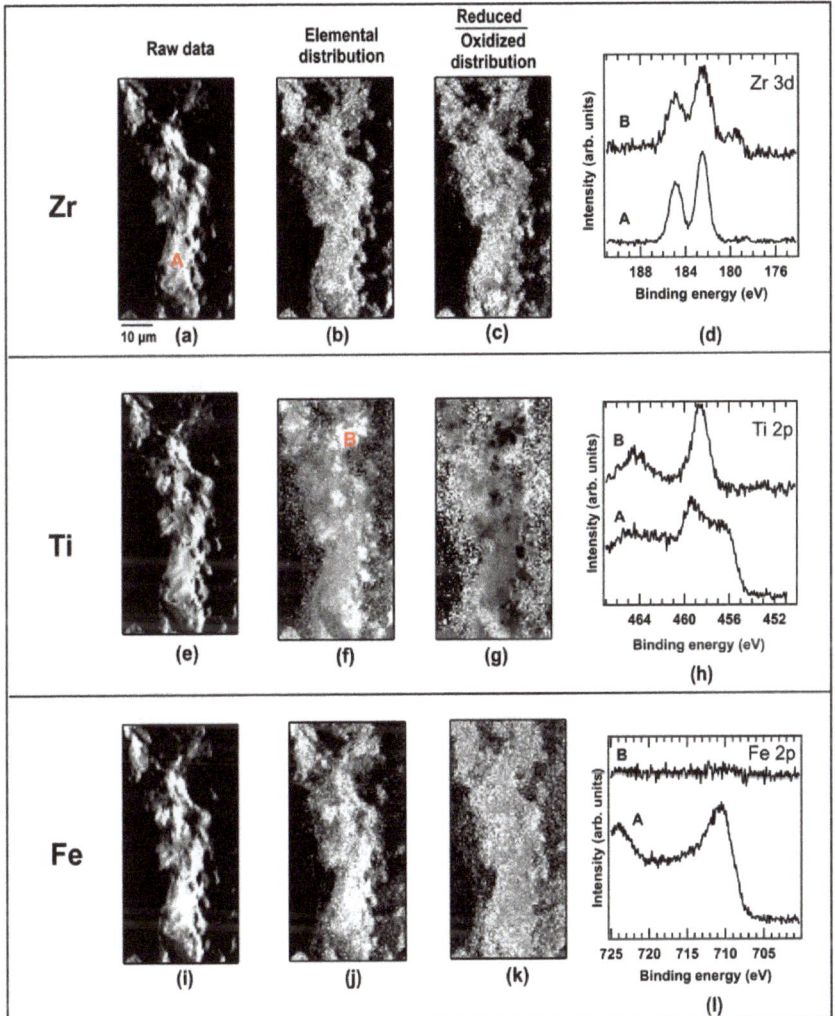

Figure 4. Element distribution and spectra of Ti, Fe and Zr in as-cast TiFe + 4 wt.% Zr. Zr (a–d), Ti (e–h), Fe (i–l). Raw data: (a,e,i), Elemental distribution (b,f,j). Reduced/oxidized distribution (c,g,k). Spectra (d,h,l).

Point A shows the presence of all three elements. This could be associated with TiFe alloy with a small amount of Zr as was seen in a previous investigation. In the case of point B, only Zr and Ti were detected. This is a somewhat puzzling observation because a Zr-Ti phase has not been detected in previous investigations of TiFe + x wt.% Zr. However, as Ti and Zr are totally miscible, this is possible. As Fe could not enter in solid solution in Zr but Ti is miscible, it is reasonable to consider that the Zr-rich region may have more Ti than Fe, especially at the very edge of the secondary phase.

The map in panel (b) shows an uneven distribution of the Zr; the change of the signal intensity between the darker and brighter areas is about 25%. The Zr 3d spectra acquired at positions A and B, labelled in panels (a) and (f), show two different surfaces chemical compositions. The spectrum at position A appears as a single typical Zr 3d spin-orbit doublet with the Zr $3d_{5/2}$ peak centered at 182.5 eV binding energy (BE) while the spectrum at position B is the convolution of the same component at 182.5 eV BE with a less intense one with its Zr $3d_{5/2}$ peak centered at 179 eV BE. Due to the presence of well-separated convoluted components, the peaks of spectrum B are broader. The peak at lower binding energy can be associated with reduced Zr in the 0/+1 valence state while the main component corresponds to Zr in the +3/+4 valence state typical of chemically stable zirconium oxide phases [24].

The Zr 3d spectra acquired at positions A and B, labelled in panels (a) and (f), show two different surfaces chemical compositions. As the reduced and oxidized components of Zr were well separated in energy, it was possible to select the energy windows of the two chemical states and calculate the ratio of the corresponding images. The map in panel (c) shows the ratio between reduced and oxidized Zr; despite the Zr is everywhere mostly fully oxidized, brighter/darker areas indicate a lower/higher presence of a weak reduced Zr, as shown by the Zr 3d core level spectra in panel (d).

The background subtracted map of Ti 2p shown in panel (f) evidences large changes in the concentration of titanium; the intensity in the brightest areas is twice that in the darker ones. To some extent, the Ti and Zr maps (panels (f) and (b), respectively) are complementary, i.e., regions with a higher Zr content show less Ti (e.g., region B). The Ti 2p core level spectra acquired at points A and B show two chemically different environments; spectrum B has the typical line shape of the TiO_2 with the Ti $2p_{3/2}$ centered at 458.6 eV in good agreement with literature references [25]. The spectrum acquired at position A has an additional shoulder at 455.8 eV binding energy compatible with a +2 valence state [26]. As done for the Zr in panel (c), the image corresponding to the ratio between reduced and oxidized Ti is shown in panel (g). The changes in the contrast suggest an uneven distribution of the ratio, with a higher local amount of Ti resulting in a higher contribution of the +4 oxidation state.

Finally, the same analysis is proposed for Fe. Panel (j) shows the distribution of this element on the surface of the probed area. As also evidenced by the two spectra acquired in points A and B, reported in panel (l), regions with and without Fe are present. Spectrum A is dominated by a peak centered at 710.8 eV binding energy corresponding to Fe_2O_3 [27]. The image with the ratio of the reduced and oxidized Fe is shown in panel (k) indicating a rather uniform contrast except in the regions where there is no iron.

Taking into account that the beam penetration is only of the order of 1 or 2 nm for Zr and Ti and even less for Fe the absence of Fe may be due to small migration of Fe upon oxidation of the edge of the secondary phase.

2.3. Hydrided Sample

Figure 5 shows the elemental distribution and spectrum of Ti, Fe and Zr on a fully hydrogenated sample. Before discussing these results, we should point out that the sample was kept under hydrogen until close to the measurement in order to keep its hydrided state. However, the time delay between hydrogenation and the measurement and also the fact that the measurement is performed under high vacuum under intense beam means that some phases of the sample may no longer be in a hydrided state. Therefore, caution should be exercised in the analysis of these results.

Three different points have been probed. Point A consists only of Ti and Fe atoms. This is in agreement with the main TiFe phase. The spectra of Ti and Fe do not show any shift in energy. This is consistent with the fact that this phase is most probably dehydrided. Point B only shows Zr and Fe signals. The Zr spectrum is fully oxidized but there is a small bump in the Fe spectrum at 717 eV. This may be an indication of the formation of a Zr-Fe-H phase. Raj et al. have shown that the high temperature phase Zr_2Fe could form the intermetallic hydride Zr_2FeH_x hydride at a temperature as low as 250 K under hydrogen pressure less than 0.5 atm [28]. Oxygen-containing compounds could also be hydrided as demonstrated by Zavaliy et al. on the hydrogenation of for example $Ti_{4-x}Zr_xFe_2O_y$ [29]. Regarding point C, the three elements are present, but their spectra show an oxidation state.

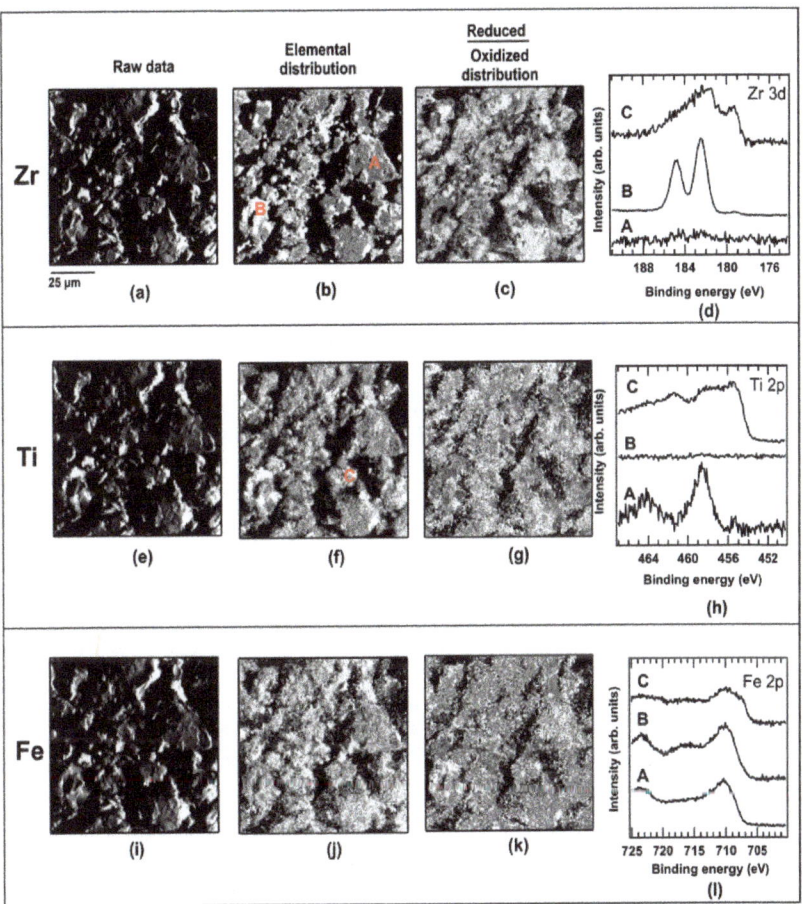

Figure 5. Element distribution and spectra of Ti, Fe and Zr in hydrogenated TiFe + 4 wt.% Zr. Zr (a–d), Ti (e–h), Fe (i–l). Raw data: (a,e,i), Elemental distribution (b,f,j). Reduced/oxidized distribution (c,g,k). Spectra (d,h,l).

2.4. Dehydrided Sample

The dehydrided elemental distribution and spectra are shown in Figure 6. Point A is TiFe alloy with basically no Zr present. This is associated with the main TiFe phase. The Ti and Fe spectra are mostly oxidized. Point B is Ti-Zr phase as seen in the as-cast alloy, however there are main differences. In the as-cast alloy both Ti and Zr in this phase showed some state of oxidation but here, oxidation is apparent only in the Zr spectrum. The Ti spectrum of points A and B indicates a TiO_2 phase. Finally, point C is only Zr precipitate that is also oxidized.

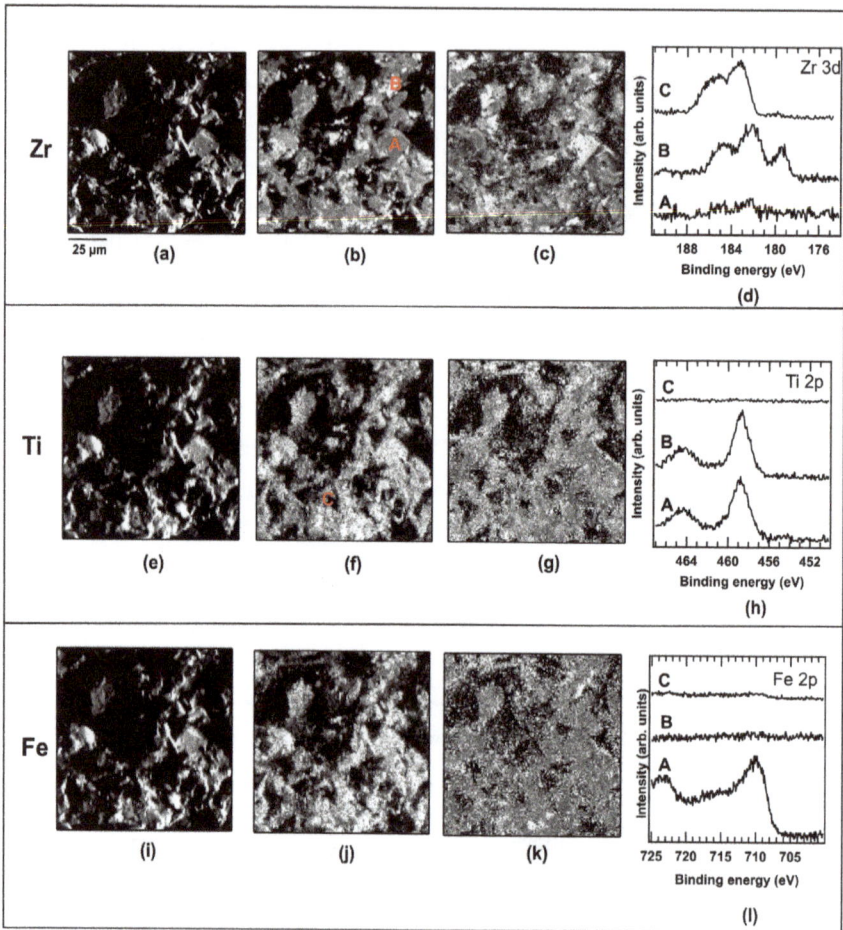

Figure 6. Element distribution and spectra of Ti, Fe and Zr in dehydrogenated TiFe + 4 wt.% Zr. Zr (a–d), Ti (e–h), Fe (i–l). Raw data: (a,e,i), Elemental distribution (b,f,j). Reduced/oxidized distribution (c,g,k). Spectra (d,h,l).

3. Materials and Methods

The as-cast sample of composition TiFeZr0.05 (TiFe + 4 wt.% Zr) was synthesized by arc melting. Details are given in ref. [12]. After synthesis, a part of the alloy was cut and polished for the as-cast investigations. The other part was crushed in an argon-filled glove box. The crushed powder was then hydrogenated at room temperature under 20 bars of hydrogen pressure using a home-made Seivert apparatus. After hydrogenation a part of the sample was taken out of the reactor. This fully hydrided sample is thereafter identified as hydrided hydrogenated TiFe + 4 wt.% Zr. The remaining part was desorbed at room temperature under dynamic vacuum for a period of two hours. This sample is identified as dehydrided TiFe + 4 wt.% Zr.

Photoemission measurements were performed by using the scanning photoemission microscope (SPEM) hosted at the ESCA Microscopy beamline at Elettra synchrotron facility (Trieste, Italy) [30]. A SPEM uses a direct approach to obtain spatial resolution consisting in the focalization of the incoming X-ray beam with Fresnel Zone Plate (ZP) optics. The X-ray microprobe can reach a dimension of 120 nm. The optical setup is completed by an

Order Sorting Aperture (OSA) positioned between the sample and the ZP selecting the first diffraction order and removing the undesired orders. Photoelectrons are collected by a SPECS PHOIBOS 100 hemispherical electron analyzer provided with a 48-channels electron detector.

The SPEM can operate in two modes: imaging and submicron-spectroscopy. In the first mode, the distribution of an atomic element or chemical state is mapped by synchronized scanning the sample with respect to the focused photon beam and collecting photoelectrons within a selected electron kinetic energy window, corresponding to the electron energy level of the element of interest. The multichannel electron detector, where each channel measures electrons with a specific kinetic energy, defined by the selected energy window, allows the simultaneous collection of 48 images. This adds a spectro-imaging option, i.e., the reconstruction of the spectrum corresponding to the covered energy window from a selected area from subdomains of the main map [31]. The second mode is the microprobe spectroscopy providing spectroscopic information from the selected spot onto the sample with a higher spectral resolution.

Contrast in the photoemission maps reflect the intensity variations of the number of photoelectrons detected at each specific position in the image and for an 8 eV binding energy window centered around the given energy. Such number depends on the number of atoms associated with the specific core level acquired in the illuminated spot, their chemical state and the surface topography. The photoemission yield increases/decreases if the probed surface is more normal/grazing to the electron analyzer axis. In order to remove the topographical contribution from the maps, an algorithm was applied consisting in dividing the map acquired at a specific core level by that collected at its higher kinetic energy background. The resulting so-called "peak/background" map will show only the chemical contribution [31]. Exceptions are the areas completely shadowed where the low count rates generate fake values.

Measurements were carried out using 650 eV photon energy and an overall spectral energy resolution of 220 meV in microspot spectroscopy and 490 meV in imaging mode. More details on the multi-channel data acquisition and processing can be found in ref. [31].

4. Conclusions

In this paper the microstructure and microchemistry of TiFe + 4 wt.% Zr was investigated in detail. The photoemission map of the as-cast alloy confirmed the higher abundance of zirconium at the edge of the zirconium-rich secondary phase. The edge is also the site of a Zr-Ti phase that was not detected using conventional SEM. The reduced/oxidized ratio was determined for all elements. It was found that, for zirconium, this ratio is higher in the zirconium-rich region while for titanium it is higher in the region with low content of titanium. In the case of iron, the reduced/oxidized ratio is constant.

The hydrided sample showed the presence of a possible Zr-Fe hydride phase. There is also a Zr-Fe-O phase that could be a hydride. In the dehydrided sample the TiFe phase was not oxidized, and in the Ti-Zr phase, oxidation is only on Zr.

The general conclusion is that in the TiFe + 4 wt.%Zr alloy there is a wide range of different oxides. As the nature of the various oxide phases change upon hydrogenation/dehydrogenation, it may indicate that these oxides play a more important role in the hydrogenation than what was previously believed.

Author Contributions: Conceptualization, S.S. and J.H.; methodology, M.A. and L.G.; validation, S.S., M.A., L.G. and J.H.; formal analysis, S.S., M.A., L.G. and J.H.; investigation, S.S., M.A., L.G. and J.H.; writing—original draft preparation, S.S. and J.H.; writing—review and editing, M.A., L.G., N.K. and E.H.J. All authors have read and agreed to the published version of the manuscript.

Funding: This research was funded in part by a Discovery grant from the Natural Sciences and Engineering Research Council of Canada (NSERC). S.S. acknowledges the financial support from the European Union's Horizon 2020 research and innovation program, under Grant Agreement No. 951815.

Data Availability Statement: Not applicable.

Acknowledgments: We acknowledge Elettra Sincrotrone Trieste for providing access to its synchrotron radiation facilities. The authors would like to thank Patrick Zeller for help in data acquisition.

Conflicts of Interest: The authors declare no conflict of interest.

References

1. Endo, N.; Suzuki, S.; Goshome, K.; Maeda, T. Operation of a bench-scale TiFe-based alloy tank under mild conditions for low-cost stationary hydrogen storage. *Int. J. Hydrogen Energy* **2017**, *42*, 5246–5251. [CrossRef]
2. HyCARE focuses on large-scale, solid-state hydrogen storage. *Fuel Cells Bull.* **2019**, *2019*, 11. [CrossRef]
3. Endo, N.; Shimoda, E.; Goshome, K.; Yamane, T.; Nozu, T.; Maeda, T. Operation of a stationary hydrogen energy system using TiFe-based alloy tanks under various weather conditions. *Int. J. Hydrogen Energy* **2020**, *45*, 207–215. [CrossRef]
4. Endo, N.; Saita, I.; Nakamura, Y.; Saitoh, H.; Machida, A. Hydrogenation of a TiFe-based alloy at high pressures and temperatures. *Int. J. Hydrogen Energy* **2015**, *40*, 3283–3287. [CrossRef]
5. Reilly, J.J.; Wiswall, R.H. Formation and Properties of Iron Titanium Hydride. *Inorg. Chem.* **1974**, *13*, 218–222. [CrossRef]
6. Vega, L.E.R.; Leiva, D.R.; Lean Neto, R.M.; Silva, W.B.; Silva, R.A.; Ishikawa, T.T.; Kiminami, C.S.; Botta, W.J. Mechanical activation of TiFe for hydrogen storage by cold rolling under inert atmosphere. *Int. J. Hydrogen Energy* **2018**, *43*, 2913–2918. [CrossRef]
7. Edalati, K.; Matsuda, J.; Iwaoka, H.; Toh, S.; Akiba, E.; Horita, Z. High-pressure torsion of TiFe intermetallics for activation of hydrogen storage at room temperature with heterogeneous nanostructure. *Int. J. Hydrogen Energy* **2013**, *38*, 4622–4627. [CrossRef]
8. Edalati, K.; Matsuda, J.; Yanagida, A.; Akiba, E.; Horita, Z. Activation of TiFe for hydrogen storage by plastic deformation using groove rolling and high-pressure torsion: Similarities and differences. *Int. J. Hydrogen Energy* **2014**, *39*, 15589–15594. [CrossRef]
9. Emami, H.; Edalati, K.; Matsuda, J.; Akiba, E.; Horita, Z. Hydrogen storage performance of TiFe after processing by ball milling. *Acta Mater.* **2015**, *88*, 190–195. [CrossRef]
10. Romero, G.; Lv, P.; Huot, J. Effect of ball milling on the first hydrogenation of TiFe alloy doped with 4 wt% (Zr + 2Mn) additive. *J. Mater. Sci.* **2018**, *53*, 13751–13757. [CrossRef]
11. Dematteis, E.M.; Dreistadt, D.M.; Capurso, G.; Jepsen, J.; Cuevas, F.; Latroche, M. Fundamental hydrogen storage properties of TiFe-alloy with partial substitution of Fe by Ti and Mn. *J. Alloys Compd.* **2021**, *874*, 12. [CrossRef]
12. Lv, P.; Guzik, M.N.; Sartori, S.; Huot, J. Effect of ball milling and cryomilling on the microstructure and first hydrogenation properties of TiFe+4 wt.% Zr alloy. *J. Mater. Res. Technol.* **2019**, *8*, 1828–1834. [CrossRef]
13. Dematteis, E.M.; Berti, N.; Cuevas, F.; Latroche, M.; Baricco, M. Substitutional effects in TiFe for hydrogen storage: A comprehensive review. *Mater. Adv.* **2021**, *2*, 2524–2560. [CrossRef]
14. Zeaiter, A.; Chapelle, D.; Cuevas, F.; Maynadier, A.; Latroche, M. Milling effect on the microstructural and hydrogenation properties of TiFe$_{0.9}$Mn$_{0.1}$ alloy. *Powder Technol.* **2018**, *339*, 903–910. [CrossRef]
15. Lv, P.; Huot, J. Hydrogenation improvement of TiFe by adding ZrMn$_2$. *Energy* **2017**, *138*, 375–382. [CrossRef]
16. Patel, A.K.; Duguay, A.; Tougas, B.; Neumann, B.; Schade, C.; Sharma, P.; Huot, J. Study of the Microstructural and First Hydrogenation Properties of TiFe Alloy with Zr, Mn and V as Additives. *Processes* **2021**, *9*, 1217. [CrossRef]
17. Gosselin, C.; Huot, J. First Hydrogenation Enhancement in TiFe Alloys for Hydrogen Storage Doped with Yttrium. *Metals* **2019**, *9*, 242. [CrossRef]
18. Jain, P.; Gosselin, C.; Huot, J. Effect of Zr, Ni and Zr$_7$Ni$_{10}$ alloy on hydrogen storage characteristics of TiFe alloy. *Int. J. Hydrogen Energy* **2015**, *40*, 16921–16927. [CrossRef]
19. Lv, P.; Huot, J. Hydrogen storage properties of Ti$_{0.95}$FeZr$_{0.05}$, TiFe$_{0.95}$Zr$_{0.05}$ and TiFeZr$_{0.05}$ alloys. *Int. J. Hydrogen Energy* **2016**, *41*, 22128–22133. [CrossRef]
20. Gosselin, C.; Huot, J. Hydrogenation Properties of TiFe Doped with Zirconium. *Materials* **2015**, *8*, 7864–7872. [CrossRef]
21. Patel, A.K.; Tougas, B.; Sharma, P.; Huot, J. Effect of cooling rate on the microstructure and hydrogen storage properties of TiFe with 4 wt% Zr as an additive. *J. Mater. Res. Technol.* **2019**, *8*, 5623–5630. [CrossRef]
22. Gosselin, C.; Santos, D.; Huot, J. First hydrogenation enhancement in TiFe alloys for hydrogen storage. *J. Phys. D Appl. Phys.* **2017**, *50*, 375303. [CrossRef]
23. Patel, A.K.; Sharma, P.; Huot, J. Effect of annealing on microstructure and hydrogenation properties of TiFe + X wt% Zr (X = 4, 8). *Int. J. Hydrogen Energy* **2018**, *43*, 6238–6243. [CrossRef]
24. Azdad, Z.; Marot, L.; Moser, L.; Steiner, R.; Meyer, E. Valence band behaviour of zirconium oxide, Photoelectron and Auger spectroscopy study. *Sci. Rep.* **2018**, *8*, 16251. [CrossRef] [PubMed]
25. Bharti, B.; Kumar, S.; Lee, H.N.; Kumar, R. Formation of oxygen vacancies and Ti(3+) state in TiO$_2$ thin film and enhanced optical properties by air plasma treatment. *Sci. Rep.* **2016**, *6*, 32355. [CrossRef] [PubMed]
26. Biesinger, M.C.; Lau, L.W.M.; Gerson, A.R.; Smart, R.S.C. Resolving surface chemical states in XPS analysis of first row transition metals, oxides and hydroxides: Sc, Ti, V, Cu and Zn. *Appl. Surf. Sci.* **2010**, *257*, 887–898. [CrossRef]
27. Yang, H.H.; Shi, B.F.; Wang, S.L. Fe Oxides Loaded on Carbon Cloth by Hydrothermal Process as an Effective and Reusable Heterogenous Fenton Catalyst. *Catalysts* **2018**, *8*, 207. [CrossRef]
28. Raj, P.; Suryanarayana, P.; Sathyamoorthy, A.; Shashikala, K.; Iyer, R.M. Zr$_2$FeH$_x$ System Hydrided at Low-Temperatures—Structural Aspects by Mossbauer and X-ray-Diffraction Studies. *J. Alloys Compd.* **1992**, *178*, 393–401. [CrossRef]

29. Zavaliy, I.Y.; Denys, R.V.; Koval'Chuck, I.V.; Riabov, A.B.; Delaplane, R.G. Hydrogenation of $Ti_{4-x}Zr_xFe_2O_y$ alloys and crystal structure analysis of their deuterides. *Chem. Met. Alloy.* **2009**, *2*, 59–67. [CrossRef]
30. Abyaneh, M.K.; Gregoratti, L.; Amati, M.; Dalmiglio, M.; Kiskinova, M. Scanning Photoelectron Microscopy: A Powerful Technique for Probing Micro and Nano-Structures. *e-J. Surf. Sci. Nanotec.* **2011**, *9*, 158–162. [CrossRef]
31. Gregoratti, L.; Barinov, A.; Benfatto, E.; Cautero, G.; Fava, C.; Lacovig, P.; Lonza, D.; Kiskinova, M.; Tommasini, R.; Mahl, S.; et al. 48-Channel electron detector for photoemission spectroscopy and microscopy. *Rev. Sci. Instrum.* **2004**, *75*, 64–68. [CrossRef]

Disclaimer/Publisher's Note: The statements, opinions and data contained in all publications are solely those of the individual author(s) and contributor(s) and not of MDPI and/or the editor(s). MDPI and/or the editor(s) disclaim responsibility for any injury to people or property resulting from any ideas, methods, instructions or products referred to in the content.

Article

Synthesis, Structure and Mg^{2+} Ionic Conductivity of Isopropylamine Magnesium Borohydride

Lasse G. Kristensen [1], Mads B. Amdisen [1], Mie Andersen [2] and Torben R. Jensen [1,*]

[1] Interdisciplinary Nanoscience Center (iNANO), Department of Chemistry, University of Aarhus, Langelandsgade 140, DK-8000 Aarhus C, Denmark
[2] Interdisciplinary Nanoscience Center (iNANO), Aarhus Institute of Advanced Studies (AIAS), Department of Physics and Astronomy, Center for Interstellar Catalysis, University of Aarhus, DK-8000 Aarhus C, Denmark
* Correspondence: trj@chem.au.dk

Abstract: The discovery of new inorganic magnesium electrolytes may act as a foundation for the rational design of novel types of solid-state batteries. Here we investigated a new type of organic-inorganic metal hydride, isopropylamine magnesium borohydride, Mg(BH$_4$)$_2$·(CH$_3$)$_2$CHNH$_2$, with hydrophobic domains in the solid state, which appear to promote fast Mg^{2+} ionic conductivity. A new synthetic strategy was designed by combination of solvent-based methods and mechanochemistry. The orthorhombic structure of Mg(BH$_4$)$_2$·(CH$_3$)$_2$CHNH$_2$ was solved *ab initio* by the Rietveld refinement of synchrotron X-ray powder diffraction data and density functional theory (DFT) structural optimization in space group $I2_12_12_1$ (unit cell, a = 9.8019(1) Å, b = 12.1799(2) Å and c = 17.3386(2) Å). The DFT calculations reveal that the three-dimensional structure may be stabilized by weak dispersive interactions between apolar moieties and that these may be disordered. Nanoparticles and heat treatment (at T > 56 °C) produce a highly conductive composite, σ(Mg^{2+}) = 2.86 × 10^{-7}, and 2.85 × 10^{-5} S cm^{-1} at −10 and 40 °C, respectively, with a low activation energy, E_a = 0.65 eV. Nanoparticles stabilize the partially eutectic molten state and prevent recrystallization even at low temperatures and provide a high mechanical stability of the composite.

Keywords: structure; ionic conductivity; complex borohydride; mechanochemistry

Citation: Kristensen, L.G.; Amdisen, M.B.; Andersen, M.; Jensen, T.R. Synthesis, Structure and Mg^{2+} Ionic Conductivity of Isopropylamine Magnesium Borohydride. *Inorganics* **2023**, *11*, 17. https://doi.org/10.3390/inorganics11010017

Academic Editor: Rainer Niewa

Received: 18 November 2022
Revised: 6 December 2022
Accepted: 15 December 2022
Published: 30 December 2022

Copyright: © 2022 by the authors. Licensee MDPI, Basel, Switzerland. This article is an open access article distributed under the terms and conditions of the Creative Commons Attribution (CC BY) license (https://creativecommons.org/licenses/by/4.0/).

1. Introduction

Due to the increased electrification of society, new ways of storing energy need to be developed, e.g., with significant improvement of the gravimetric and volumetric capacity as well as the safety of batteries, while simultaneously keeping the costs down. In recent years, the increase in capacity in Li-ion batteries using carbon anodes has stagnated and is approaching the theoretical limit [1]. The development of "post-Li-ion" batteries is therefore of critical importance [2]. There are increasing efforts towards sustainability, and the use of more abundant elements, such as sodium and magnesium, could cut the cost of the raw materials both economically and environmentally [3]. The all-solid-state battery is a new promising technology, which is expected to provide several advantages, such as easier assembling and production, as well as an improved safety profile and lifetime. This is partly related to the solid electrolyte, which is expected to be made thinner, provide faster charging and discharging rates, help to avoid short-circuits in the battery, and increase thermal stability during cooling, as the compounds are already solid when compared to traditional liquid electrolytes [4]. The Achilles heel appears to be in creating fast ionic conductivity in the solid state, in particular for divalent cations, such as Mg^{2+}.

Magnesium appears to be less prone to forming dendrites when compared to lithium, however, obtaining fast ionic conductivity in the solid state of divalent cations is very challenging. This is particularly true at relevant ambient temperatures where very few solid-state magnesium-based electrolyte materials have been reported, as compared to monovalent cations such as Li$^+$ and Na$^+$. In recent years, novel magnesium borohydrides have provided high Mg^{2+} conductivities even at low temperatures ($\approx 10^{-4}$ S cm^{-1} at 40 °C), making this class of materials very interesting [5–9]. These new compounds are built using ionic and covalent bonds and therefore have very low electronic conductivity. Furthermore, metal borohydrides have the benefit of being compatible with metal anodes, which increase both the gravimetric and volumetric densities in a final cell [10–12]. However, cationic conductivity still needs to be further increased for battery applications and challenges, and contact and electrochemical stability issues must be addressed [13]. Replacing lithium is challenging, since lithium is light, mobile in the solid state and since compatible electrodes often operate with a large electrochemical stability window [2].

Detailed knowledge of the phenomena that are responsible for fast cation conductivity in the solid state may lead to rational design of novel materials. Recently, borohydrides have shown greatly increased ionic conductivities when coordinated to a neutral ligand [7–9,14]. Furthermore, thermal and mechanical properties are also altered, allowing for malleable compounds or even liquid-like electrolytes [8,15]. Here we present the synthesis, structure, and ionic conductivity of a new type of monoisopropylamine magnesium borohydride, $Mg(BH_4)_2 \cdot (CH_3)_2CHNH_2$, and incorporation of this complex into a nanocomposite.

2. Results

Initial Characterization. A new synthesis method combining solvent-based techniques to prepare the reactants, and mechanochemistry to form the product has been developed [16,17]. Magnesium borohydride, α-$Mg(BH_4)_2$ (denoted, **s1**), was synthesized by a solvothermal method and then dissolved in isopropylamine (IPA), $(CH_3)_2CHNH_2$. A white solid was filtered off and dried in vacuum (25 °C, 40 min) to form diisopropylamine magnesium borohydride, $Mg(BH_4)_2 \cdot 2(CH_3)_2CHNH_2$ (**s4**). A new crystalline compound was formed by mechanochemical treatment of α-$Mg(BH_4)_2$ and $Mg(BH_4)_2 \cdot 2(CH_3)_2CHNH_2$ in the molar ratio 1:1 at 350 rpm with a total milling time of 120 min. Mechanochemical treatment at a shorter time or with lower ball-to-powder ratio provides partly reacted products (sample **s2**). This is illustrated in Figure 1, which provides diffraction patterns of the reactants α-$Mg(BH_4)_2$ (**s1**) and $Mg(BH_4)_2 \cdot 2(CH_3)_2CHNH_2$ (**s4**) and the new compound (**s3**). The diffraction pattern of the sample **s3** is assigned to a single crystalline phase and all observed Bragg reflections can be accounted for by an orthorhombic unit cell, $a = 9.8019(1)$ Å, $b = 12.1799(2)$ Å and $c = 17.3386(2)$ Å, which is similar to previously proposed data for the suggested composition, $Mg(BH_4)_2 \cdot (CH_3)_2CHNH_2$ [6].

Structural analysis. The successful synthesis of single-phase monoisopropylamine magnesium borohydride, $Mg(BH_4)_2 \cdot (CH_3)_2CHNH_2$, allowed for the measurement of high-quality synchrotron radiation powder X-ray diffraction (SR-PXD) data (see Figure 1). The structure was solved ab initio, in the orthorhombic space group $I2_12_12_1$, using direct space methods (implemented in the program FOX). Several other structural models were investigated but rejected due to an unsatisfactory fit to the experimental diffraction data, unrealistic coordination and/or instability when optimised by density functional theory (DFT), see supporting information. Several cycles of DFT structural optimization and Rietveld refinement were conducted to develop the final structural model. The structure is composed of magnesium in two different coordination environments. In the first environment, Mg^{2+} is coordinated to four BH_4^-, similar to the structure of α-$Mg(BH_4)_2$, however the coordination is planar, whereas it is tetrahedral in α-$Mg(BH_4)_2$. In the second environment, magnesium is tetrahedrally coordinated to two BH_4^- and two $(CH_3)_2CHNH_2$ molecules. These structural units are bridged by BH_4^- to form a zigzag 1D chain propagating along the a-axis, forming a 'flat helix' as shown in Figure 2. From DFT it was found that bridging tetrahydridoborohydride (BH_4^-) complexes, Mg–BH_4^-–Mg, have a bidentate κ^2 coordination to Mg in $[Mg(BH_4)_4]$ and a tridentate κ^3 coordination to Mg in $[Mg(BH_4)_2(NH_2CH(CH_3)_2)_2]$. Terminal BH_4^- was found to have a bidentate κ^2 coordination to Mg. Mg–N (2.08 Å) and terminal Mg–B (2.20 Å) distances are similar to what have been reported previously [6,7,18–20].

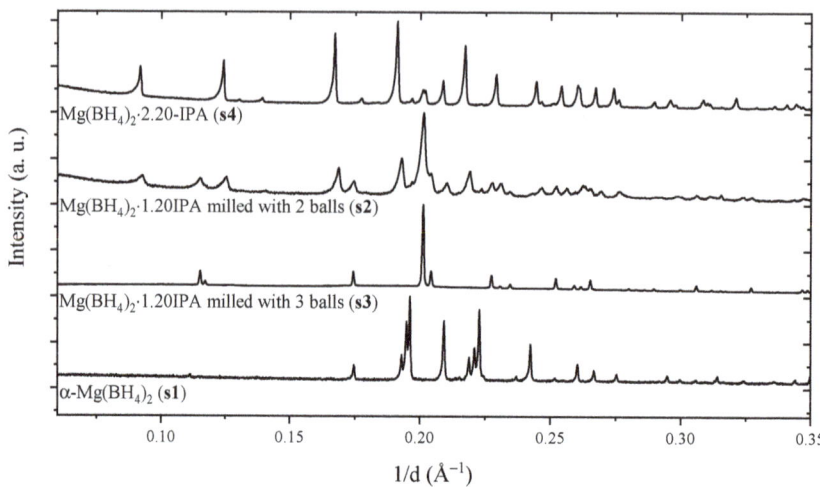

Figure 1. Powder X-ray diffraction patterns of samples **s1** to **s4** measured at the I11 beamline at the Diamond light source, Oxford (UK) (**s3**, λ = 0.826366(3) Å) and in-house (**s1**, **s2**, **s4**, λ = 1.5406 Å). The pattern of sample **s3** is distinct to that of the reactants, α-Mg(BH$_4$)$_2$ (**s1**) and Mg(BH$_4$)$_2$·2IPA (**s4**), and is assigned to a new compound, Mg(BH$_4$)$_2$·(CH$_3$)$_2$CHNH$_2$. Sample **s2** was treated in a less intense manner by mechanochemistry, which resulted in an incomplete reaction.

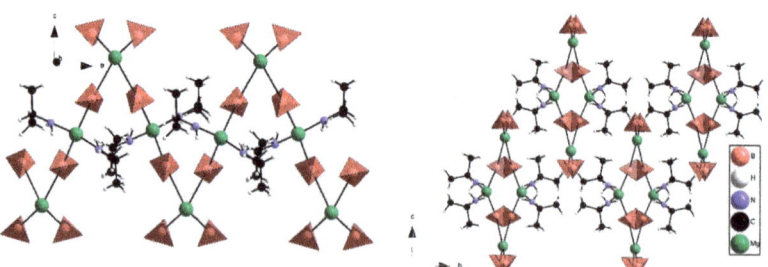

Figure 2. Experimental structure of Mg(BH$_4$)$_2$·(CH$_3$)$_2$CHNH$_2$ viewed in the *a-c* plane (**left**) and *b-c* plane (**right**). In the *a-c* plane the chain-like structure with alternating Mg^{2+} environments is visible. The *b-c* plane shows how the chains arrange to form flat polar helixes, which are surrounded by apolar regions. Atoms: magnesium (green), carbon (black), nitrogen (blue), BH$_4^-$ (red tetrahedron) and hydrogen (white).

The bridging BH$_4^-$ group is not placed in the center between Mg$_{(1)}$ (Mg in [Mg(BH$_4$)$_4$]) and Mg$_{(2)}$ (Mg in [Mg(BH$_4$)$_2$(NH$_2$CH(CH$_3$)$_2$)$_2$]). The distances are Mg$_{(1)}$–B = 3.14 Å and Mg$_{(2)}$–B = 2.57 Å for the experimental structure and Mg$_{(1)}$–B = 2.60 Å and Mg$_{(2)}$–B = 2.33 Å for the DFT-optimized structure, respectively. This suggests that the bridging BH$_4^-$ group is more tightly bound to Mg$_2$, as compared to Mg$_1$. The apolar regions in the structure formed by CH$_3$ groups in neighboring IPA molecules are close (C–C, 2.5–2.7 Å). As further discussed below, weak dispersive interactions between the IPA molecules may play a role in stabilizing the 3D structure in the *b,c*-plane, as seen in Figure 2 (right). Furthermore, the DFT results suggest a certain degree of disorder of the (CH$_3$)$_2$CH– moiety, which could explain the discrepancy between the observed and calculated diffraction patterns as seen in Figure 3.

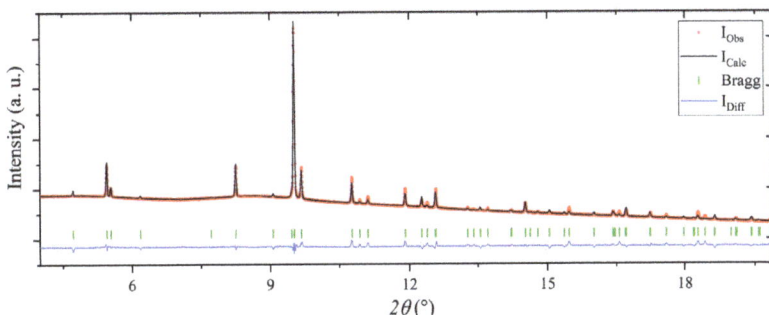

Figure 3. Rietveld refinement of the structural model of $Mg(BH_4)_2 \cdot (CH_3)_2CHNH_2$ in the space group $I2_12_12_1$, using the powder X-ray diffraction pattern of $Mg(BH_4)_2 \cdot 1.20IPA$ (**s3**). R-factors: $R_p = 1.26$, $R_{wp} = 2.47$, conventional R-factors: $R_p = 37.6$ and $R_{wp} = 23.2$ and $R_{Bragg} = 20.3$.

The DFT structural optimization was done with two different exchange–correlation functionals with and without accounting for van-der-Waals (vdW) interactions, beginning from the experimental structure shown in Figure 2. Independently of the used functions, it was found that the experimental structure is metastable and that it becomes increasingly disordered upon further refinement. This indicates a 'flat energy landscape' between the experimental structure and various disordered structural models (see Figures S1–S4 and Table S1 in the supporting information). In all DFT optimizations (with and without vdW), the distance between CH_3 groups in neighboring IPA molecules increases to about 3.5–3.9 Å. This indicates that the interactions between the CH_3 groups are weak. Calculations for the butane dimer reveal typical vdW C–C distances of around 4 Å [21]. However, the chemical bonding scheme in all the different structural models, both experimental and theoretical, is the same but with different degrees of structural distortion. The distortion primarily occurs for the $(CH_3)_2CH–$ moieties, which indicates that more than one conformation of these could exist, resulting in static or dynamic disorder to the structure. The DFT results also reveal that some H atoms in the neighboring borohydride and amine groups are close ($d_{H-H} = 2.5–2.70$ Å), which suggests that di-hydrogen bonds, $B-H^{\delta -} \cdots ^{+\delta}H-N$, bind the 1-D chains together.

Thermal analysis. Figure 4 shows *in situ* temperature-resolved synchrotron radiation powder X-ray diffraction of $Mg(BH_4)_2 \cdot 1.20IPA$ (**s3**) in the temperature range from 20 to 75 °C. Initially, the sample contained one crystalline compound, $Mg(BH_4)_2 \cdot (CH_3)_2CHNH_2$. However, during heating, a new set of diffraction peaks appeared at 31 °C, which are assigned to the crystallization of $Mg(BH_4)_2 \cdot 2(CH_3)_2CHNH_2$. Sample analysis using liquid-state ^1H NMR reveals that **s3** contains 1.20 IPA per $Mg(BH_4)_2$ (see Table 1). Therefore amorphous $Mg(BH_4)_2 \cdot 2(CH_3)_2CHNH_2$ may be recrystallized upon heating or surface-adsorbed IPA may react with monoisopropylamine magnesium borohydride through a gas–solid reaction. Notice that the boiling point of IPA is 32 °C. At 58 °C, the diffraction from $Mg(BH_4)_2 \cdot 2(CH_3)_2CHNH_2$ disappears, which is at a significantly lower temperature than previously reported (87 °C) [6]. This may indicate eutectic melting of the composite $Mg(BH_4)_2 \cdot (CH_3)_2CHNH_2$-$Mg(BH_4)_2 \cdot 2(CH_3)_2CHNH_2$ since the diffracted intensity from both compounds decreases above 51 °C. After the disappearance of diffraction from $Mg(BH_4)_2 \cdot 2(CH_3)_2CHNH_2$ at 58 °C, the remaining diffracted intensity from $Mg(BH_4)_2 \cdot (CH_3)_2CHNH_2$ further decreases in intensity and disappears at 73 °C. After storage for 4 months at RT, the diffraction data of **s3** measured at RT revealed the crystallization of minor amounts of $Mg(BH_4)_2 \cdot 2(CH_3)_2CHNH_2$.

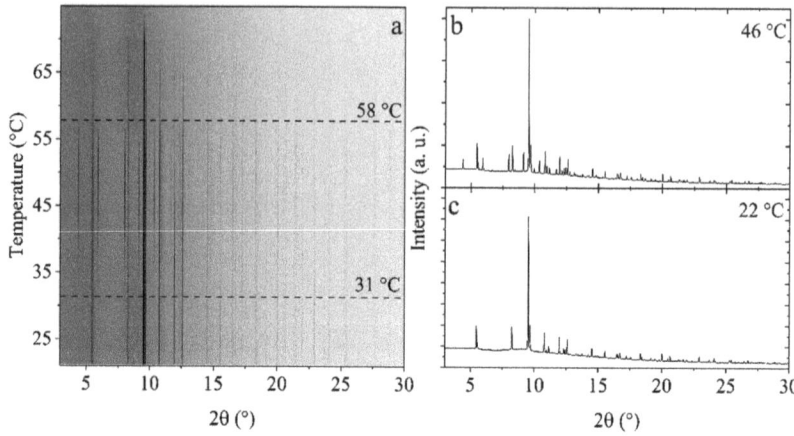

Figure 4. (**a**) Temperature-resolved *in situ* synchrotron radiation powder X-ray diffraction of Mg(BH$_4$)$_2$·1.20IPA (**s3**) from 20 to 75 °C (heating rate of 2 °C/min). (**b,c**) Selected diffraction patterns at 46 and 22 °C, respectively. The data were acquired at the I11 beamline of the Diamond light source (λ = 0.826366 Å).

Table 1. Overview of the investigated samples including synthesis method, molar reactant ratios, crystalline compounds observed in the product using powder X-ray diffraction, and the (CH$_3$)$_2$CHNH$_2$ content as measured by liquid-state ^{1}H nuclear magnetic resonance.

Sample	Synthesis Method	Reactants	Reactant Ratio (s3/(s1 + s3))	Crystalline Compounds	IPA Content (^{1}H NMR)
s1	Solvent-based	-	-	α-Mg(BH$_4$)$_2$	-
s2	Mechanochem.	s1 & s4	0.44	Mg(BH$_4$)$_2$·(CH$_3$)$_2$CHNH$_2$	1.20
s3	Mechanochem.	s1 & s4	0.44	Mg(BH$_4$)$_2$·(CH$_3$)$_2$CHNH$_2$	1.20
s4	Solvent-based	-	-	Mg(BH$_4$)$_2$·2(CH$_3$)$_2$CHNH$_2$	2.20
s5	Mechanochem.	s3 & Al$_2$O$_3$	-	Mg(BH$_4$)$_2$·(CH$_3$)$_2$CHNH$_2$ + Al$_2$O$_3$ (50 wt%)	1.20

Figure 5 displays the thermal analysis of Mg(BH$_4$)$_2$·1.20(CH$_3$)$_2$CHNH$_2$ (**s3**), i.e., thermogravimetric analysis (TGA), differential scanning calorimetry (DSC) and mass spectrometry (MS) measured simultaneously for the same sample. A thermal DSC event was observed at 56 °C, in accordance with the disappearance of diffraction from Mg(BH$_4$)$_2$·2(CH$_3$)$_2$CHNH$_2$, and assigned to eutectic melting of a fraction of the sample. This event is accompanied by a weak indication of the release of hydrogen observed by MS. In the temperature range of 108 to 138 °C, a mass loss of $\Delta m/m$ = 6.59 wt% corresponding to the release of 0.14 molecules of IPA per formula unit was observed, and the remaining sample composition is ~1 IPA per formula unit. Release of IPA in this temperature range, 108 to 138 °C, was also detected by MS, accompanied by a minor release of hydrogen. In the temperature range of 138 to 225 °C a major mass loss of $\Delta m/m$ = 42.61 wt% was observed, corresponding to a loss of 0.9 IPA. Mass spectroscopy reveals an increasing release of IPA and H$_2$ with a maximum release rate around 200 °C. The release of hydrogen suggests a chemical reaction between the organic moiety (IPA) and the borohydride complex. A total mass loss of 49.20 wt% was observed in the temperature range of RT to 225 °C.

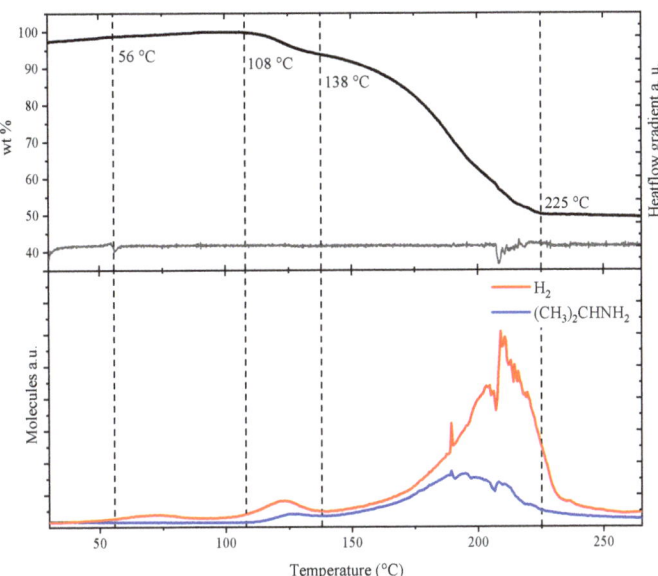

Figure 5. Simultaneous measurement of thermogravimetry (TGA), differential scanning calorimetry (DSC) and mass spectroscopy (MS) of Mg(BH$_4$)$_2$·IPA (**s3**). (**Top**) TGA data and the gradient of the differential scanning calorimetry data. (**Bottom**) MS analysis of hydrogen and isopropylamine release.

Magnesium ionic conductivity. The Mg^{2+} ionic conductivity of Mg(BH$_4$)$_2$·1.20(CH$_3$)$_2$-CHNH$_2$ (**s3**) was measured by electrochemical impedance spectroscopy (EIS) during heating from −7.9 to 40 °C, see Figure 6. The same sample (**s3**) was then heated to 40 °C and cooled to 30 °C three times in order to convert amorphous material to Mg(BH$_4$)$_2$·2(CH$_3$)$_2$CHNH$_2$ and to further stabilize the tablet prior to the second EIS measurement. Activation energies, E_A, were extracted from a plot of log(σT) versus $1/T$ as described in the experimental section. The electronic conductivity of this class of compounds is negligible [5,8,9].

The Mg^{2+} ionic conductivity of Mg(BH$_4$)$_2$·1.20IPA (**s3**) is high and increases exponentially in the temperature range of −10 to 40 °C, see Figure 6. The conductivity is slightly higher after thermal treatment, i.e., of the composite (~0.8)Mg(BH$_4$)$_2$·(CH$_3$)$_2$CHNH$_2$ − (~0.2)Mg(BH$_4$)$_2$·2(CH$_3$)$_2$CHNH$_2$. Noteworthy, a low activation energy was observed for the first measurement, E_A = 0.93 eV of **s3**, which further decreased to E_A = 0.86 eV after heat treatment, i.e., of the composite (~0.8)Mg(BH$_4$)$_2$·(CH$_3$)$_2$CHNH$_2$−(~0.2)Mg(BH$_4$)$_2$·2(CH$_3$)$_2$CHNH$_2$.

Previous investigations reveal a significant increase in the Mg^{2+} conductivity of composites containing different crystalline isopropylamine magnesium borohydride compounds, which was also observed here. Furthermore, the compounds were able to strip and plate magnesium with an oxidative stability of 1.2 V vs. Mg/Mg^{2+} [6]. Adding 50 wt% Al$_2$O$_3$ (13 nm) to the sample (**s3**) resulted in sample **s5**. Initially, the conductivity was similar to the original sample; however, as the sample was heated to above the melting point of the Mg(BH$_4$)$_2$·2(CH$_3$)$_2$CHNH$_2$ (at 56 °C) (see Figure 4), the conductivity significantly increased (see Figure S4). The sample deformed during this transition and became thinner and wider. This was accounted for during calculation of the conductivity. During cooling, this highly conductive state with low activation energy (0.60 eV) was maintained to 30 °C. The highly conductive state shows a similar (but lower) conductivity to that of Mg(BH$_4$)$_2$·1.5(CH$_3$)$_2$CH–NH$_2$ containing 50 nm MgO particles [6]. This difference is due to the lower content of the eutectic molten state in Mg(BH$_4$)$_2$·1.20IPA-Al$_2$O$_3$ (**s5**) as compared to Mg(BH$_4$)$_2$·1.5-MgO. As the sample deformed during heating, a new pellet was made by applying a pressure of 2.5 Gpa for 30 s, releasing the pressure and then heating to 65 °C while in the press. Finally, a pressure of 2.5 Gpa was applied to the sample for 30 s at RT. This resulted in a highly viscous sample, likely due to the melt stabilization effect mentioned in refs. [5,8]. Due to the soft nature of the sample, the sample was relaxed onto the electrodes at RT for 16 h before measurements to ensure optimal contact. The conductivity of Mg(BH$_4$)$_2$·1.20IPA−Al$_2$O$_3$ (50 wt%, **s5**) is shown in Figure 6.

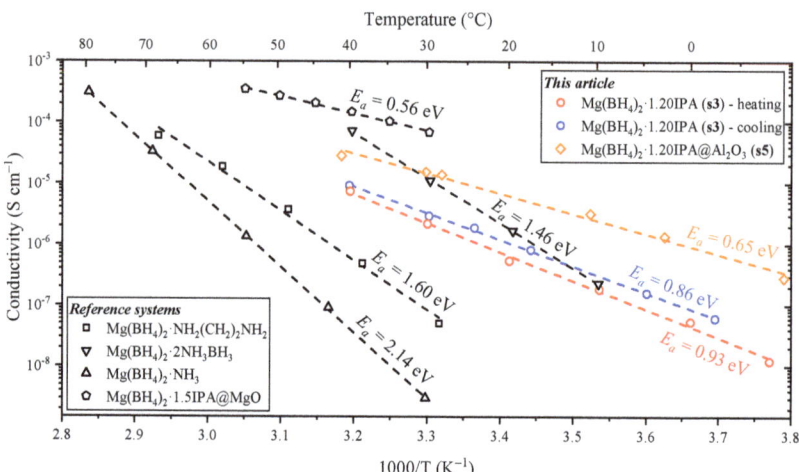

Figure 6. Magnesium (Mg^{2+}) ionic conductivity as a function of temperature measured by electrochemical impedance spectroscopy (EIS) of Mg(BH$_4$)$_2$·(CH$_3$)$_2$CHNH$_2$ (**s3**) during heating from −7.9 to 40 °C (data shown as circles). The same sample (**s3**) was then heated and cooled three times between 30 and 40 °C before being cooled to −2.6 °C and measured by EIS during heating to 40 °C (open spheres). Nyquist plots of Mg(BH$_4$)$_2$-1.20IPA-Al$_2$O$_3$ (**s5**) are provided in Figure S6. Magnesium ionic conductivity data and activation energies of Mg(BH$_4$)$_2$·NH$_2$(CH$_2$)$_2$NH$_2$, Mg(BH$_4$)$_2$·NH$_3$ and Mg(BH$_4$)$_2$·2NH$_3$BH$_3$, and of Mg(BH$_4$)$_2$·1.5IPA−MgO are also included for comparison (data from refs. [6,7,9,22]).

The "paste-like" nature of sample **s5** was maintained even after storage at −18 °C for 24 h, indicating that recrystallization was inhibited by the presence of nanoparticles. The thermal stability of Mg(BH$_4$)$_2$·1.20(CH$_3$)$_2$CHNH$_2$ (**s3**) was limited by the presence of ~20% Mg(BH$_4$)$_2$·2(CH$_3$)$_2$CHNH$_2$, which represents eutectic melting with a fraction of the sample at T = 56 °C. Thus, this sample (**s3**) has low thermal stability as compared to other similar compounds, i.e., Mg(BH$_4$)$_2$·2NH$_3$BH$_3$ (47 °C) [9], Mg(BH$_4$)$_2$·NH$_2$(CH$_2$)$_2$NH$_2$ (75 °C) [22] and Mg(BH$_4$)$_2$·NH$_3$ (90 °C) [7]. However, the activation energy for Mg^{2+} cationic conductivity is significantly lower for Mg(BH$_4$)$_2$·(CH$_3$)$_2$CHNH$_2$ (**s3**), E_A = 0.86 eV, as compared to similar compounds presented in Figure 6 This advantage is even more pronounced for the nanocomposite Mg(BH$_4$)$_2$·(CH$_3$)$_2$CHNH$_2$−Al$_2$O$_3$ (**s5**), with an activation energy of 0.65 eV. Thus, the advantage of this new material is a moderate temperature dependence of the cationic conductivity, which has an exceptionally high value at low temperatures. Furthermore, the soft nature of the sample may improve contact with the electrodes, which is a major challenge in these all-solid-state systems.

3. Discussion

Recently, the Mg^{2+} conductivity mechanism of Mg(BH$_4$)$_2$·NH$_3$ was investigated using diffraction, structure refinements and DFT. The structure of Mg(BH$_4$)$_2$·NH$_3$ was found to be very flexible owing to a three-dimensional network of di-hydrogen bonds, B−H$^{\delta−}$···$^{+\delta}$H−N, but also to the BH$_4^-$−Mg^{2+} coordination, which varies from in edge to corner coordination (κ^1 to κ^3). Furthermore, the migration of Mg^{2+} cations is also assisted by a neutral molecule, NH$_3$, which is exchanged between the framework and interstitial magnesium [7,23]. The IPA analogue, $Mg(BH_4)_2 \cdot (CH_3)_2CHNH_2$, investigated here, resembles the above-mentioned compound by having a one-dimensional chain-like structure and that the organic and inorganic moieties are interconnected by weak interactions.

In the following section we discuss and compare the experimental and DFT-optimized structural models, as well as the role of weak dispersive interactions. In the experimental structure obtained from the Rietveld refinement, see Figure 2, the isopropyl groups point towards each other: C−H···H−C (d_{H-H} = 1.9–3.4 Å). From DFT we find that this ordered structure is metastable compared to more disordered structures. The metastable DFT structure and the experimental structure are very similar, except for the terminal BH_4^- which is closer to the isopropyl groups in the experimental structure, as seen in Figure S2. This could hint at a B−H$^{\delta-}$···H−C interaction, which is absent in the DFT structure. In the more disordered (and more stable) DFT structures, we found a distortion of the helixes; see Figures S1 and S3. While the helixes remain, terminal BH_4^- moves from a planar coordination to a tetrahedral coordination. Terminal $Mg_{(1)}$–B distances increased from 2.2 Å to 2.4 Å and BH_4^- has a bidentate (κ^2) coordination compared to the tridentate (κ^3) coordination found from the Rietveld refinement. For bridging $Mg_{(1)}$–B–$Mg_{(2)}$, the BH_4^- group is still not placed in the center between $Mg_{(1)}$ and $Mg_{(2)}$ (B–((IPA)$_2Mg_{(2)}$)–B = 2.57 Å and B–((BH_4^-)$_2Mg_{(1)}$)–B = 2.31 Å) with a bidentate (κ^2) coordination to $Mg_{(1)}$ and a tridentate (κ^3) coordination to $Mg_{(2)}$. Both the experimental and theoretical models found a coordination number of eight for Mg^{2+}. However, experimental hydrogen positions obtained by the Rietveld refinement are a result of anti-bump restraints in the ab initio structural solution process and are not refined.

In the more disordered DFT models, the C−H···H−C distances significantly increased (d_{H-H} = 2.5–3.7 Å) compared to the experimental structure. However, in all structures, the neighbouring IPA molecules are rather close and within distances that are typical for weak dispersive interactions [21]. It therefore seems plausible that such interactions are important for stabilizing the 3D structure. We did not observe any significant difference between DFT structures optimized with and without account for van der-Waals interactions, but this may be because the unit cell is kept fixed at the experimentally measured size during the relaxation, which limits how far the different groups are able to move with respect to each other. In Figure S3, several DFT structures that lie between the metastable and most stable structure in energy can be seen. The existence of these shallow local minima on the potential energy surface (and likely many more) supports the claim that the sample is highly disordered due to the flat energy landscape. The disordered DFT-optimized structural models often have a slightly poorer fit to the diffraction data, which relates to the very different structural descriptions. DFT provides an atomic scale 'momentary' view of the unit cell, whereas diffraction provides an average structure over time and space. Furthermore, DFT describes the 0 K potential energy surface, whereas the powder pattern was measured at room temperature. Failure to converge to a refined model has also been observed for ammine rare earth borohydrides $RE(BH_4)_3 \cdot 4NH_3$ (RE^{3+} = La, Ce, Pr, Nd) [24,25].

Thus, both DFT and diffraction reveal a new composition and structure of the compound investigated here, which has some degree of structural disorder and is held together by weak interactions. The high Mg^{2+} ionic conductivity and moderate activation energy for cation migration are assigned to these structural properties. Composites of crystalline materials, $Mg(BH_4)_2 \cdot x(CH_3)_2CHNH_2$, x = 1 or 2, have higher conductivity and eutectic melting. The molten state can be stabilized by nanoparticles to form a mechanically and thermally more stable nanocomposite. This nanocomposite has higher Mg^{2+} conductivity and lower activation energy, which has previously been assigned to surface effects, e.g., the wetting of nanoparticles by thin layers of eutectic molten liquid. These new phenomena for the rational design of functional battery materials are demonstrated in this work.

4. Materials and Methods

4.1. Synthesis

Magnesium borohydride, $Mg(BH_4)_2$, was synthesized as described in refs. [26,27]. Anhydrous toluene (purity 99.8%, 40 mL) and anhydrous dimethylsulphide borane, $(CH_3)_2SBH_3$ (90% in toluene, 10 mL) were added to a round-bottomed flask. While stirring, di-*n*-butylmagnesium $Mg(C_4H_9)_2$ (1.0 M in heptane with up to 1 wt% triethylaluminum, 28 mL) was slowly added within 2 min, ensuring an excess of $(CH_3)_2SBH_3$ at all times [26]. A white precipitate immediately formed upon addition of $Mg(C_4H_9)_2$ and the reaction continued for 20 h with stirring at room temperature. The product, $Mg(BH_4)_2 \cdot \frac{1}{2}S(CH_3)_2$, was washed with toluene and heated to 143 °C for 4.5 h in a evacuated Schlenk tube to form α-$Mg(BH_4)_2$, denoted sample **s1**.

Diisopropylamine magnesium borohydride, $Mg(BH_4)_2 \cdot 2(CH_3)_2CHNH_2$ was synthesized as described in ref. [6], by dissolving finely ground α-$Mg(BH_4)_2$ (300 mg) in 1.2 mL isopropylamine, $(CH_3)_2CHNH_2$, IPA) while stirring. The reaction was allowed to continue for 30 min in an ice-bath, after which the product was dried for 40 min in vacuum at 25 °C and denoted **s4**.

Monoisopropylamine magnesium borohydride, $Mg(BH_4)_2 \cdot (CH_3)_2CHNH_2$ was mechanochemically synthesized using $Mg(BH_4)_2 \cdot 2(CH_3)_2CHNH_2$ (**s4**, 277.9 mg, 1.5 mmol) and α-$Mg(BH_4)_2$ (**s1**, 101.9 mg, 1.9 mmol). The reactants were ball-milled in a WC vial using three 10 mm WC balls for 60 repetitions of 2 min, with the milling intervened by 2 min of pause, i.e., total milling time 120 min and a ball-to-sample ratio of ~54/1. This sample is denoted **s3**.

A nanocomposite was prepared by adding nano particulate, 50 wt% Al_2O_3 (13 nm), to sample **s3** mechanochemically. Al_2O_3 (100.1 mg) and **s3** (100.5 mg) were milled at 350 rpm for 2 min intervened by 2 min of pause, i.e., total milling time 120 min and a ball-to-sample ratio of ~68/1.

4.2. Characterization

Powder X-ray diffraction (PXD) data were obtained using a Rigaku Smartlab diffractometer equipped with a monochromatic rotating Cu source (λ = 1.54056 Å). Sample preparation was done in an argon environment where samples were packed in 0.5 mm (outer diameter) borosilicate capillaries and sealed with grease to avoid air exposure.

In situ synchrotron radiation powder X-ray diffraction (SR-PXD) data were acquired at the I11 beamline at the Diamond light source, Oxford, UK (λ = 0.826366 Å) [28]. The samples were measured in the temperature range from 20 to 75 °C at a heating rate of 2 °C per min.

The software FOX was used for indexing of the unit cell whereafter the structure was solved by ab initio structure determination [29,30]. Subsequently, atomic positions were refined by Rietveld refinements with the software Fullprof, treating BH_4^- and $(CH_3)_2CHNH_2$ as rigid bodies [30,31].

Density functional theory (DFT) structural optimization was carried out using the GPAW code v. 21.1.0 with a plane wave basis set [32,33], the Atomic Simulation Environment (ASE) software package [34] and the exchange–correlation functionals PBE [35] and BEEF-vdW [36]. Several structure optimization runs were carried out using different plane-wave cutoff energies between 340 eV and 550 eV and either (1·1·1) or (2·2·2) k point sampling. The optimizations were initiated from the experimental structure as shown in Figure 2 and terminated when the maximum force on any atom fell below 0.01 eV/Å. The small changes to the numerical settings caused the optimization runs to find different local minima on the potential energy surface, corresponding to a structure that resembled the experimental one, as well as several more disordered structures. During the optimizations, all atoms were allowed to relax, and the unit cell was kept fixed to the experimentally determined size. The structures and energies presented in the supporting information were obtained from a final structural optimization run of the obtained minimum energy structures using a cutoff energy of 500 eV and (1·1·1) k point sampling.

Nuclear magnetic resonance (NMR) measurements were performed on a Bruker Ascend 400 MHz spectrometer equipped with a ^{1}H–^{13}C–^{15}N 5 mm TXI liquid state probe. Samples were dissolved in deuterated dimethylsulfoxide in NMR tubes before measuring. The integrated intensities of ^{1}H on BH_4^- and isopropylamine were used to calculate the sample composition. The integral of these peaks was normalized to their abundance in the sample as described in ref. [6].

Thermogravimetric analysis (TGA) and differential scanning calorimetry (DSC) were measured using a PerkinElmer STA 6000 coupled with a mass spectrometer (MS) (Hiden Analytical HPR-20 QMS sampling system). Approximately 2 mg was placed in a closed Al_2O_3 crucible and was heated from 30 to 300 °C ($\Delta T/\Delta t$ = 2 °C/min) with an argon purge rate of 30 mL/min. The lid had a small hole for outlet gas, which was examined using mass spectrometry for hydrogen (m/z = 2) and isopropylamine (m/z = 44).

Electrochemical impedance spectroscopy (EIS) measurements were performed from $7 \cdot 10^7$ Hz to 1 Hz using a Biologic MTZ-35 impedance analyzer with a symmetrical molybdenum sample holder equipped with a 4 probe setup. Samples were pressed in a hydraulic press at 1 tonne at RT for 1 min. Samples were heated using a custom-made furnace at 2 °C/min and kept at a constant temperature for one minute when the set temperature was reached. Data were fitted using an $Q_1/(R_1+Q_2)$ equivalent circuit, where R_1 represents the charge transfer resistance and Q_1 and Q_2 are constant phase elements. Q_1 accounts for the depressed semicircles of real systems and Q_2 is used to represent the mass transfer, as the observed mass transfer cannot be described by standard capacitors or Warburg elements. When fitted, Q_1 will act as a capacitor (α~0.97), essentially creating an RC circuit with R_1 and Q_2 acting as mass transfer elements (α~0.7). From the charge transfer resistance, ionic conductivity is calculated as $d/A/R_1$, where d is the thickness of the pellet and A is the area. Activation energies (E_a) were extracted from linear fits to $\log(\sigma_i)$ versus $1/T$, where σ_i is the ionic conductivity. Because of the 4 probe setup, we assumed the resistance of the setup to be negligible. First, the sample was cooled to -7.9 °C using an ethanol and dry ice bath where it was held for 15 min to ensure the temperature had stabilized. Using a heat blower, the mixture was brought up to 0 °C where dry ice was added to stabilize the temperature. Above RT, the custom-made oven was used instead. The sample was subjected to temperature increases between 30 and 40 °C three times before being cooled to -2.6 °C. Due to sample s2 becoming soft, EIS measurements were only conducted up to 40 °C.

5. Conclusions

The compound $Mg(BH_4)_2 \cdot (CH_3)_2CHNH_2$ was synthesized using a mechanochemical method. From SR-PXD it was found to have a polar helical 1D structure consisting of alternating magnesium environments ($[Mg(BH_4)_4]$ and $[Mg(BH_4)_2(NH_2CH(CH_3)_2)_2]$). These are surrounded by weakly interacting apolar layers to form a 3D structure. The structure was solved in $I2_12_12_1$ with the unit cell parameters a = 9.8019(1) Å, b = 12.1799(2) Å and c = 17.3386(2) Å. Discrepancies between the refined model and collected data were attributed to disorder, especially from the weakly interacting isopropyl groups $(CH_3)_2CH-$. From DFT optimization it was found that the structure is metastable and that more stable distorted structures could exist. This could also explain the discrepancies between the refined model and the data. From SR-PXD, the compound was found to have a melting point of 73 °C while $Mg(BH_4)_2 \cdot 2(CH_3)_2CHNH_2$ formed during heating and melted at 56 °C, which is lower than reported previously. This is assigned to eutectic melting. From TGA-DSC-MS investigations, IPA release first occurs at 108 °C however this might be due to surface-coordinated IPA being released. The compound was found to have a very low activation energy of 0.85 eV, which is likely caused by weak hydrophobic interactions in interstitial sites. The composite (~0.8)$Mg(BH_4)_2 \cdot (CH_3)_2CHNH_2-$(~0.2)$Mg(BH_4)_2 \cdot 2(CH_3)_2CHNH_2$ was melted onto 13 nm Al_2O_3 nanoparticles at 60 °C while in a compressed state after applying a pressure of 2.5 GPa. This formed a paste-like material which lowered the activation energy to 0.65 eV and increased conductivity at 40 °C by a factor of three and at $-10\ °C$ by a factor of ten. This paste-like state remains stable for more than 24 h and at temperatures lower than $-10\ °C$.

Supplementary Materials: The following supporting information can be downloaded at: https://www.mdpi.com/article/10.3390/inorganics11010017/s1, Figure S1: Most stable structure of $Mg(BH_4)_2·(CH_3)_2CHNH_2$, Figure S2: Comparison between the metastable structures, Figure S3: Structures gained from DFT, Figure S4: Rejected model found from ab inito structural solution, Figure S5: Ionic conductivity of $Mg(BH_4)_2·IPA@Al_2O_3$ (s5) without preheating, Figure S6: Nyquist plot of $Mg(BH_4)_2·1.2IPA-Al_2O_3$; Table S1: DFT relative potential energies.

Author Contributions: Formal analysis, investigation, data curation, visualization, L.G.K.; Formal analysis, investigation, data curation, M.B.A.; methodology, software, validation, resources, M.A.; Conceptualization, supervision, project administration, funding acquisition T.R.J.; All authors contributed to writing—original draft preparation and to review and editing; All authors have read and agreed to the published version of the manuscript.

Funding: The work was supported by the Danish Council for Independent Research, Technology and Production Solid-State Magnesium Batteries—SOS-MagBat (9041-00226B) and the Danish Natural Science Research Councils (DanScatt). Funding from the Danish Ministry of Higher Education and Science through the SMART Lighthouse is gratefully acknowledged.

Data Availability Statement: SR-PXD data as well as EIS data will be made available on DOI: 10.6084/m9.figshare.20579484.

Acknowledgments: Affiliation with the Center for Integrated Materials Research (iMAT) at Aarhus University is gratefully acknowledged. The synchrotron data used in this study were collected at the I11 beamline at the Diamond light source. We thank Stephen Thompson and Eamonn Connolly for setting up the experiment.

Conflicts of Interest: The authors declare no conflict of interest.

References

1. Wu, F.; Maier, J.; Yu, Y. Guidelines and Trends for Next-Generation Rechargeable Lithium and Lithium-Ion Batteries. *Chem. Soc. Rev.* **2020**, *49*, 1569–1614. [CrossRef] [PubMed]
2. Kulova, T.L.; Fateev, V.N.; Seregina, E.A.; Grigoriev, A.S. A Brief Review of Post-Lithium-Ion Batteries. *Int. J. Electrochem. Sci.* **2020**, *15*, 7242–7259. [CrossRef]
3. U.S. Geological Survey. *Mineral Commodity Summaries 2022*; Mineral Commodity Summaries: Reston, VA, USA, 2022; Volume 2022, p. 202. [CrossRef]
4. Famprikis, T.; Canepa, P.; Dawson, J.A.; Islam, M.S.; Masquelier, C. Fundamentals of Inorganic Solid-State Electrolytes for Batteries. *Nat. Mater.* **2019**, *18*, 1278–1291. [CrossRef] [PubMed]
5. Skov, L.N.; Grinderslev, J.B.; Rosenkranz, A.; Lee, Y.-S.; Jensen, T.R. Towards Solid-State Magnesium Batteries: Ligand-Assisted Superionic Conductivity. *Batter. Supercaps* **2022**, *5*, e202200163. [CrossRef]
6. Kristensen, L.G.; Amdisen, M.B.; Skov, L.N.; Jensen, T.R. Fast Magnesium Ion Conducting Isopropylamine Magnesium Borohydride Enhanced by Hydrophobic Interactions. *Phys. Chem. Chem. Phys.* **2022**, *24*, 18185–18197. [CrossRef]
7. Yan, Y.; Dononelli, W.; Jørgensen, M.; Grinderslev, J.B.; Lee, Y.-S.; Cho, Y.W.; Černý, R.; Hammer, B.; Jensen, T.R. The Mechanism of Mg2+ Conduction in Ammine Magnesium Borohydride Promoted by a Neutral Molecule. *Phys. Chem. Chem. Phys.* **2020**, *22*, 9204–9209. [CrossRef]
8. Yan, Y.; Grinderslev, J.B.; Jørgensen, M.; Skov, L.N.; Skibsted, J.; Jensen, T.R. Ammine Magnesium Borohydride Nanocomposites for All-Solid-State Magnesium Batteries. *ACS Appl. Energy Mater.* **2020**, *3*, 9264–9270. [CrossRef]
9. Kisu, K.; Kim, S.; Inukai, M.; Oguchi, H.; Takagi, S.; Orimo, S. Magnesium Borohydride Ammonia Borane as a Magnesium Ionic Conductor. *ACS Appl. Energy Mater.* **2020**, *3*, 3174–3179. [CrossRef]
10. Xiao, N.; McCulloch, W.D.; Wu, Y. Reversible Dendrite-Free Potassium Plating and Stripping Electrochemistry for Potassium Secondary Batteries. *J. Am. Chem. Soc.* **2017**, *139*, 9475–9478. [CrossRef]
11. Wang, Z.; Hu, J.; Han, L.; Wang, Z.; Wang, H.; Zhao, Q.; Liu, J.; Pan, F. A MOF-Based Single-Ion Zn2+ Solid Electrolyte Leading to Dendrite-Free Rechargeable Zn Batteries. *Nano Energy* **2019**, *56*, 92–99. [CrossRef]
12. Cao, C.; Li, Y.; Feng, Y.; Peng, C.; Li, Z.; Feng, W. A Solid-State Single-Ion Polymer Electrolyte with Ultrahigh Ionic Conductivity for Dendrite-Free Lithium Metal Batteries. *Energy Storage Mater.* **2019**, *19*, 401–407. [CrossRef]
13. Xia, S.; Wu, X.; Zhang, Z.; Cui, Y.; Liu, W. Practical Challenges and Future Perspectives of All-Solid-State Lithium-Metal Batteries. *Chem* **2019**, *5*, 753–785. [CrossRef]
14. Yan, Y.; Grinderslev, J.B.; Lee, Y.-S.; Jørgensen, M.; Cho, Y.W.; Černý, R.; Jensen, T.R. Ammonia-Assisted Fast Li-Ion Conductivity in a New Hemiammine Lithium Borohydride, LiBH4·1/2NH3. *Chem. Commun.* **2020**, *56*, 3971–3974. [CrossRef]

15. Blanchard, D.; Nale, A.; Sveinbjörnsson, D.; Eggenhuisen, T.M.; Verkuijlen, M.H.W.; Suwarno; Vegge, T.; Kentgens, A.P.M.; Jongh, P.E. de Nanoconfined LiBH4 as a Fast Lithium Ion Conductor. *Adv. Funct. Mater.* **2015**, *25*, 184–192. [CrossRef]
16. Huot, J.; Cuevas, F.; Deledda, S.; Edalati, K.; Filinchuk, Y.; Grosdidier, T.; Hauback, B.C.; Heere, M.; Jensen, T.R.; Latroche, M.; et al. Mechanochemistry of Metal Hydrides: Recent Advances. *Materials* **2019**, *12*, 2778. [CrossRef] [PubMed]
17. Richter, B.; Grinderslev, J.B.; Møller, K.T.; Paskevicius, M.; Jensen, T.R. From Metal Hydrides to Metal Borohydrides. *Inorg. Chem.* **2018**, *57*, 10768–10780. [CrossRef] [PubMed]
18. Černý, R.; Filinchuk, Y.; Hagemann, H.; Yvon, K. Magnesium Borohydride: Synthesis and Crystal Structure. *Angew. Chem. Int. Ed.* **2007**, *46*, 5765–5767. [CrossRef]
19. Jepsen, L.H.; Ban, V.; Møller, K.T.; Lee, Y.-S.; Cho, Y.W.; Besenbacher, F.; Filinchuk, Y.; Skibsted, J.; Jensen, T.R. Synthesis, Crystal Structure, Thermal Decomposition, and 11B MAS NMR Characterization of Mg(BH4)2(NH3BH3)2. *J. Phys. Chem. C* **2014**, *118*, 12141–12153. [CrossRef]
20. Jepsen, L.H.; Ley, M.B.; Filinchuk, Y.; Besenbacher, F.; Jensen, T.R. Tailoring the Properties of Ammine Metal Borohydrides for Solid-State Hydrogen Storage. *ChemSusChem* **2015**, *8*, 1452–1463. [CrossRef]
21. Goursot, A.; Mineva, T.; Kevorkyants, R.; Talbi, D. Interaction between N-Alkane Chains: Applicability of the Empirically Corrected Density Functional Theory for Van Der Waals Complexes. *J. Chem. Theory Comput.* **2007**, *3*, 755–763. [CrossRef]
22. Roedern, E.; Kühnel, R.-S.; Remhof, A.; Battaglia, C. Magnesium Ethylenediamine Borohydride as Solid-State Electrolyte for Magnesium Batteries. *Sci. Rep.* **2017**, *7*, 46189. [CrossRef] [PubMed]
23. Grinderslev, J.B.; Skov, L.N.; Andreasen, J.G.; Ghorwal, S.; Skibsted, J.; Jensen, T.R. Methylamine Lithium Borohydride as Electrolyte for All-Solid-State Batteries. *Angew. Chem. Int. Ed.* **2022**, *61*, e202203484. [CrossRef] [PubMed]
24. Grinderslev, J.B.; Ley, M.B.; Lee, Y.-S.; Jepsen, L.H.; Jørgensen, M.; Cho, Y.W.; Skibsted, J.; Jensen, T.R. Ammine Lanthanum and Cerium Borohydrides, M(BH4)3·nNH3; Trends in Synthesis, Structures, and Thermal Properties. *Inorg. Chem.* **2020**, *59*, 7768–7778. [CrossRef]
25. Grinderslev, J.B.; Jensen, T.R. Trends in the Series of Ammine Rare-Earth-Metal Borohydrides: Relating Structural and Thermal Properties. *Inorg. Chem.* **2021**, *60*, 2573–2589. [CrossRef] [PubMed]
26. Zanella, P.; Crociani, L.; Masciocchi, N.; Giunchi, G. Facile High-Yield Synthesis of Pure, Crystalline Mg(BH4)2. *Inorg. Chem.* **2007**, *46*, 9039–9041. [CrossRef]
27. Filinchuk, Y.; Richter, B.; Jensen, T.R.; Dmitriev, V.; Chernyshov, D.; Hagemann, H. Porous and Dense Magnesium Borohydride Frameworks: Synthesis, Stability, and Reversible Absorption of Guest Species. *Angew. Chem. Int. Ed. Engl.* **2011**, *50*, 11162–11166. [CrossRef] [PubMed]
28. Thompson, S.P.; Parker, J.E.; Potter, J.; Hill, T.P.; Birt, A.; Cobb, T.M.; Yuan, F.; Tang, C.C. Beamline I11 at Diamond: A New Instrument for High Resolution Powder Diffraction. *Rev. Sci. Instrum.* **2009**, *80*, 075107. [CrossRef]
29. Ravnsbæk, D.B.; Sørensen, L.H.; Filinchuk, Y.; Besenbacher, F.; Jensen, T.R. Screening of Metal Borohydrides by Mechanochemistry and Diffraction. *Angew. Chem., Int. Ed.* **2012**, *51*, 3582–3586. [CrossRef]
30. Favre-Nicolin, V.; Černý, R. FOX, 'free Objects for Crystallography': A Modular Approach Toab Initiostructure Determination from Powder Diffraction. *J. Appl. Crystallogr.* **2002**, *35*, 734–743. [CrossRef]
31. Favre-Nicolin, V. Free Objects for Crystallography. Available online: https://sourceforge.net/projects/objcryst/ (accessed on 18 November 2022).
32. Mortensen, J.J.; Hansen, L.B.; Jacobsen, K.W. Real-Space Grid Implementation of the Projector Augmented Wave Method. *Phys. Rev. B* **2005**, *71*, 035109. [CrossRef]
33. Enkovaara, J.; Rostgaard, C.; Mortensen, J.J.; Chen, J.; Dułak, M.; Ferrighi, L.; Gavnholt, J.; Glinsvad, C.; Haikola, V.; Hansen, H.A.; et al. Electronic Structure Calculations with GPAW: A Real-Space Implementation of the Projector Augmented-Wave Method. *J. Phys. Condens. Matter* **2010**, *22*, 253202. [CrossRef]
34. Larsen, A.H.; Mortensen, J.J.; Blomqvist, J.; Castelli, I.E.; Christensen, R.; Dułak, M.; Friis, J.; Groves, M.N.; Hammer, B.; Hargus, C.; et al. The Atomic Simulation Environment—a Python Library for Working with Atoms. *J. Phys. Condens. Matter* **2017**, *29*, 273002. [CrossRef] [PubMed]
35. Perdew, J.P.; Burke, K.; Ernzerhof, M. Generalized Gradient Approximation Made Simple. *Phys. Rev. Lett.* **1996**, *77*, 3865–3868. [CrossRef] [PubMed]
36. Wellendorff, J.; Lundgaard, K.T.; Møgelhøj, A.; Petzold, V.; Landis, D.D.; Nørskov, J.K.; Bligaard, T.; Jacobsen, K.W. Density Functionals for Surface Science: Exchange-Correlation Model Development with Bayesian Error Estimation. *Phys. Rev. B* **2012**, *85*, 235149. [CrossRef]

Disclaimer/Publisher's Note: The statements, opinions and data contained in all publications are solely those of the individual author(s) and contributor(s) and not of MDPI and/or the editor(s). MDPI and/or the editor(s) disclaim responsibility for any injury to people or property resulting from any ideas, methods, instructions or products referred to in the content.

Article

Electrolytes in Multiple-Phase Hydrogen Storage Reactions

John J. Vajo [1], Jasim Uddin [2], Son-Jong Hwang [3] and Jason Graetz [1,*]

[1] HRL Laboratories, LLC, 3011 Malibu Canyon Road, Malibu, CA 90265, USA; jjvajo@hrl.com
[2] Liox Power, Inc., 129 North Hill Avenue, Suite 707, Pasadena, CA 91103, USA; jasimuddin@gmail.com
[3] Division of Chemistry and Chemical Engineering, California Institute of Technology, Pasadena, CA 91125, USA; sjhwang@caltech.edu
* Correspondence: jgraetz@hrl.com

Abstract: Multiple-phase hydrogen storage materials such as metal alanates and borohydrides, and destabilized systems offer the possibility of high hydrogen storage capacity with favorable thermodynamics. However, the multiphase nature of these materials intrinsically limits the kinetics due to the required transport of species between phases, which are typically in dry powder form. To address this limitation, the influence of added electrolytes is explored. This approach is motivated by analogy with similar multiphase battery reactions that show reduced kinetic limitations while necessarily containing electrolytes. Previous experimental results showing improved kinetics for MgH_2/Sn (using a $LiBH_4/KBH_4$ eutectic electrolyte) and $NaAlH_4$ (using a diglyme electrolyte) are further analyzed in terms of this analogy. The results show that the analogy is useful and rate constants are increased. Importantly, the inclusion of an electrolyte also appears to alleviate the continuously decreasing rates with the extent of reaction, which is characteristic of many multiphase hydrides. Instead, reaction rates are approximately constant until near completion. Together, these effects can lead to >10× shorter overall reaction times. In addition, new results are presented for the hydrogenation of MgB_2 using $Li/K/CsI$ and $Li/K/CsCl$ eutectic electrolytes, where >60% conversion to $Mg(BH_4)_2$ was demonstrated at 350 bar.

Keywords: hydrogen storage material; complex hydride; destabilized hydride; solid-state reaction; electrolyte; eutectic

Citation: Vajo, J.J.; Uddin, J.; Hwang, S.-J.; Graetz, J. Electrolytes in Multiple-Phase Hydrogen Storage Reactions. *Inorganics* **2023**, *11*, 267. https://doi.org/10.3390/inorganics11070267

Academic Editors: Craig Buckley, Mark Paskevicius, Torben R. Jensen and Terry Humphries

Received: 15 May 2023
Revised: 21 June 2023
Accepted: 22 June 2023
Published: 24 June 2023

Copyright: © 2023 by the authors. Licensee MDPI, Basel, Switzerland. This article is an open access article distributed under the terms and conditions of the Creative Commons Attribution (CC BY) license (https://creativecommons.org/licenses/by/4.0/).

1. Introduction

Many candidate hydrogen storage materials with high capacities and thermodynamics appropriate for proton exchange membrane (PEM) fuel cells used in transportation applications contain multiple solid phases that must nucleate, grow, and be consumed as hydrogen is released and stored. The presence of multiple solid phases in these materials hinders the kinetics of the solid phase transformations that occur as hydrogen is exchanged because solid–solid reactions can only occur where particles of different phases are in physical contact at the atomic scale. This contact is difficult given the typical mixed powder form of these materials and the irregular shape of powder particles at the atomic scale. Overall, this limits the rate at which hydrogen can be released and stored. As a result, most multiple solid-phase hydrogen storage materials are not practical for commercial use.

In our previous work, an investigation of electrolyte-assisted hydrogen storage reactions in destabilized hydrides (MgH_2/Sn) and complex hydrides ($LiAlH_4$, $NaAlH_4$, $Mg(BH_4)_2$) revealed significantly reduced reaction times for hydrogen desorption and uptake in the presence of an electrolyte [1,2]. In this work, we provide motivation and background for the use of electrolytes by describing the kinetic limitations in terms of subjective kinetic temperatures and excess free energies. Using these descriptors, we compare multiphase hydrogen storage reactions with analogous multiphase battery reactions, in which electrolytes are necessarily always present due to the electrochemical form of the reactions [3]. Within this context, we summarize the influence of electrolytes on MgH_2/Sn

(using a LiBH$_4$/KBH$_4$ eutectic electrolyte), NaAlH$_4$ (using a diglyme electrolyte) and also give further results for the hydrogenation of MgB$_2$ (using Li/K/CsI and Li/K/CsCl eutectic electrolytes). We found that, in addition to the expected increase in reaction rates, electrolytes enable reactions to proceed at a constant rate, independent of the extent of the reaction.

2. Motivation and Background

To describe the kinetic limitations for thermochemical reactions (in dry powder form) and compare them with those for electrochemical reactions (containing liquid electrolytes) we first discuss the MgH$_2$/Si destabilized hydride reaction and then the Mg/Sn battery reaction.

2.1. Kinetic Limitations and Excess Free Energy

The MgH$_2$/Si system is based on MgH$_2$, which has a high gravimetric hydrogen capacity of 7.6 wt% but is thermodynamically too stable for practical use. Specifically, the enthalpy (ΔH) for dehydrogenation is 74.5 kJ/mol-H$_2$ and the entropy (ΔS) is 135 J/K-mol-H$_2$ [4] (thermodynamic values for the reactions discussed in this work are compiled in Table 1). These values give a temperature for an equilibrium hydrogen pressure of 1 bar (T_{1bar}) of 280 °C, using $T_{1bar} = \Delta H/\Delta S$, which is too high for transportation applications. However, the thermodynamic properties of MgH$_2$ can be tuned by using silicon as a destabilizing additive [5,6]. MgH$_2$ and Si react according to the reaction 2 MgH$_2$ + Si → Mg$_2$Si + 2 H$_2$, which contains 5.0 wt% hydrogen and has a much lower T_{1bar} of ~15 °C (Table 1), due to the stability of Mg$_2$Si relative to pure Mg. For this reaction to proceed, MgH$_2$ in a single-step reaction, or Mg metal in a two-step reaction, must react with Si to form Mg$_2$Si. This requires atomic scale contact between the MgH$_2$ and Si solid phases, which is difficult to achieve. As a result, effectively no reaction occurs between MgH$_2$ and Si at 15 °C. As shown in Figure 1a for a typical formulation of a mechanically milled mixture of 2 MgH$_2$ + Si undergoing a constant heating ramp (2 °C/min), the onset of the desorption reaction occurs at ~270 °C. Following Ref. [7] (p. 4554), we define this temperature as the kinetic temperature, T_K = 270 °C. This temperature is a subjective assessment of where a reaction begins to occur at a practical rate. For this system, the rate at 270 °C is 0.28 wt%-H$_2$/h. The difference between T_{1bar} and T_K illustrates the kinetic limitation, i.e., T_K = ~270 °C >> T_{1bar} = ~15 °C.

Table 1. Thermodynamic values for various hydrogen storage and battery reactions.

Reaction	ΔG^0 (kJ/mol-H$_2$ or Mg)	ΔH (kJ/mol-H$_2$)	ΔS (J/K-mol-H$_2$)	T_{1bar} (°C)
MgH$_2$	37.0	74.5	135	280
MgH$_2$/Si	0.7	36.8	128	15
MgH$_2$/Sn	2.3	39.0	125	39
Mg/Si	−37.5			
Mg/Sn	−34.7			
LiBH$_4$/MgH$_2$	15.4	45.8	104	170
Mg(BH$_4$)$_2$ (1st step)		39.2	99.8	20
NaAlH$_4$ (1st step)		39.1	127	33

Using T_K, we can further define the kinetic limitation in terms of an excess kinetic free energy (ΔG_K) using the expression $\Delta G_K = (T_K - T_{1bar}) \bullet \Delta S$. We note that ΔG_K is not a true thermodynamic quantity; rather it is subjective based on T_K chosen at a practical rate of dehydrogenation. Although reactions are thermodynamically spontaneous for $\Delta G < 0$, we express ΔG_K as a positive quantity to represent the free energy required to drive a kinetically hindered reaction that is in excess of the thermodynamic driving force. The values above give ΔG_K = 34 kJ/mol-H$_2$ for the MgH$_2$/Si system (values for ΔG_K are tabulated in Table 2). Considerable effort has been devoted to improving (i.e., lowering) T_K and ΔG_K. Using catalysts and nanoscale particles to increase interfacial area, T_K = 200 °C with ΔG_K = 24 kJ/mol-H_2 has been achieved [8]. Although improved, these values are still too high for practical use.

Other hydrogen storage materials display similar excess kinetic free energies. For example, the hydrogen storage reaction 2 $LiBH_4$ + MgH_2 → 2 LiH + MgB_2 + 2 H_2 contains 11.4 wt% hydrogen with T_{1bar} = 170 °C (Table 1), while experimentally, T_K = 380 °C [7]. Thus, ΔG_K = 22 kJ/mol-H_2. Another example is $Mg(BH_4)_2$, which has 14.8 wt% hydrogen and dehydrogenates according to the overall reaction $Mg(BH_4)_2$ → MgB_2 + 4 H_2. In detail, this reaction actually occurs in 3 or 4 steps beginning with the formation of $MgB_{12}H_{12}$ at T_K = 250 °C [7,9,10]. The enthalpies and entropies for these reactions have not been determined experimentally, although density functional theory calculations give T_{1bar} = 20 °C (Table 1) for the 1st step [11]. This gives, ΔG_K = 23 kJ/mol-H_2. These examples are further evidence that dehydrogenation reactions across solid/solid interfaces are kinetically limited and typically require ≥20 kJ/mol-H_2 of excess kinetic free energy to initiate the reaction.

2.2. Analogy with Multiphase Electrochemical Battery Reactions

In contrast to the extremely poor kinetics of the multiphase hydrogen storage reactions, similar multiphase battery reactions (known as "alloy" and "conversion" reactions in the battery research community) can operate at or near room temperature [12]. Examples include Li/SnO_2 [13], which forms Li_2O irreversibly and Li_nSn alloys reversibly, Li/FeF_2 [14], which forms LiF + Fe reversibly, and Mg/Sn [15], which forms Mg_2Sn reversibly.

To illustrate quantitatively the difference between a battery reaction and an analogous hydrogen storage reaction, we consider here the free energy driving forces (ΔG) for the Mg/Sn alloying battery reaction and then relate these to the MgH_2/Si hydrogen storage reaction. Discharge of a Mg/Sn battery, given by the reaction Mg + 0.5 Sn → 0.5 Mg_2Sn, is exothermic with a calculated free energy ΔG = −34.7 kJ/mol-Mg, which for this electrochemical reaction corresponds to a reversible potential E_{rev} = 0.18 V (using ΔG = −$n \bullet F \bullet E$ with n = 2 and F = 96,485 C/mol). As shown in Figure 1b, a complete reaction between Mg and Sn has been observed at room temperature in a battery with an electrolyte appropriate for Mg^{2+} [15]. The discharge reaction occurs at 0.16 V (ΔG = −30.9 kJ/mol-Mg), which is lower than E_{rev} indicating some kinetic limitation. This is typically called an overpotential. However, to make a connection to hydrogen storage reactions, we use free energy. From the differences in free energy with −34.7 kJ/mol-Mg possible but only −30.9 kJ/mol-Mg obtained, the excess free energy is (as above, expressed as a positive quantity) ΔG_K = 3.8 kJ/mol-Mg. For an electrochemical reaction in general, the excess kinetic free energy can be given by ΔG_K = | −$n \bullet F \bullet (E_{dis/recharge} - E_{rev})$ |, where $E_{dis/recharge}$ is the actual measured potential for discharge or recharge and the absolute value is use to express ΔG_K as a positive quantity. More importantly, the reverse recharging reaction which is endothermic (ΔG = +34.7 kJ/mol-Mg) was also observed at room temperature by applying a voltage of 0.21 V (−40.5 kJ/mol-Mg), giving ΔG_K = 5.8 kJ/mol-Mg. Thus, in the battery environment, this multiphase alloying reaction was reversibly driven electrochemically at room temperature at practical rates with an excess kinetic free energy ΔG_K ~ 5 kJ/mol.

The chemistry of Mg reacting (alloying) with Sn and Si is similar. For example, with Si the battery reaction Mg + 0.5 Si → 0.5 Mg_2Si has ΔG = −37.5 kJ/mol-Mg (versus −34.7 kJ/mol-Mg for Sn). However, the reaction rates are much different for the analogous MgH_2/Si hydrogen storage reaction where endothermic dehydrogenation is not observed until ≥200 °C [8], at which temperature the excess free energy is 24 kJ/mol-H_2. A comparison of these analogous reactions reveals that a ΔG_K = 5 kJ is sufficient to drive the battery reaction (*electro*chemically), while ΔG_K = 24 kJ is needed to drive a similar hydrogen storage reaction (*thermo*chemically). Furthermore, despite being exothermic, the reverse hydrogenation reaction of Mg_2Si has not been observed even under 1850 bar H_2 [16], which is equivalent to an excess free energy of ~20 kJ at room temperature.

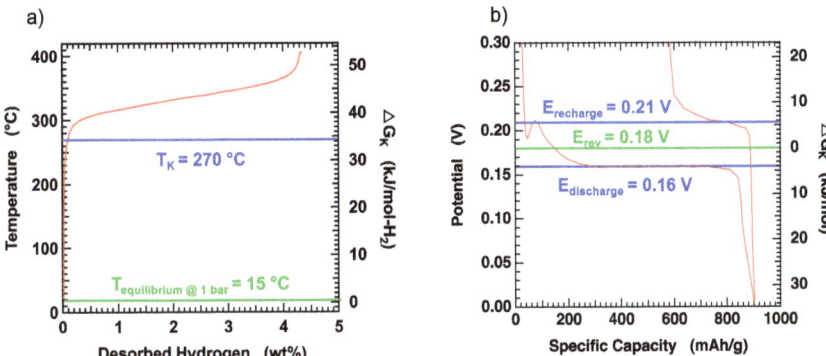

Figure 1. The analogy between thermochemical hydrogen storage reactions and electrochemical battery reactions. (**a**) MgH$_2$/Si destabilized the hydride system. Following the convention used for batteries, the extent of the hydrogen desorption reaction (during heating ramp of 2 °C/min) is shown on the x-axis (given by the amount of desorbed hydrogen in wt%) with the driving force for the reaction (given by the temperature) on the left y-axis. Although the equilibrium temperature at 1 bar is calculated to be 15 °C, no significant hydrogen desorption occurs until the temperature is increased to 270 °C. The corresponding excess kinetic free energy is given on the right y-axis. (**b**) Mg/Sn battery reaction (based on data obtained from Ref. [15]). A conventional depiction is shown with the extent of reaction (given by the specific capacity in mAh/g) on the x-axis and the driving force (given for an electrochemical reaction by the potential in V) on the left y-axis. The capacity is shown as increasing during the discharge reaction (forming Mg$_2$Sn) and decreasing during the recharge reaction (reforming Mg + Sn). The calculated reversible, i.e., equilibrium, potential is 0.18 V. Discharge is seen to occur at a lower potential of 0.16 V while the recharge occurs at a higher potential of 0.21 V. The corresponding excess kinetic free energy is given on the right y-axis with positive values below 0.18 V pertaining to the discharge reaction and positive values above 0.18 V pertaining to the recharge reaction.

We contend that the lower excess free energy required to drive reactions electrochemically (as opposed to thermochemically) is due to the presence of an electrolyte [17]. In batteries, the ion transport species from one electrode is *necessarily* solubilized in a liquid solvent as an ion (e.g., the cation Mg^{2+} for the Mg/Sn example) because the electrons travel through an external circuit providing the electrical work of the battery [3]. Solubilization in a liquid solvent facilitates transport between different electrodes and enables the direct, atomic scale contact between reacting species over the full electrolyte-wetted surface area of the Sn electrode. Once transported to the Sn electrode, the electrons and Mg^{2+} ions must still react in a solid/solid reaction, with Mg and Sn inter-diffusing and alloying to form Mg$_2$Sn. However, now the reaction is occurring within a single particle. Given that the relatively low $\Delta G_K \sim 5$ kJ in the battery reaction reflects the overall reaction, including both transport through the electrolyte and interdiffusion within the Sn electrode, the much higher $\Delta G_K \sim >20$ kJ in hydrogen storage reactions indicate additional limitations in the transport between particles. This suggests that the excess free energy required to drive multiphase hydrogen storage reactions may be significantly reduced (i.e., lowering the reaction temperature) with appropriate electrolytes to improve the transport rates of the various species involved in the reaction.

Even with an electrolyte, a hydrogen storage reaction would still be thermochemical, with the term "electrolyte" here meaning a liquid phase containing or consisting of dissociated mobile cations and anions. The addition of an electrolyte would be equivalent to chemically short-circuiting a battery by mixing both electrode materials together. In this case, as solubilized ions form (e.g., Mg^{2+}) and enter the electrolyte, the electrons must still travel through atomic scale contacts between different powder particle phases. However, due to its lighter mass and therefore more delocalized quantum mechanical nature, electron transport is typically much faster than ion transport.

Table 2. Excess kinetic free energies for various hydrogen storage reactions and for the Mg/Sn battery reaction.

Reaction	Conditions	ΔG_K (kJ/mol-H_2 or Mg)
MgH_2/Si	No electrolyte	34
MgH_2/Si [8]	Nanoscale/catalyzed/no electrolyte	24
MgH_2/Sn	No electrolyte	22
MgH_2/Sn	With $LiBH_4$/KBH_4 eutectic	20
Mg/Sn discharge [15]	With battery electrolyte	3.8
Mg/Sn charge [15]	With battery electrolyte	5.8
$LiBH_4$/MgH_2	Catalyzed/no electrolyte	22
$Mg(BH_4)_2$ (1st step) [7]	No electrolyte	23
$NaAlH_4$ (1st step)	No electrolyte	19
$NaAlH_4$ + $TiCl_3$ (1st step)	Catalyzed/no electrolyte	4.7
$NaAlH_4$ + $TiCl_3$ (1st step)	Catalyzed/with glyme electrolyte	2.8

3. Results and Discussion

Based on a comparison of thermochemical and electrochemical multiphase reactions, we have investigated the use of liquid electrolytes in the form of eutectics, ionic liquids, or solvents containing dissolved salts, to facilitate the transport of atoms such as Li, Mg, B, and Al, between phases and thereby increase reaction rates (or reduce desorption temperatures and hydrogenation pressures) in complex and destabilized hydride materials. Here we further analyze previous results for MgH_2/Sn and $NaAlH_4$ and then present new results for the hydrogenation of MgB_2.

3.1. Analysis of the Electrolyte-Assisted MgH_2/Sn Hydrogen Storage Reaction Using ΔG_K

The MgH_2/Sn destabilized hydride system was studied using a eutectic electrolyte with a composition of 0.725 $LiBH_4$/0.275 KBH_4 [1]. This composition was chosen to have a melting point (~110 °C) [18] well below the melting point of Sn (232 °C). Straightforward measurement of T_K during a heating ramp (as shown in Figure 1a for MgH_2/Si) was not observed due to the extremely slow reaction rates even with the eutectic. Instead, rates were measured over extended times at fixed temperatures of 150 °C, 175 °C, and 200 °C. Although slow, the rate of dehydrogenation at 150 °C increased 12× from 0.0008 wt%-H_2/h without eutectic to 0.01 wt%-H_2/h with the eutectic. At 200 °C, the rate with eutectic is 0.23 wt%-H_2/h, which is similar to the rate described above for MgH_2/Si (without electrolyte). Thus, we chose T_K = 200 °C, which gives ΔG_K = 20 kJ/mol-H_2 (based on data in Table 1). By using the measured rates and extrapolating, we estimate that the same rate (0.23 wt%-H_2/h) would occur without the eutectic at 215 °C. This gives ΔG_K = 22 kJ/mol-H_2, so there is a modest reduction of 2 kJ/mol-H_2 upon the addition of the eutectic electrolyte.

In addition to improving dehydrogenation, the 0.725 $LiBH_4$/0.275 KBH_4 electrolyte also enabled full (re)hydrogenation of Mg_2Sn back to MgH_2/Sn using 1000 bar H_2 at 215 °C–175 °C [1]. Without the eutectic, essentially no hydrogenation was observed. This was the first time that significant hydrogenation was observed in the Mg_2Si or Mg_2Sn systems. A similar attempt using an eutectic electrolyte was made with Mg_2Si but no hydrogenation was seen. The thermodynamics of the two systems are similar. The difference may have been a more inert oxide layer on the Mg_2Si and/or more mobility in the Sn-based system due to the relatively low melting points for Sn (232 °C) and Mg_2Sn (778 °C) compared to Si (1400 °C) and Mg_2Si (1100 °C).

3.2. Analysis of Electrolyte-Assisted NaAlH$_4$ Dehydrogenation Using the Avrami-Erofe'ev Model

The influence of electrolytes was also explored for the dehydrogenation of LiAlH$_4$ and NaAlH$_4$ [2]. These hydrides dehydrogenate in two steps beginning from a single phase. Thus, the multiphase reaction transport limitations between reacting phases described for MgH$_2$ + Si or Sn might be considered to not apply. However, there are three solid phase products when MAlH$_4$ dehydrogenates, M$_3$AlH$_6$ + Al for the 1st step and MH + Al for the 2nd step. These products must form at 3-phase boundaries that could present transport limitations. Indeed, pure NaAlH$_4$ does not decompose until it melts at ~180 °C, despite a thermodynamic T_{1bar} = 33 °C (Table 1) for the 1st step. This gives ΔG_K = 19 kJ/mol-H$_2$, which is somewhat lower than other hydrogen storage reactions but still much higher than the Mg/Sn battery reaction.

For NaAlH$_4$ catalyzed with 3 mol% TiCl$_3$, dehydrogenation was compared without an electrolyte and with 50 wt% diglyme [2]. No additional electrolyte salt was included with the diglyme. Rather, the solubility of NaAlH$_4$ or NaCl (formed during milling with TiCl$_3$) was relied upon to possibly provide mobile [AlH$_4$]$^-$ or Na$^+$ ions. The results are shown in Figure 2 following the format in Figure 1. The two dehydrogenation steps are clearly seen. Without electrolyte, the T_K for the 1st step is 70 °C, giving ΔG_K = 4.7 kJ/mol-H$_2$. This is a significant decrease from pure NaAlH$_4$ and similar to that for the Mg/Sn battery reaction. Thus, as is well known, catalyzed NaAlH$_4$ dehydrogenates with rates that are, or are close to, practical. Including diglyme further decreases T_K by 15 °C to 55 °C, ΔG_K = 2.8 kJ/mol-H$_2$, a decrease of ~2 kJ/mol-H$_2$. However, perhaps more significantly, is how the electrolyte influences the rates as the reactions proceed. For the 1st step, the rates at a given wt% with and without electrolytes are similar, except that the reaction occurs ~15 °C lower with electrolytes. A more distinct difference is seen in the 2nd step where the dependence of the rate on the extent of reaction is very different. Specifically, with diglyme the rate increased reaching ~3.5 wt%-H$_2$/h at 4 wt% desorbed hydrogen while without electrolyte the rate simply decreases starting from ~0.5 wt%-H$_2$/h at 3 wt%.

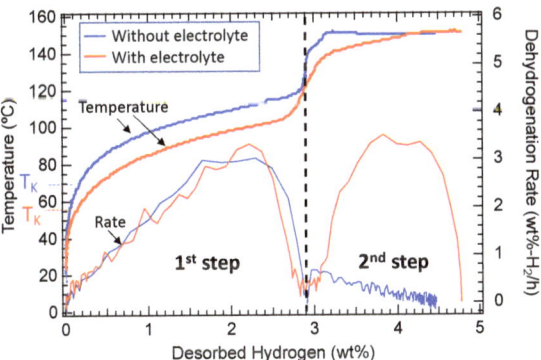

Figure 2. Dehydrogenation of NaAlH$_4$ without and with an electrolyte. Following Figure 1, the extent of the reaction is given on the x-axis with the driving force (temperature) on the left y-axis. The dehydrogenation rate is given on the right y-axis. Without electrolyte (blue), T_K = 70 °C. With 50 wt% diglyme (red), T_K = 55 °C. For both samples, 3 mol% TiCl$_3$ was added as a catalyst by milling, and the temperature was ramped to 150 °C at 0.5 °C/min. The wt% desorbed hydrogen is given with respect to the NaAlH$_4$ + 0.03 TiCl$_3$ mass.

To understand this further, a full kinetic analysis was performed using the Avrami-Erofe'ev (A-E) model:

$$\alpha = 1 - \exp(-(kt)^n), \qquad (1)$$

which is typical for systems with rates driven by nucleation and growth, where α is the extent of reaction, k is a temperature-dependent rate constant, and n is a growth parameter (with typical values between ~1 and 4) that is related to nucleation and growth of the product phases [19]. A series of A-E calculations were performed at a fixed rate constant to illustrate how the growth parameter (n) affects the shape of the desorption curve (Figure 3). Simulated curves are shown in Figure 3a at various values of n (all other parameters were fixed). Similarly, Figure 3b,c show the same data plotted as the rate vs. time and rate vs. fractional decomposition, respectively. At $n = 1$, the A-E equation is equivalent to a traditional first-order (homogeneous) rate equation where the reaction starts at the highest rate and decays as the reaction proceeds. For systems with $2 \leq n \leq 4$ the fractional decomposition curve takes on a sigmoidal shape with an initial induction period (nucleation), followed by an acceleratory period (growth of nuclei), and finally, a deceleration period (growth with overlap). For these systems, the n value determines the growth geometry and the maximum rate (during the acceleratory period) increases with n.

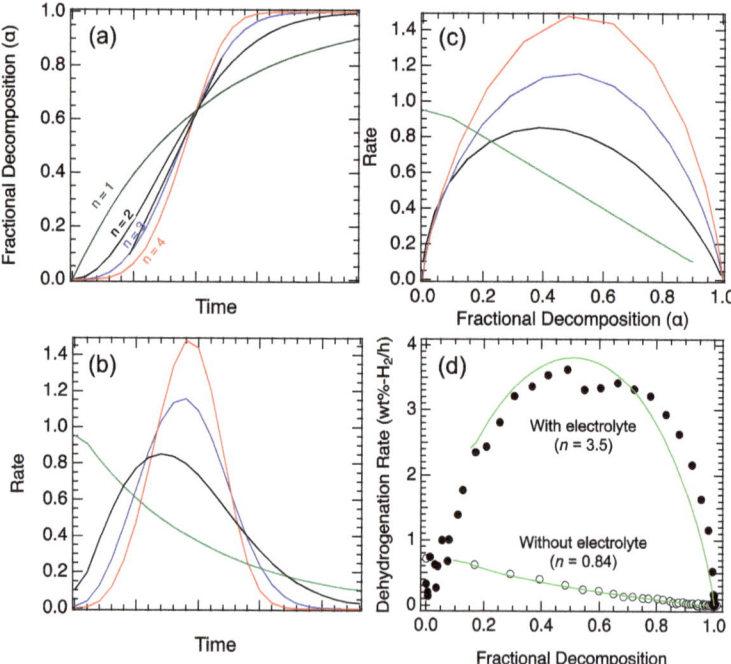

Figure 3. Avrami-Erofe'ev kinetic analysis of hydride decomposition. Desorption curves were generated using the A-E equation ($\alpha = 1 - \exp(-(kt)^n)$) using a constant rate (k) with $n = 1-4$ showing (**a**) fractional decomposition (extent of reaction) vs. time (arbitrary units), (**b**) rate (derivative of fractional decomposition) vs. time (arbitrary units) and (**c**) rate vs. fractional decomposition. (**d**) shows the results from the 2nd desorption step of NaAlH$_4$ (Na$_3$AlH$_6$ + Al $\rightarrow$ 3 NaH + Al + 3/2 H$_2$) [2] with (filled circles) and without (open circles) electrolyte along with the A-E fits (green traces).

Using data from Figure 2, the rate of dehydrogenation from the 2nd step (Na$_3$AlH$_6$ + Al $\rightarrow$ 3 NaH + Al + 3/2 H$_2$) with and without the electrolyte is shown in Figure 3d where the fractional decomposition (extent of reaction) is plotted on the x-axis and the temperature is constant (150 °C). As previously mentioned, the first step (not shown) is first order (with $n \sim 1$) with and without electrolytes. The 2nd desorption step without the electrolyte is clearly characteristic of a first-order reaction ($n \sim 1$) with a continuously decreasing rate with the extent of the reaction (open circles in Figure 3d). A very different shape is observed with the sample containing an electrolyte, where the rate increases initially, plateaus (constant rate), and then decreases at the very end of the reaction (filled circles). This behavior is consistent with A-E with $n = 3.5$. The increase in the growth parameter from $n \sim 1$ to $n = 3.5$ coupled with a $\sim 5\times$ increase in the rate constant leads to the observed $\sim 10\times$ decrease in overall reaction time.

The unusual kinetic behavior observed in the 2nd desorption step of NaAlH$_4$ was also seen with an electrolyte in MgH$_2$/Sn [1] and LiAlH$_4$ 2nd step [2]. In multiphase hydrogen storage reactions (and more generally, solid-state reactions) without an electrolyte, the majority of the reaction can be treated as the first order, where the rate continuously decreases or the temperature must be continuously increased to maintain a constant rate. The difference with an electrolyte suggests that in addition to improving *inter*particle transport, the electrolyte also enables *intra*particle transport, presumably over the particle surface (assuming the electrolyte does not penetrate or break up a multiphase particle). For a solid-state reaction that generates two new solid phases (such as dehydrogenation of M$_3$AlH$_6$ to MH + Al), a reaction occurs along a triple-phase boundary (assuming the product phases are not continuously nucleated). These boundaries are spatially very confined; locally, they are one-dimensional. This confinement will favor possible alternative transport pathways such as surface transport linking the different phases. This transport could be facilitated by an electrolyte, which effectively increases the velocity of the advancing reactant-product interface, and decreases the activation energy. The concept of expanding the growth geometry (in addition to the rate constants) is supported by the kinetic analysis which shows an increase in the growth parameters from $n = 1$ with no electrolyte to $n = 3$ with electrolyte.

3.3. Influence of Electrolytes on MgB$_2$ Hydrogenation

The hydrogenation of MgB$_2$ was also studied [1]. Similar to the dehydrogenation of LiAlH$_4$ and NaAlH$_4$, an influence of an electrolyte might not be expected for MgB$_2$ because hydrogenation begins from a single phase (and for MgB$_2$ also ends in a single phase, Mg(BH$_4$)$_2$). However, it is thought that there are multiple intermediate phases including MgH$_2$, Mg, and MgB$_{12}$H$_{12}$ [7] that perhaps could be contained within a single particle. As described above, the reaction at the confined multiphase boundaries could be improved by providing surface transport through an electrolyte.

Significant hydrogenation of MgB$_2$ was shown using a 0.33 LiI/0.33 KI/0.33 CsI eutectic electrolyte at 53 wt% with hydrogen treatment at 1000 bar and 320 °C for 50 h. [1]. As shown in Table 3, from a starting boron-based composition of 96% B in MgB$_2$ and 4% B in (impurity) boron oxides (Table 3, row 1) using ^{11}B nuclear magnetic resonance (NMR), the hydrogen treatment with eutectic resulted in a final composition with 72% of the boron as [BH$_4$]$^-$ (Table 3, row 3). In contrast, without the eutectic, only 3% of the B was hydrogenated to [BH$_4$]$^-$ (Table 3, row 2). The portion of the ^{11}B NMR spectrum focused on the borohydride region is shown in Figure 4. Comparison with literature spectra indicates the formation of Mg(BH$_4$)$_2$ together with CsBH$_4$ and LiBH$_4$. Although the ^{11}B chemical shift can be influenced by the local environment and mixed cation solid solution phases are possible, an approximate composition (based on areas between the minima between the peaks) is 12%BH$_4$ in CsBH$_4$; 74%BH$_4$ in Mg(BH$_4$)$_2$; 14%BH$_4$ in LiBH$_4$. The extremely high pressure used (1000 bar) clearly resulted in some hydrogenation of the eutectic.

In this work, we present new results where lower pressures were explored. Reducing the pressure to 700 bar and (unintentionally) the temperature to 300 °C (Table 3, row 4) reduced the [BH$_4$]$^-$ fraction drastically, from 72% to 13%. The inclusion of 3 at% TiF$_3$ as a catalyst (by milling) increased the [BH$_4$]$^-$ fraction to 26% but also increased the fraction of non-[BH$_4$]$^-$ boron hydride species (BH$_x$) from 4% to 9% (Table 3, row 5). These results indicate that well-known catalysts for complex hydrides (such as TiF$_3$) can still operate in an electrolyte environment. This is not surprising if we consider that the catalyst facilitates reaction on the surface of and within individual powder particles, while the eutectic facilitates transport between particles.

Table 3. ^{11}B NMR analysis of hydrogenated MgB$_2$ formulations.

	Composition	Eutectic	Hydrogenation Conditions (50 h)	MgB$_2$	B-O	BH$_x$	[BH$_4$]$^-$
1	MgB$_2$	Li/K/CsI 53 wt%	No treatment	0.96	0.04	0	0
2	MgB$_2$	none	1000 bar 320 °C	0.93	0.04	0	0.03
3	MgB$_2$	Li/K/CsI 53 wt%	1000 bar 320 °C	0.21	0.04	0.04	0.72
4	MgB$_2$	Li/K/CsI 50 wt%	700 bar 300 °C	0.79	0.04	0.04	0.13
5	MgB$_2$ + 3 at% TiF$_3$	Li/K/CsI 50 wt%	700 bar 300 °C	0.61	0.04	0.09	0.26
6	MgB$_2$ + 3 at% TiF$_3$	Li/K/CsI 50 wt%	350 bar 320 °C	0.68	0.05	0.11	0.16
7	MgB$_2$ + 3 at% TiCl$_3$	Li/K/CsI 50 wt%	350 bar 310 °C	0.65	0.07	0.13	0.14
8	MgB$_2$ + 3 at% TiCl$_3$	Li/K/CsCl 50 wt%	350 bar 310 °C	0.35	0.05	0.11	0.49
9	MgB$_2$ + 0.2 LiH + 3 at% TiCl$_3$	Li/K/CsCl 50 wt%	350 bar 310 °C	0.21	0.03	0.03	0.73

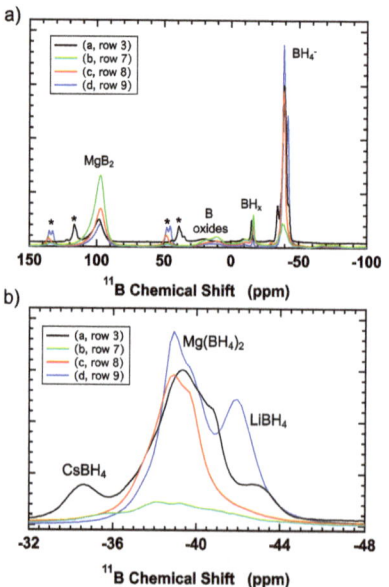

Figure 4. ^{11}B NMR spectra following hydrogenation of MgB$_2$/electrolyte formulations. (**a**) Full spectrum showing the MgB$_2$, boron oxides, BH$_x$, and [BH$_4$]$^-$ regions; * indicate spinning sidebands. (**b**) [BH$_4$]$^-$ region. Curve a (black): MgB$_2$ with 53 wt% LiKCsI electrolyte at 1000 bar, 320 °C, 50 h. Curve b (green): MgB$_2$ + 3 at% TiCl$_3$ with 50 wt% LiKCsI electrolyte at 350 bar, 310 °C, 50 h. Curve c (red): MgB$_2$ + 3 at% TiCl$_3$ with 50 wt% LiKCsI electrolyte at 350 bar, 310 °C, 50 h. Curve d (blue): MgB$_2$ + 0.2 LiH + 3 at% TiCl$_3$ with 50 wt% LiKCsCl electrolyte at 350 bar, 310 °C, 50 h. Row designations refer to Table 3.

Further reducing the pressure to 350 bar (at 320 °C) reduced the $[BH_4]^-$ fraction to 16% but actually increased the BH_x fraction to 11% (Table 3, row 6). Switching the catalyst from TiF_3 to $TiCl_3$ (and using a different high-pressure hydrogenation apparatus, at 310 °C) gave similar results with 14% $[BH_4]^-$ and 13% BH_x (Table 3, row 7). The increased BH_x fraction at lower pressure suggests that these species are intermediates on the way to formation of the fully hydrogenated $[BH_4]^-$ anions and that lower pressure favors these partially hydrogenated species. While the yields are much greater than the yields possible without a eutectic, the total yield of hydrogenated MgB_2 was only ~30% with only ~50% of that yield being $[BH_4]^-$. Although the yield is low, the hydrogenation appears to give predominately $Mg(BH_4)_2$ as shown in Figure 4, curve b.

Next, to try and improve the yields, the electrolyte was switched from the Li/K/Cs iodide-based eutectic to a Li/K/Cs chloride-based eutectic, with the specific composition of 0.575 LiCl/0.165 KCl/0.26 CsCl [20,21]. Significant improvement was seen with the $[BH_4]^-$ fraction increasing from 0.14 (with the iodide) to 0.49 (with the chloride), while the BH_x fraction actually decreased slightly (Table 3, row 8). The NMR spectrum (Figure 4, curve c) suggests the formation of relatively pure $Mg(BH_4)_2$. Compared with the sample hydrogenated at 1000 bar, the peaks associated with $CsBH_4$ and $LiBH_4$ are not clearly visible.

The chloride eutectic has a melting point of ~266 °C (higher than the iodide, ~210 °C) but still below the temperature for these hydrogenations, >300 °C. During hydrogenation, the higher melting point may lead to greater viscosity and therefore reduced transport properties. However, counteracting the viscosity and perhaps more important, the molar mass of the chloride eutectic is lower; 80.4 g/mol-Cl compared to 186.5 g/mol-I. This difference means that at 50 wt%, the eutectic: MgB_2 molar ratio increases from 0.25:1 for the iodide eutectic to 0.57:1 for the chloride eutectic, a factor of ~2.3×. Thus, during hydrogenation, there are 2.3× more liquid-state anions (Cl^-) and total cations (individually 4 × Li^+, 1.15 × K^+, and 1.8 × Cs^+) to possibly (depending on solubility) facilitate the transport of Mg^{2+} cations and $[BH_4]^-$ or other intermediate anions within the reacting mixture. Testing other eutectic compositions with suitable melting temperatures may reveal the relative importance of the Cl^- anions compared to the different cations in facilitating hydrogenation.

A final variation involved the addition of 0.2 LiH per mole of MgB_2. The 2 LiH + MgB_2 system is a well-known destabilized hydride system, which can be nearly completely hydrogenated to 2 $LiBH_4$ + MgH_2 at ~100 bar [7]. Thus at 350 bar, the addition of 0.2 LiH should result in the facile formation of 0.2 $LiBH_4$, reacting 10% of the MgB_2 and leaving the remaining 90% for possible hydrogenation to $Mg(BH_4)_2$. As shown in Table 3 (row 9), hydrogenation at 350 bar of a MgB_2 + 0.2 LiH + 3 at% $TiCl_3$ mixture with 50 wt% LiKCsCl eutectic resulted in 73% of the boron as $[BH_4]^-$ with only 3% as BH_x. From the NMR spectrum (Figure 4, curve d), the $[BH_4]^-$ is a mixture of $Mg(BH_4)_2$ and $LiBH_4$. Using the area of the spectrum (divided using the minimum between the peaks) indicates that ~25% of the $[BH_4]^-$ is present as $LiBH_4$. This is more than the expected amount of ~14%, assuming 10% of the $[BH_4]^-$ boron was $LiBH_4$ and the remaining 63% was in $Mg(BH_4)_2$. The origin is this difference is not understood. One possibility is that the simple area estimate is inaccurate. Another possibility is that some of the Li from the chloride eutectic is forming $LiBH_4$ although, there was no clear indication of $LiBH_4$ in the sample without added LiH. Despite clearly being a mixture, the utility of this formulation lies in the dehydrogenation behavior, i.e., if it cycles hydrogen well with high capacity, it is not particularly important whether pure $Mg(BH_4)_2$ or a mixture with $LiBH_4$ is formed. Thus far, the dehydrogenation has not been tested. Overall, the total amount of hydrogenation (73% $[BH_4]^-$) in the MgB_2 + 0.2 LiH sample hydrogenated at 350 bar is comparable to that originally achieved at 1000 bar. Including the complete formulation with the 50 wt% electrolytes, the uptake is 5.6 wt% hydrogen. If similar uptake could be achieved with an electrolyte at 28 wt% (or 10.5 wt%), the uptake would be 8 (or 10) wt% hydrogen.

4. Materials and Methods

The materials and methods including the Sieverts apparatus, the 1000 bar hydrogenation system, and the ^{11}B NMR setup (Bruker, Billerica, MA, USA) used for the MgH_2/Sn, MgB_2, $LiAlH_4$, and $NaAlH_4$ experiments have been described completely in Refs. [1,2]. For the additional MgB_2 hydrogenation experiments, a custom high-pressure manifold was constructed using a 25 mL Series 4740 pressure vessel from Parr Instruments (Moline, IL, USA) with valves and fittings from High-Pressure Equipment (Erie, PA, USA). The pressure was limited to ~400 bar using burst disks. To reach 350 bar from a standard hydrogen tank and regulator, the manifold contained a coiled $\frac{1}{4}$ inch diameter tubing volume that could be immersed in liquid nitrogen. Based on the manifold volume including the Parr vessel, the volume of the coil was chosen so that when the system was pressurized to 50 bar with the coil in liquid nitrogen, the pressure increased to 175 bar when the coil was warmed to room temperature (with the H_2 tank sealed off). Finally, heating the Parr vessel to 310 °C further increased the pressure to 350 bar.

5. Conclusions

The use of electrolytes to improve the kinetics of thermochemical multiphase hydrogen storage reactions has been explored based on an analogy with electrochemical multiphase battery reactions. A subjective excess free energy, based on a subjective practical reaction temperature, was used to relate the hydrogen storage and battery reactions. The difference between electrolyte-based and solid-state diffusion-based interparticle atomic transport between the reacting phases was discussed. Using this analogy, the hydrogen storage reaction kinetics with and without added electrolytes for Mg_2/Sn, $NaAlH_4$, and MgB_2 were analyzed. While the hypothesized increases in reaction rates were observed, the kinetics displayed more complex behavior. Perhaps more important than a simple reduction in reaction temperature and contrary to most multiphase solid-state reactions, the kinetics showed reaction rates that were often independent of the extent of the reaction. The behavior suggests that electrolytes also improve intraparticle transport, likely over the surface of a particle. This transport alleviates the spatial restrictions of multiple phase boundaries inherent in solid-state reactions.

The practical aspects of using electrolytes for hydrogen storage reactions were not discussed in detail here. For most of the reactions studied here, compositions containing ~50 wt% electrolytes were used. To be practical, likely <~25 wt% is needed. However, by a final analogy with batteries, where electrolyte loadings are typical ~15 wt%, we feel that optimized commercially viable compositions are feasible.

6. Patents

U.S. Patent with the number US-11050075-B1 resulted from this work.

Author Contributions: Conceptualization, J.J.V. and J.G.; methodology, J.J.V. and J.G.; validation, J.J.V., J.G., J.U. and S.-J.H.; writing—original draft preparation, J.J.V. and J. G.; writing, reviewing and editing, J.J.V., J.G., J.U. and S.-J.H.; funding acquisition, J.J.V.; Eutectic preparations and characterization, and high temperature and pressure hydrogenation, J.U.; NMR data acquisition and analysis, S.-J.H. All authors have read and agreed to the published version of the manuscript.

Funding: This research was funded by the U.S. Department of Energy, contract number DE-EE0007849.

Data Availability Statement: The data for this work is available from the corresponding author.

Acknowledgments: We thank Cullen Quine for design and construction of the high-pressure hydrogenation system and Dan Addison for discussions throughout this work.

Conflicts of Interest: The authors declare no conflict of interest. The funders had no role in the design of the study; in the collection, analyses, or interpretation of data; in the writing of the manuscript; or in the decision to publish the results.

References

1. Vajo, J.J.; Tan, H.; Ahn, C.C.; Addison, D.; Hwang, S.-J.; White, J.L.; Wang, T.C.; Stavila, V.; Graetz, J. Electrolyte-assisted hydrogen storage reactions. *J. Phys. Chem. C* **2018**, *122*, 26845–26850. [CrossRef]
2. Graetz, J.; Vajo, J.J. Electrolyte-assisted hydrogen cycling in lithium and sodium alanates at low pressures and temperatures. *Energies* **2020**, *13*, 5868. [CrossRef]
3. Huggins, R.A. *Advanced Batteries Material Science Aspects*; Springer: New York, NY, USA, 2009; pp. 1–6. [CrossRef]
4. Yang, X.; Li, W.; Zhang, J.; Hou, Q. Hydrogen storage performance of Mg/MgH$_2$ and its improvement measures: Research progress and trends. *Materials* **2023**, *16*, 1587. [CrossRef] [PubMed]
5. Shang, Y.; Pistidda, C.; Gizer, G.; Klassen, T.; Dornheim, M. Mg-based materials for hydrogen storage. *J. Magnes. Alloy.* **2021**, *9*, 1837–1860. [CrossRef]
6. Tan, X.F.; Kim, M.; Yasuda, K.; Nogita, K. Strategies to enhance hydrogen storage performances in bulk Mg-based hydrides. *J. Mater. Sci. Technol.* **2023**, *153*, 139–158. [CrossRef]
7. Klebanoff, L.E.; Keller, J.O. 5 Years of hydrogen storage research in the U.S. DOE Metal Hydride Center of Excellence (MHCoE). *Int. J. Hydrog. Energy* **2013**, *38*, 4533–4576. [CrossRef]
8. Polanski, M.; Bystrzycki, J. The influence of different additives on the solid-state reaction of magnesium hydride (MgH$_2$) with Si. *Int. J. Hydrog. Energy* **2009**, *34*, 7692–7699. [CrossRef]
9. Li, X.; Yan, Y.; Jensen, T.R.; Filinchuk, Y.; Dovgaliuk, I.; Chernyshov, D.; He, L.; Li, Y.; Li, H.-W. Magnesium borohydride Mg(BH$_4$)$_2$ for energy applications: A review. *J. Mater. Sci. Technol.* **2023**, *161*, 170–179. [CrossRef]
10. Li, H.-W.; Yan, Y.; Orimo, S.; Zuttel, A.; Jensen, C.M. Recent progress in metal borohydrides for hydrogen storage. *Energies* **2011**, *4*, 185–214. [CrossRef]
11. Ozolins, V.; Majzoub, E.H.; Wolverton, C. First-principles prediction of thermodynamically reversible hydrogen storage reactions in the Li-Mg-Ca-B-H system. *J. Am. Chem. Soc.* **2009**, *131*, 230–237. [CrossRef] [PubMed]
12. Palacin, M.R. Recent advances in rechargeable battery materials: A chemist's perspective. *Chem. Soc. Rev.* **2009**, *38*, 2565–2575. [CrossRef] [PubMed]
13. Lan, X.; Xiong, X.; Liu, J.; Yuan, B.; Hu, R.; Zhu, M. Insight into reversible conversion reactions in SnO$_2$-based anodes for lithium storage: A review. *Small* **2022**, *18*, 2201110. [CrossRef] [PubMed]
14. Olbrich, L.F.; Xiao, A.W.; Pasta, M. Conversion-type fluoride cathodes: Current state of the art. *Curr. Opin. Electrochem.* **2021**, *30*, 100779. [CrossRef]
15. Singh, N.; Arthur, T.S.; Ling, C.; Matsui, M.; Mizuno, F. A high energy-density tin anode for rechargeable magnesium-ion batteries. *Chem Comm.* **2013**, *49*, 149–151. [CrossRef] [PubMed]
16. Paskevicius, M.; Sheppard, D.A.; Chaudhary, A.-L.; Webb, C.J.; Gray, E.; Mac, A.; Tian, H.Y.; Peterson, V.K.; Buckley, C.E. Kinetic limitations in the Mg–Si–H system. *Int. J. Hydrog. Energy* **2011**, *36*, 10779–10786. [CrossRef]
17. Comanescu, C. Paving the way to the fuel of the future–Nanostructured complex hydrides. *Int. J. Mol. Sci.* **2023**, *24*, 143. [CrossRef] [PubMed]
18. Ley, M.B.; Roedern, E.; Jensen, T.R. Eutectic melting of LiBH$_4$-KBI$_4$. *Phys. Chem. Chem Phys.* **2014**, *16*, 24194–24199. [CrossRef] [PubMed]
19. Brown, W.E.; Dollimore, D.; Galwey, A.K. Theory of solid state reaction kinetics. In *Comprehensive Chemical Kinetics*; Bamford, C.H., Tipper, C.F.H., Eds.; Elsevier: New York, NY, USA, 1980; Volume 22, pp. 41–109. ISBN 0444418075.
20. Redkin, A.; Korzun, I.; Yaroslavtseva, T.; Reznitskikh, O.; Zaikov, Y. Isobaric heat capacity of molten halide eutectics. *J. Therm. Anal. Calorim.* **2017**, *128*, 621–626. [CrossRef]
21. Murakami, T.; Nohira, T.; Ogata, Y.H.; Ito, Y. Electrochemical window of a LiCl–KCl–CsCl melt. *Electrochem. Solid State Lett.* **2005**, *8*, E1–E3. [CrossRef]

Disclaimer/Publisher's Note: The statements, opinions and data contained in all publications are solely those of the individual author(s) and contributor(s) and not of MDPI and/or the editor(s). MDPI and/or the editor(s) disclaim responsibility for any injury to people or property resulting from any ideas, methods, instructions or products referred to in the content.

Collectable Single Pure-Pd Metal Membrane with High Strength and Flexibility Prepared through Electroplating for Hydrogen Purification

Naruki Endo [1,*], Yumi Kaneko [2], Norikazu Dezawa [2], Yasuhiro Komo [2] and Masanobu Higuchi [2]

1. Renewable Energy Research Center, National Institute of Advanced Industrial Science and Technology (AIST), 2-2-9 Machiikedai, Koriyama 963-0298, Japan
2. Sanno Co., Ltd., 5-8-8, Tsunashima-Higashi, Kouhoku-ku, Yokohama 223-0052, Japan
* Correspondence: naruki.endo@aist.go.jp

Abstract: Among the various film preparation methods, electroplating is one of the simplest and most economical methods. However, it is challenging to collect a dense single Pd film through plating, owing to the accumulation of stress in the film during the process. Therefore, the characteristics of a single plated film have not been clearly identified, although pure Pd is widely used in metallic-hydrogen-purification membranes. In this study, stress concentration in film during preparation was reduced by optimizing the plating process, and a dense single flat film was successfully collected. No impurities were detected. Thus, a high-purity Pd film was prepared. Its surface texture was found to be significantly different from that of the rolled film, and several approximately 5 μm sized aggregates were observed on the surface. The plated film is reported to have mechanical properties superior to those of the rolled film, with twice the displacement and four times the breaking point strength. The hydrogen permeabilities of the plated film (5.4×10^{-9}–1.1×10^{-8} mol·m^{-1}·s^{-1}·Pa$^{-1/2}$ at 250–450 °C) were comparable to those of the rolled and reported films, indicating that the surface texture does not have a strong effect on hydrogen permeability. The results of this study promote the practical use of Pd-based membranes through electroplating.

Keywords: palladium; Pd-based membrane; electroplating; hydrogen purification

1. Introduction

Hydrogen has been gaining increased significance in the global trend toward the realization of a carbon-neutral society by 2050–2060. As most hydrogen is currently produced via reforming reactions [1], purification processes are necessary to obtain pure hydrogen with high added value. Hydrogen purification using metallic-hydrogen-separation membranes yields 100% pure hydrogen in a one-step process [2]. Thus, the practical applications of such membranes are being explored [1,2].

There are two major categories of metallic-hydrogen-purification/separation membranes, namely Pd-based [3–16] and non-Pd-based [17–20]. The former has been the subject of extensive research and development [6–16]; such membranes prepared using the rolling method are now in practical use, albeit on a small scale. However, in addition to the high price of Pd itself, the membrane preparation process becomes more expensive as the rolled membrane becomes thinner, which is a major issue. Other dry-process methods include vapor deposition, such as physical vapor deposition (PVD) [21,22] and chemical vapor deposition (CVD) [23,24]. While PVD and CVD enable strict film thickness and composition control, they are not suitable for mass production due to high equipment costs and strict depositing conditions.

There are two wet-process methods, namely electroless plating (ELP) [25–28] and electroplating [29,30]. ELP is one of the most promising methods because it can deposit on various types of supports, although it has disadvantages of complicated pretreatment and difficult thickness control. Gade et al. prepared single pure Pd and PdCu films using ELP to evaluate the permeabilities of Pd-based films and reported that it was comparable to those of rolled films and previously reported values [31]. However, it is difficult to control ELP thickness because ELP conducts film deposition through spontaneous reaction; thus, single defect-free films of <7.2 μm could not be collected. The mechanical property of the single film has not been investigated at all.

Recently, we reported the preparation of a Pd alloy through electroplating to reduce expenses [32,33]. Electroplating offers advantages of low-cost equipment, mass-production suitability, easy control of thickness, and preparation of high-purity membranes [4,5]. In our previous studies, we plated a PdCu alloy in a one-step process and found that the plated film had higher strength and was more flexible than a rolled PdCu film [32,33]. However, the composition of PdCu alloys significantly impacted hydrogen permeability [31,34]. This makes it particularly difficult to control alloy composition during plating. However, we started our work with a PdCu alloy because the alloy has the advantage of containing more than half Cu (at%), which prevents the accumulation of stress in the film and makes it as easy to collect as a single film.

In general, a pure Pd film absorbs hydrogen during the plating process and becomes hard and brittle [35,36]. This makes it difficult to obtain a single flat film (see Figure 1a). Although pure Pd is the most typical metallic-hydrogen-purification membrane, there is no report on the preparation and characterization of a dense single Pd film through electroplating (<10 μm thickness). Further, the mechanical and hydrogen-permeable properties of a pure-Pd-plated film are not well understood. Moreover, for other metals, the mechanical properties change significantly, owing to hydrogen absorption during electroplating [37].

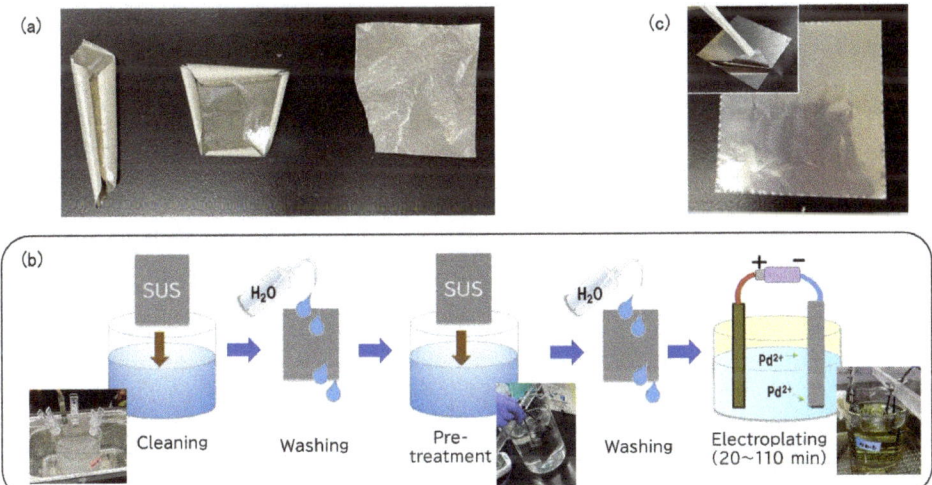

Figure 1. (**a**) Photograph of the plated Pd films prepared under conventional plating conditions. The film was neither a single flat one, due to the stress caused by hydrogen absorption during the plating process, nor was it broken when peeled from an SUS (stainless steel) substrate. (**b**) Schematic of each process involved in electroplating. No special processes were used; only general-purpose equipment was used. (**c**) Collected single flat pure-Pd-plated film. The inset depicts the process of peeling the plated film from a substrate.

In this short communication, we report the characterization of a single pure-Pd-plated film itself. This was conducted by applying the know-how of single film collection, cultivated during the PdCu one-step plating process to achieve pure Pd plating, and we succeeded in recovering dense single flat films with a thickness of 3–15 μm. Mechanical properties of thin films, with a minimum thickness of 3 µm, and their hydrogen permeability were evaluated. The main objective is to show that a single pure Pd film prepared by electroplating has desirable properties for practical use.

2. Materials and Methods

As it is essential to prepare pure Pd films at a low cost, all the equipment used for electroplating, including the plating solution and substrates, were general-purpose products. The electroplating process is depicted in Figure 1b. An SUS304 (non-porous stainless steel) plate (60 mm × 60 mm) was used as the substrate, which was electrolytically degreased (6 V, 10–120 s) using an alkaline degreasing solution (pH 12–13). Next, the surfaces were washed first with distilled water and then with hydrochloric acid (1–6 N; by immersion for 10–120 s). Thereafter, the surfaces were washed with distilled water again and electroplated under the following conditions. PALLABRIGHT SST (Japan Pure Chemical Co., Ltd., Tokyo, Japan) was used as the Pd-plating solution. A potentiostat/galvanostat HA-151B (Hokuto Denko Co., Ltd., Tokyo, Japan) was used as the power supply for plating. The plating was performed for 20–110 min for a thickness of 3–15 µm at a current density of 0.5–2.0 A/dm^2 and a temperature of 40–65 °C. These conditions were set to achieve a Pd film with a thickness of 3–15 µm. The key here is to not clean the substrate too much so that the plated film can be removed from the substrate, but rather to plate relatively slowly, thereby preventing hydrogen absorption on the Pd film during electroplating. By doing so, as depicted in Figure 1c, stress is not accumulated in the film, and the film becomes flexible so that it can be removed as a single flat film.

Rolled Pd films (99.9% Pd, Tanaka Kikinzoku Kogyo Co., Ltd., Tokyo, Japan) were prepared using the cold rolling process. Pd ingots were rolled to thicknesses of 15 and 50 µm. We obtained rolled films of diameters 12.1 and 20 mm. However, rolled films with thicknesses of <15 µm were not available because such films are impossible to prepare.

The prepared films were characterized through scanning electron microscopy (SEM), energy-dispersive X-ray (EDX) spectrometry, and X-ray diffraction (XRD). The surface morphology of all films was observed through field-emission scanning electron microscopy (FE-SEM; FEI QUANTA FEG250). The concentrations of Pd and other elements were analyzed through an EDAX AMETEK OCTAN PRO EDX spectrometer. Its lower detection limits were several wt% for light elements (Be, B, C, N, O and F) and 0.2–1 wt% for other elements. FE-SEM and EDX spectrometry were performed under vacuum at room temperature (~20 °C). The thickness of each film sample was determined from cross-sectional SEM observations. XRD was performed using a Rigaku Smartlab X-ray diffractometer with Cu Kα radiation. The scanned 2θ angles ranged from 20° to 100°.

The mechanical strength measurements were carried out using SHIMAZU EZ-TEST series EZ-SX. The sample films with 20 mm diameters were cut from a 500 mm × 500 mm Pd film. The plated and rolled films were 3.2–15.1 and 15 µm thick, respectively. The sample was held in place using a holder and pressed using a spherical jig (ϕ7 mm). The crosshead speed was 1 mm/min. The mechanical strength tests were measured three times for each sample of the same film thickness and the displacement against load force was measured within a measurement error of ±5%. These mechanical measurements were carried out using our proposed method [32,38,39].

The hydrogen permeation tests were performed using a conventional gas-permeation apparatus; the details of these tests are described in [32,40]. The samples used for comparison were plated Pd (3.2–10.2 µm thickness) and rolled Pd (50 µm thickness). Each sample was cut to ϕ12.1 mm and sealed using Cu gaskets (with outer and inner diameters of ϕ12.1 mm and ϕ5.2 mm, respectively). The diameter used to calculate hydrogen permeability was ϕ7.1 mm, which was the diameter of the Cu gaskets tightened by Swagelok VCR fittings. Hydrogen gas (99.9999%) was introduced on the upstream side (p_1 = 400–200 kPa), and the downstream side was kept under atmospheric pressure (p_2 = 101 kPa). Thereafter, the temperature was decreased incrementally by 50 °C from 450 °C and measured after 30 min. The hydrogen gas permeating the membranes was measured using a mass flow meter (HORIBA MODEL SEF-E40; 50 mL/min). A helium (He) leak test was performed before/after each hydrogen permeation test, and no He flow rate was observed. This shows that each plated film was defect-free and dense.

3. Results and Discussion

Electroplated Pd could be easily delaminated from the substrate by cutting it around the edges (see Figure 1c), yielding a 60 mm × 60 mm film. The surface observation results obtained through FE-SEM are presented in Figure 2a–c. The SEM images of the surfaces of the pure-Pd-plated film before and after the hydrogen-permeation test are presented in (a) and (b), respectively. The SEM image of the rolled film is depicted in Figure 2c for comparison; the image is before performing the test because there was almost no change in its microstructure before and after the hydrogen-permeation test. The surface texture after plating, depicted in Figure 2a, indicates an aggregate structure with a thickness of approximately 3–5 µm and a rough surface. This structure has a cauliflower-like texture, which is similar to that reported in previously published papers [31,35]. After the test (Figure 2b), the surface structure was altered due to the high temperature, and the grain boundaries of the aggregates became more distinct; therefore, the surface roughness increased. In contrast, the rolled film had rolled lines and no aggregates, which were seen in the plated film, and its surface structure was smoother than that of the plated film (see Figure 2a–c).

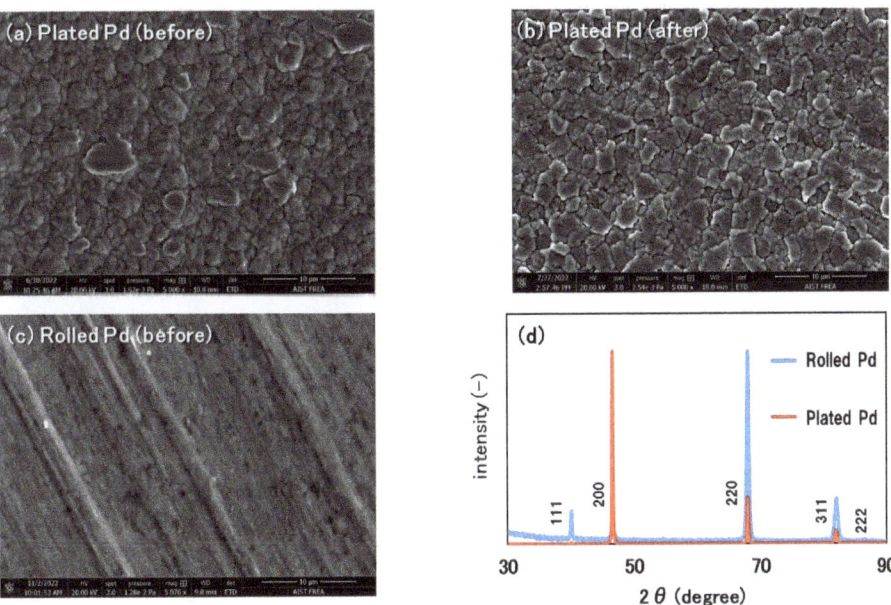

Figure 2. SEM images of the surfaces of the pure-Pd-plated film (**a**) before and (**b**) after the hydrogen-permeation tests; (**c**) rolled film. (**a,c**) Depiction of the as-plated and as-rolled states. (**d**) XRD profiles of the plated (red line) and rolled (blue line) Pd films before the hydrogen-permeation tests. Both samples had different peak intensities due to grain orientation, but they had the same FCC structure. The 111 peak of the plated film is very low but appears at the same angle as the rolled film.

Figure 2d presents the XRD results for the plated and rolled films. Both samples are depicted after preparation. Although the peak intensities of both samples are different due to the orientation of the crystal grains during each preparation process, the peak positions for both are identical, indicating an FCC structure. No peaks indicating any other structure were observed, and the lattice parameter obtained from this result was

$a = 3.8910 \pm 0.0002$ Å, which is a slightly larger value than previously reported for pure Pd because of hydrogen absorption [41]. Compositional analysis through EDX was also performed. The spectrum obtained for the plated film was the same as that obtained for the rolled film, and no elements except Pd were detected. From these results, it can be concluded that a single pure Pd film can be plated.

Further, the mechanical strength of the prepared pure Pd film was investigated. The results of the measurement are presented in Figure 3a. The horizontal and vertical axes represent the displacement (mm) and load force (N), respectively. The plated films were 3.2–15.1 µm thick, and the rolled film was 15 µm. The peaks in the curves indicate the points of film breakage. The results for the plated films indicate that the thicker the film, the larger the displacement and the higher the load at the points of film breakage. The displacement of the 15 µm thick rolled film was smaller than that of the 5.5 µm thick plated film, and the breaking point of the rolled film appeared between 9.5 and 5.5 µm. Figure 3b presents the plots of the film thickness (µm) on the horizontal axis and breaking load force (N) on the vertical axis. The plots indicate that the breaking load increases with an increase in film thickness. The breaking load of the rolled film is equivalent to that of the 7.5 µm thick plated film; this indicates that the plated film with half the thickness has the same strength as that of the rolled film. Comparing both films with the same thickness of 15 µm, it can be seen that the breaking load for the plated film is approximately four times higher than that of the rolled film. These results indicate that the plated films are stronger and more flexible than the rolled films.

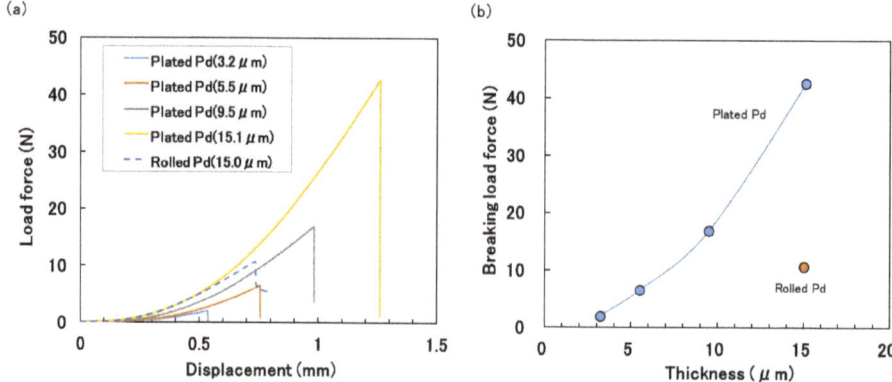

Figure 3. (**a**) Plots of load force (N) vs. displacement (mm) for the as-plated Pd films (3.2–15.1 µm thickness) and as-rolled one (15 µm thickness). (**b**) Plots of breaking load force (in N) vs. thickness (µm) of the plated and rolled pure Pd films.

The plated film is more flexible than the rolled film because the latter unavoidably undergoes work hardening, during which point defects and dislocations are introduced into the film during the rolling process [32]. In the former case, it is considered that our plating process restricts the absorption of hydrogen to the maximum extent possible, such that stress does not accumulate easily in the film and film ductility is maintained. In contrast, the plated films demonstrate a much higher breaking strength than the rolled film because pure Pd prepared through plating is generally known to have higher hardness than bulk Pd [35,36]. The high strength and flexibility of the plated film indicate that the amount of used Pd can be reduced, and material handling can be simplified when rolled film is replaced by plated film. These are significant advantages in practical applications.

The results of the hydrogen-permeation tests are presented in Figure 4a,b. The vertical axes in both figures represent the permeated hydrogen flow rate, whereas the horizontal axes in (a) and (b) represent the film thickness and reciprocal of the film thickness (1/thickness), respectively. For each film thickness, the permeated hydrogen flow rate was higher at higher temperatures; the thinner the film, the larger the difference in flow rate. Compared with the permeated hydrogen flow rate at the same temperature, the flow rate for the film with 1/2 or 1/3 thickness increased by almost two or three times. Figure 4b indicates that the plots for all temperatures are approximately on a straight line passing through the origin. This indicates that hydrogen permeation through the plated films with thicknesses of approximately 3–10 μm was mostly diffusion limitation. This also implies that the quality of the films with different thicknesses was similar. It is generally understood that hydrogen permeation changes from diffusion limitation to surface molecular dissociation limitation at a thickness of ≤5 μm [4]; however, our pure Pd film indicated a diffusion limitation even at a thickness of 3.2 μm. We are currently evaluating thinner films to preliminarily observe that flow dependence tends to deviate from linearity at thicknesses of approximately 1.5–2.0 μm.

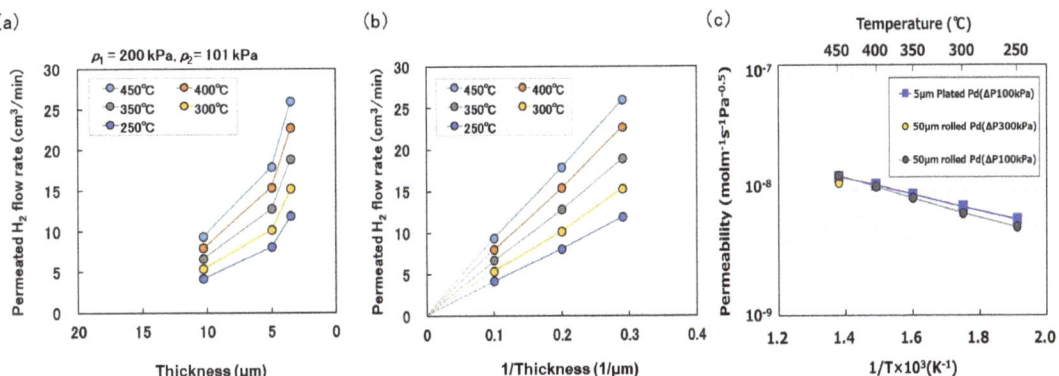

Figure 4. (a) Dependence of the permeated hydrogen flow rate on film thickness at various temperatures. (b) Dependence on the thickness (1/thickness) of permeated hydrogen flow rate at each film thickness. (c) Temperature dependence of the hydrogen permeabilities of plated Pd (5 μm thickness) and rolled Pd (50 μm thickness) at 250–450 °C. The experimental conditions are indicated in the figure.

The temperature dependence of hydrogen permeability of the plated film (5 μm thickness) is depicted in Figure 4c, along with the results for the rolled film (50 μm thickness) for comparison. Hydrogen permeability gradually decreased with a decrease in temperature. A similar result was obtained for the rolled film, although the plated film exhibited a slightly higher value, suggesting the contribution of increased surface roughness. The hydrogen permeability of the plated film was the same for a thickness of 10 μm (not shown in Figure 4c). These permeability values were comparable to the values reported in the literature [42]. The results obtained indicate that although the surface structures of the plated and rolled films were significantly different (see Figure 2a–c), their effect on hydrogen permeability was not so strong. This is significant knowledge pertaining to the development of Pd-based hydrogen-permeable membranes through electroplating in the future.

4. Conclusions

In this communication, pure Pd film was deposited through electroplating and successfully recovered as a dense single flat film. The collected film exhibited an FCC crystal structure, did not contain any impurities, and comprised highly pure Pd. Unlike conventional Pd films, the plated film prepared in this study demonstrated low stress accumulation; the displacement and breaking strength were twice and four times higher, respectively, than those of the rolled films. Although the surface texture of the plated film was significantly different from that of the rolled film, the hydrogen permeability of the former was comparable to that of the latter and previously reported films. Such plated films can be prepared easily and at a low cost; they also demonstrate excellent mechanical properties. Therefore, the Pd-plated film is favorable for practical applications. Detailed investigations of the plating process and relationship between its microstructure and mechanical strength are currently underway.

Author Contributions: N.E. conceived the experimental design. N.E. and N.D. co-wrote the paper and analyzed the data. Y.K. (Yumi Kaneko) and N.D. performed the electroplating. N.D. measured the mechanical properties and hydrogen permeability. Y.K. (Yasuhiro Komo) and M.H. conceived and contributed to the overall project planning. All authors have read and agreed to the published version of the manuscript.

Funding: This work was partially supported by the Fukushima Prefecture Subsidy.

Data Availability Statement: The data obtained in this study are referenced in the main body of the paper and are available from the corresponding author upon reasonable request.

Acknowledgments: The authors acknowledge Narita, Oouchi, and Nakamura for technical support. N.E. acknowledges Jung for his helpful comments.

Conflicts of Interest: The authors declare no conflict of interest.

References

1. Shu, J.; Grandjean, B.P.A.; Van Neste, A.V.; Kaliaguine, S. Catalytic Palladium-Based Membrane Reactors: A Review. *Can. J. Chem. Eng.* **1991**, *69*, 1036–1060. [CrossRef]
2. Basile, A.; Gallucci, F.; Tosti, S. Synthesis, Characterization, and Applications of Palladium Membranes. *Membr. Sci. Technol.* **2008**, *13*, 255–323. [CrossRef]
3. Paglieri, S.N.; Way, J.D. Innovations in Palladium Membrane Research. *Sep. Purif. Methods* **2002**, *31*, 1–169. [CrossRef]
4. Yun, S.; Ted Oyama, S.T. Correlations in Palladium Membranes for Hydrogen Separation: A Review. *J. Membr. Sci.* **2011**, *375*, 28–45. [CrossRef]
5. Conde, J.J.; Maroño, M.; Sánchez-Hervás, J.M. Pd-Based Membranes for Hydrogen Separation: Review of Alloying Elements and Their Influence on Membrane Properties. *Sep. Purif. Rev.* **2017**, *46*, 152–177. [CrossRef]
6. Zhang, Y.; Komaki, M.; Nishimura, C. Morphological Study of Supported Thin Pd and Pd–25Ag Membranes upon Hydrogen Permeation. *J. Membr. Sci.* **2005**, *246*, 173–180. [CrossRef]
7. Kikuchi, E. Membrane Reactor Application to Hydrogen Production. *Catal. Today* **2000**, *56*, 97–101. [CrossRef]
8. Sonwane, C.G.; Wilcox, J.; Ma, Y.H. Solubility of Hydrogen in PdAg and PdAu Binary Alloys Using Density Functional Theory. *J. Phys. Chem. B* **2006**, *110*, 24549–24558. [CrossRef]
9. Shi, L.; Goldbach, A.; Zeng, G.; Xu, H. Preparation and Performance of Thin-Layered PdAu/Ceramic Composite Membranes. *Int. J. Hydrogen Energy* **2010**, *35*, 4201–4208. [CrossRef]
10. Zeng, G.; Goldbach, A.; Shi, L.; Xu, H. On Alloying and Low-Temperature Stability of Thin, Supported PdAg Membranes. *Int. J. Hydrogen Energy* **2012**, *37*, 6012–6019. [CrossRef]
11. Millet, P.; Ngameni, R.; Decaux, C.; Grigoriev, S.A. Hydrogen Sorption by $Pd_{77}Ag_{23}$ Metallic Membranes. Role of Hydrogen Content, Temperature and Sample Microstructure. *Int. J. Hydrogen Energy* **2011**, *36*, 4262–4269. [CrossRef]
12. Montesinos, H.; Julián, I.; Herguido, J.; Menéndez, M. Effect of the Presence of Light Hydrocarbon Mixtures on Hydrogen Permeance through Pd–Ag Alloyed Membranes. *Int. J. Hydrogen Energy* **2015**, *40*, 3462–3471. [CrossRef]
13. Kamakoti, P.; Morreale, B.D.; Ciocco, M.V.; Howard, B.H.; Killmeyer, R.P.; Cugini, A.V.; Sholl, D.S. Prediction of Hydrogen Flux through Sulfur-Tolerant Binary Alloy Membranes. *Science* **2005**, *307*, 569–573. [CrossRef]
14. Decaux, C.; Ngameni, R.; Solas, D.; Grigoriev, S.; Millet, P. Time and Frequency Domain Analysis of Hydrogen Permeation across PdCu Metallic Membranes for Hydrogen Purification. *Int. J. Hydrogen Energy* **2010**, *35*, 4883–4892. [CrossRef]
15. Zhang, K.; Way, J.D. Palladium-Copper Membranes for Hydrogen Separation. *Sep. Purif. Technol.* **2017**, *186*, 39–44. [CrossRef]
16. Zhao, C.; Goldbach, A.; Xu, H. Low-Temperature Stability of Body-Centered Cubic PdCu Membranes. *J. Membr. Sci.* **2017**, *542*, 60–67. [CrossRef]
17. Phair, J.W.; Donelson, R. Developments and Design of Novel (Non-palladium-Based) Metal Membranes for Hydrogen Separation. *Ind. Eng. Chem. Res.* **2006**, *45*, 5657–5674. [CrossRef]
18. Dolan, M.D. Non-Pd BCC Alloy Membranes for Industrial Hydrogen Separation. *J. Membr. Sci.* **2010**, *362*, 12–28. [CrossRef]
19. Dolan, M.D.; Song, G.; Liang, D.; Kellam, M.E.; Chandra, D.; Lamb, J.H. Hydrogen Transport through $V_{85}Ni_{10}M_5$ Alloy Membranes. *J. Membr. Sci.* **2011**, *373*, 14–19. [CrossRef]
20. Buxbaum, R.E.; Marker, T.L. Hydrogen Transport through Non-porous Membranes of Palladium-Coated Niobium, Tantalum and Vanadium. *J. Membr. Sci.* **1993**, *85*, 29–38. [CrossRef]

21. Jayaraman, V.; Lin, Y.S. Synthesis and Hydrogen Permeation Properties of Ultrathin Palladium-Silver Alloy Membranes. *J. Membr. Sci.* **1995**, *104*, 251–262. [CrossRef]
22. McCool, B.; Xomeritakis, G.; Lin, Y.S. Composition Control and Hydrogen Permeation Characteristics of Sputter Deposited Palladium-Silver Membranes. *J. Membr. Sci.* **1999**, *161*, 67–76. [CrossRef]
23. Xomeritakis, G.; Lin, Y.S. Fabrication of a Thin Palladium Membrane Supported in a Porous Ceramic Substrate by Chemical Vapor Deposition. *J. Membr. Sci.* **1996**, *120*, 261–272. [CrossRef]
24. Kikuchi, E.; Nemoto, Y.; Kajiwara, M.; Uemiya, S.; Kojima, T. Steam Reforming of Methane in Membrane Reactors: Comparison of Electroless-Plating and CVD Membranes and Catalyst Packing Modes. *Catal. Today* **2000**, *56*, 75–81. [CrossRef]
25. Huang, Y.; Li, X.; Fan, Y.; Xu, N. Palladium-Based Composite Membranes: Principle, Preparation and Characterization. *Prog. Chem.* **2006**, *18*, 230–237.
26. Thoen, P.M.; Roa, F.; Way, J.D. High Flux Palladium–Copper Composite Membranes for Hydrogen Separations. *Desalination* **2006**, *193*, 224–229. [CrossRef]
27. Tosti, S. Supported and Laminated Pd-Based Metallic Membranes. *Int. J. Hydrogen Energy* **2003**, *28*, 1445–1454. [CrossRef]
28. Dittmeyer, R.; Höllein, V.; Daub, K. Membrane Reactors for Hydrogenation and Dehydrogenation Processes Based on Supported Palladium. *J. Mol. Catal. A Chem.* **2001**, *173*, 135–184. [CrossRef]
29. Tong, J.; Shirai, R.; Kashima, Y.; Matsumura, Y. Preparation of a Pinhole-Free Pd–Ag Membrane on a Porous Metal Support for Pure Hydrogen Separation. *J. Membr. Sci.* **2005**, *260*, 84–89. [CrossRef]
30. Bryden, K.J.; Ying, J.Y. Electrodeposition Synthesis and Hydrogen Absorption Properties of Nanostructured Palladium–Iron Alloys. *Nanostruct. Mater.* **1997**, *9*, 485–488. [CrossRef]
31. Gade, S.K.; Thoen, P.M.; Way, J.D. Unsupported Palladium Alloy Foil Membranes Fabricated by Electroless Plating. *J. Membr. Sci.* **2008**, *316*, 112–118. [CrossRef]
32. Endo, N.; Furukawa, Y.; Goshome, K.; Yaegashi, S.; Mashiko, K.-i.; Tetsuhiko, M. Characterization of Mechanical Strength and Hydrogen Permeability of a PdCu Alloy Film Prepared by One-Step Electroplating for Hydrogen Separation and Membrane Reactors. *Int. J. Hydrogen Energy* **2019**, *44*, 8290–8297. [CrossRef]
33. Kato, Y.; Maeda, T.; Endo, N.; Yaegashi, S.; Furukawa, Y.; Dezawa, N. Hydrogen Permeable Membranes and Their Preparation Methods. Japan Patent P6695929, 24 April 2020. (In Japanese).
34. Knapton, A.G. Palladium Alloys for Hydrogen Diffusion Membranes. *Platin. Met. Rev.* **1977**, *21*, 44–50.
35. Raub, C.J. Electroplating of Palladium for Electrical Contacts. *Platin. Met. Rev.* **1982**, *26*, 158–166.
36. Yasumura, K. Palladium and Palladium Alloy Bath. *J. Surf. Finish. Soc. Jpn.* **2004**, *55*, 640–645. [CrossRef]
37. Snavely, C.A. A Theory for the Mechanism of Chromium Plating; A Theory for the Physical Characteristics of Chromium Plate. *J. Electrochem. Soc.* **1947**, *92*, 537. [CrossRef]
38. Yaegashi, S.; Endo, N.; Kumakawa, M.; Suzuki, S.; Maeda, T. Method and Apparatus for Evaluating Flexibility of Sheet-Type Testing Materials. Japan Patent P6265196, 24 January 2018.
39. Endo, N.; Yaegashi, S.; Maehata, T.; Kumakawa, M.; Suzuki, S.; Mashiko, K.-i.; Maeda, T. High Thermal Stability and Flexibility of Thin Porous Ni Metal Support Prepared by Electroplating Deposition for Pd Alloy Membranes. *Mater. Trans.* **2017**, *58*, 1093–1096. [CrossRef]
40. Nishimura, C.; Komaki, M.; Amano, M. Hydrogen Permeation Characteristics of Vanadium-Nickel Alloys. *Mater. Trans. JIM* **1991**, *32*, 501–507. [CrossRef]
41. Arblaster, J.W. Crystallographic Properties of Palladium. *Platin. Met. Rev.* **2012**, *56*, 181–189. [CrossRef]
42. Amano, M.; Nishimura, C.; Komaki, M. Effect of High Concentration CO and CO_2 on Hydrogen Permeation through the Palladium Membrane. *Mater. Trans. JIM* **1990**, *31*, 404–408. [CrossRef]

Disclaimer/Publisher's Note: The statements, opinions and data contained in all publications are solely those of the individual author(s) and contributor(s) and not of MDPI and/or the editor(s). MDPI and/or the editor(s) disclaim responsibility for any injury to people or property resulting from any ideas, methods, instructions or products referred to in the content.

Article

Hydrogen Compression Materials with Output Hydrogen Pressure in a Wide Range of Pressures Using a Low-Potential Heat-Transfer Agent

Xu Zhang [1], Yu-Yuan Zhao [1], Bao-Quan Li [1], Mikhail Prokhorenkov [2], Elshad Movlaev [2], Jin Xu [1], Wei Xiong [1], Hui-Zhong Yan [1,*] and Sergey Mitrokhin [2,*]

[1] State Key Laboratory of Baiyunobo Rare Earth Resource Researches and Comprehensive Utilization, Baotou Research Institute of Rare Earths, Baotou 014030, China; lzhangxu@126.com (X.Z.); zhaoyuyuan1028@126.com (Y.-Y.Z.); lbqxplbq@126.com (B.-Q.L.); xujinaou123@126.com (J.X.)

[2] Chemistry Department, Lomonosov Moscow State University, Moscow 119991, Russia; mikl1995@yandex.ru (M.P.); movlaev@hydride.chem.msu.ru (E.M.)

* Correspondence: yhzmail@126.com (H.-Z.Y.); mitrokhin@hydride.chem.msu.ru (S.M.)

Citation: Zhang, X.; Zhao, Y.-Y.; Li, B.-Q.; Prokhorenkov, M.; Movlaev, E.; Xu, J.; Xiong, W.; Yan, H.-Z.; Mitrokhin, S. Hydrogen Compression Materials with Output Hydrogen Pressure in a Wide Range of Pressures Using a Low-Potential Heat-Transfer Agent. *Inorganics* **2023**, *11*, 180. https://doi.org/10.3390/inorganics11050180

Academic Editor: Craig Buckley

Received: 17 March 2023
Revised: 12 April 2023
Accepted: 13 April 2023
Published: 24 April 2023

Copyright: © 2023 by the authors. Licensee MDPI, Basel, Switzerland. This article is an open access article distributed under the terms and conditions of the Creative Commons Attribution (CC BY) license (https://creativecommons.org/licenses/by/4.0/).

Abstract: In order to meet the demand of metal hydride–hydrogen compressors (MHHC) and their hydrogen compression materials for high-pressure hydrogen filling in a hydrogen energy field, four kinds of hydrogen storage alloys with low-grade heat source (<373 K) heating outputs and different hydrogen pressures (up to 80 MPa) were developed as hydrogen compression materials. The preliminary compositions of the hydrogen storage alloys were determined by using a statistical model and research experience. The rare earth series AB_5 and Ti/Zr base AB_2 hydrogen storage alloys were prepared using a high-temperature melting method. The composition, structure, and hydrogenation/dehydrogenation plateau characteristics of the alloys were tested by an inductively coupled plasma mass spectrometer (ICP-MAS), X-ray diffractometer (XRD), and pressure–composition isothermal (PCT) tester. The median output pressures of the four-stage hydrogen storage alloys at 363 K were 8.90 MPa, 25.04 MPa, 42.97 MPa, and 84.73 MPa, respectively, which met the requirements of the 20 MPa, 35 MPa, and 70 MPa high-pressure hydrogen injections for the MHHCs. In fact, due to the tilted pressure plateau of the PCT curve, the synergy between the adjacent two alloys still needed to be adjusted.

Keywords: hydrogen compression materials; hydrogen storage alloy; metal hydride hydrogen compressors; hydrogen storage and supply; kinetics; thermodynamics; crystal structures

1. Introduction

Hydrogen energy has the advantages of having abundant resources, convenient storage, a wide application range, being clean and low-carbon, and a capacity for interconnection and collaboration [1–5]. H_2-O_2 fuel cells, as a model of hydrogen energy utilization, have entered the market introduction stage in the fields of transportation, energy storage, distributed power supply, and so on [6–9]. In particular, the power of fuel cell vehicle engines has been greatly improved, and they can run more than 700 km with a 70 MPa hydrogen storage tank [10]. Therefore, terminal hydrogenation stations will become a kind of popular public facility, but the high construction cost of these hydrogenation stations is the same dilemma faced by the global hydrogen energy industry. The use of a metal hydride–hydrogen compressor (MHHC) could reduce the current high input of these hydrogenation stations [1,5,11].

Advanced hydrogen storage and supply systems can use low-grade heat sources (such as industrial waste heat at T < 573 K) or solar heat to heat MH to dehydrogenate and convert H_2 into high-pressure and high-purity H_2. The main advantages of this are that low-grade heat sources are used instead of electricity, and the operation is simple, with no moving parts, a compact structure, safety, and reliability. It is a better choice than conventional (mechanical) and newly developed (electrochemical or ionic liquid piston) hydrogen compression methods. Since the concept of MHHCs was proposed in the United States Patent (US3516263) in 1970, the research and development work, from principle to application, has become more and more in-depth. MHHCs have been used in aerospace, hydrogen isotope processing, water pump/actuator power, and other specific applications. Norway has built a 70 MPa demonstration hydrogenation station using MHHCs [1,12–14].

The main components of MHHCs are hydrogen storage alloy materials that can form MH [1,5,14–17], such as rare earth AB_5 intermetallic compounds [18–23], titanium/zirconium base AB and AB_2 intermetallic compounds [12,24–38], and vanadium base solid solutions [39]. The hydrogen storage alloy is accompanied by electron transfer and a change in hydrogen pressure and heat while reversibly absorbing/releasing hydrogen. At the same time, its equilibrium pressure changes exponentially with a change in temperature. A variety of alloys derived from the above alloy systems offer the possibility of hydrogen pressure outputs over a wide pressure range, using water as a heat transfer medium (T < 373 K). Recently, hydrogen compression materials with a pressure range of 3.2–85 MPa have been developed in China using low-grade heat sources, which can be used to fill high-pressure hydrogen storage devices for hydrogen energy applications [5]. The advanced hydrogen compression materials developed in low-pressure ranges (<25 MPa) are AB_5 type rare earth hydrogen storage alloys, and for a high-pressure (>25 MPa) hydrogen output, the preferred hydrogen compression materials are AB_2 type titanium base hydrogen storage alloys [5]. In this paper, considering the hydrogenation requirements of a long tube trailer with 20 MPa hydrogen and an on-board hydrogen tank with 35 MPa and 70 MPa hydrogen, hydrogen storage alloy materials of a 1–4 stage hydrogen compression for MHHCs were designed and prepared, with the highest output hydrogen pressure exceeding 80 MPa. Compared with the existing advanced hydrogen compression materials, the AB_2 titanium base hydrogen storage alloy with a more adjustable pressure range was used in the low-pressure range, and the plateau characteristics of all the stages of the hydrogen storage alloy were further optimized to improve the application performance of the MHHCs.

During the development of MHHCs at home and abroad, the research work on hydrogen storage alloys has mainly focused on the simulation, testing, and evaluation of the main properties of the materials. In fact, the composition, preparation technology, and technological conditions of these materials are very important for improving their application efficiency. In this paper, the basic composition of the hydrogen storage alloy used in the MHHCs was designed by the Moscow State University of Russia (MSU), according to its theoretical research results. The China Baotou Rare Earth Research Institute (BRIRE) further optimized the composition of the alloy through experimental verification and studied the preparation technology and process conditions. The industrialization technology and application products of 1–4 stage hydrogen storage alloy materials (AB_n) have been developed.

2. Results and Discussion

The preliminary compositions of the 1–4 stage hydrogen storage alloys were determined by using a statistical model developed in MSU, which is described in the literature [40]. Based on the existing research results and experimental work, BRIRE developed the alloy product composition and an ICP analysis confirmed that it was basically consistent with the design composition. The first alloy consisted of two rare earth series AB_5 types, the second and third alloys were Laves phase $TiCr_2$ types, and the fourth alloy was a Laves phase $TiFe_2$ type. Table 1 lists the design compositions and ICP analysis results of the first-stage rare earth hydrogen storage alloys. Due to the high vapor pressure of Ca elements, the content of the Ca elements in the composition of the First-2 alloy differs greatly from the design value and causes the fluctuation of the other components. Fortunately, the First-2 alloy products with the ICP results showed the expected properties, so no further alloy products were prepared in accordance with the design composition. Table 2 lists the design compositions of the 2–4 stage hydrogen storage alloys and the ICP analysis results of the products. Table 3 lists the alloy numbers, compositions, test temperatures (T), maximum hydrogen storage capacities (C_{max}), and their corresponding hydrogenation/dehydrogenation pressures (P_a, P_d), hydrogenation/dehydrogenation plateau slope factors (S_f), hysteresis coefficients (H_f), reaction enthalpies (ΔH), and entropies (ΔS).

Table 1. Design composition and ICP analysis results of rare earth hydrogen storage alloy.

Alloys	Compositions	La	Ce	Ca	Y	Ni
First-1	Design	19.84	5.00	/	5.29	69.86
	ICP	19.20	5.04	/	5.02	70.74
First-2	Design	16.93	6.83	1.47	3.25	71.53
	ICP	16.11	6.32	0.80	2.71	74.06

Table 2. Design composition and ICP analysis results of 2–4 stage hydrogen storage alloys.

Alloys	Compositions	Ti	Zr	Cr	Mn	Cu	V	Fe
Second	Design	30.89	/	36.91	24.82	4.10	3.29	/
	ICP	29.90	/	37.44	25.72	4.22	2.72	/
Third	Design	31.09	/	47.27	7.14	/	/	14.51
	ICP	30.42	/	47.50	7.45	/	/	14.63
Fourth	Design	22.96	10.94	/	/	/	9.16	56.93
	ICP	23.03	10.45	/	/	/	9.96	56.56

Table 3. The main characteristics of 1–4 stage hydrogen storage alloys.

Alloys	Compositions	T (K)	C_{max} (H/f.u.)	P_a (MPa)	P_d (MPa)	S_f	H_f	ΔH (kJ·mol^{-1} H$_2$)	ΔS (J·mol^{-1}·K^{-1} H$_2$)
First-1	La$_{0.6}$Ce$_{0.15}$Y$_{0.25}$Ni$_{5.0}$	303	6.25	2.52	1.72	0.40	0.38	−21.74 [a]	−98.69 [a]
		323	6.23	4.44	3.21	0.40	0.32	−25.02 [d]	−106.25 [d]
		363	6.24	10.55	8.90	0.42	0.17		
First-2	La$_{0.5}$Ce$_{0.2}$Y$_{0.15}$Ca$_{0.15}$Ni$_{5.0}$	303	6.60	2.05	1.44	0.23	0.35	−22.59 [a]	−99.81 [a]
		323	6.51	3.73	2.71	0.21	0.32	−24.34 [d]	−102.63 [d]
		363	6.09	9.08	7.15	0.25	0.24		
Second	TiCr$_{1.1}$Mn$_{0.7}$V$_{0.1}$Cu$_{0.1}$	298	2.55	5.90	4.72	1.46	0.22	−20.28 [a]	−102.14 [a]
		323	2.49	11.93	11.52	1.05	0.03	−22.19 [d]	−106.78 [d]
		353	2.40	21.13	19.06	1.22	0.10		
		363	/	26.13 [c]	25.04 [c]	/	/		
Third	TiCr$_{1.4}$Mn$_{0.2}$Fe$_{0.4}$	293	2.65	16.07	15.86	0.74	0.01	−14.51 [a]	−91.89 [a]
		323	2.58	29.64	27.06	0.94	0.09	−12.337 [d]	−84.37 [d]
		353	2.59	44.05	37.38	1.59	0.16		
		298	/	17.85	17.60	/	/		
		363	/	51.59 [c]	42.97 [c]	/	/		
Fourth	Ti$_{0.8}$Zr$_{0.2}$Fe$_{1.7}$V$_{0.3}$	293	2.90	35.0	31.50	1.24	0.10	−13.11 [a]	−93.61 [a]
		323	2.63	61.50	48.00	1.38	0.24	−12.75 [d]	−91.18 [d]
		353	2.58	87.13	77.36	1.33	0.12		
		298	/	39.05	33.70	/	/		
		363	/	100.75 [c]	84.73 [c]	/	/		

[a]: The value was calculated by absorption plateau. [c]: Linear extrapolation value from the Van't Hoff diagram in Section 2.3. [d]: The value was calculated by desorption plateau.

2.1. Material Composition Design and Preparation

The application of MHHCs requires the consideration of two important technical indexes: the cycle yield and the compression ratio. The cycle yield is related to the reversible hydrogen storage capacity of hydrogen compression materials. The compression ratio is related to the hydrogenation enthalpy or plateau pressure and plateau hysteresis coefficient of hydrogen compression materials [1,5,41]. In addition, the activation properties, kinetic properties, plateau pressure coordination, flatness of the pressure plateau, and stability of the hydrogen absorption/discharge cycle need to be considered. In order to meet the MHHCs' requirements for hydrogen compression materials, the composition and preparation of these hydrogen compression materials are particularly important.

Reversible hydrogen storage capacity refers to the amount of hydrogen storage that can be released by hydrogen compression materials at a certain temperature, which is mainly related to the type of materials and the type and content of the component elements [17,38]. In general, at the same temperature, the composition of a material with a higher proportion of hydrogen-absorbing elements has a higher hydrogen storage capacity. Increasing this reversible hydrogen storage capacity is beneficial to improving the MHHC cycle yield.

The enthalpy of a hydrogenation reaction is inversely related to the plateau pressure of the material [5]. Generally speaking, the larger the cell volume of the constituent phase of the hydrogen compression material, the lower the plateau pressure and the higher the enthalpy of the hydrogenation reaction. An increase in the absolute enthalpy is beneficial for improving the compression ratio of MHHCs, but a material with a high enthalpy value has a low pressure plateau and cannot achieve the purpose of a high-pressure hydrogen output.

Plateau hysteresis reflects the difference in the degrees of hydrogenation and dehydrogenation of hydrogen compression materials, which is caused by the stress of hydride formation [42]. The hysteresis factor (H_f), as shown in Equation (1) [32], is utilized in describing this plateau hysteresis.

$$H_f = \ln P_a / P_d \tag{1}$$

The larger the cell size of the hydrogen compression material, the smaller the hysteresis of the plateau. Plateau hysteresis reduces the compression ratio of MHHCs, and at the same time, damages the cyclic stability of the material's hydrogenation/dehydrogenation [33,43].

The surface of the hydrogen compression materials needs to be in a good active state before the normal hydrogenation/dehydrogenation reaction, that is, the oxide layer that is formed on the surface by active elements such as La, Ce, Ti, and Zr in the material's composition must be removed. Doping some rare earth elements or rare earth mixtures in the material's composition can improve the activation properties of these hydrogen storage materials [44].

The kinetic properties of hydrogen compression materials are very important for improving the working efficiency of MHHCs. The contents of some of the metal elements in the material's composition, such as Ni, Cr, and Al, etc., and the state of its surface elements are related to the intrinsic kinetic properties of the materials [45].

The synergy of the plateau pressure between the two adjacent stages, i.e., the dehydrogenation pressure of the former-stage material at a high temperature is higher than the hydrogenation pressure of the latter-stage material at a low temperature, ensures the complete hydrogenation/dehydrogenation of each stage material.

The flatness of the material's pressure plateau is related to the interstitial site of the different volumes caused by the fluctuation of the material's composition [46,47], which mainly affects the synergy of the above plateau pressure and the kinetic performance. The slope of the pressure plateaus is usually characterized by slope factor S_f, which is described by Equation (2) [32]:

$$S_f = d(\ln P)/d(H\ wt\%) \tag{2}$$

where H is the mass hydrogen storage density of the alloys.

Resulting from the existence of S_f, the alloys absorb/desorb hydrogen incompletely at the midpoint of the plateau pressure.

The hydrogenation/dehydrogenation cycle stability of hydrogen compression materials during their application is related to the service life of MHHCs. The main factors leading to the cyclic stability of the materials are poisoning, disproportionation, and amorphous and lattice defects. Impurities such as CO, H_2O, and O_2 in hydrogen make the material toxic and its performance deteriorate. Water and oxygen can significantly poison Ti-based AB_2/AB-type alloys [48], while AB_5-type alloys are easily poisoned by CO at ppm amounts [49]. In comparison to AB_5-type alloys, the cycle life of AB_2-type alloys tends to be longer, which may be ascribed to the fewer defects they produce in the process of hydrogen absorption and desorption [50].

The development of all the stages of hydrogen compression materials also needs to consider the characteristics of an MHHC system operation. The hydrogen source used by

MHHCs should first adopt hydrogen produced by the electrolysis of water using renewable energy sources. At a normal temperature (293–303 K), the hydrogen output pressure is usually greater than 2.5 MPa, and the maximum can reach 4 MPa. Hydrogen produced by the electrolysis of water mainly contains impurities such as water vapor and oxygen. If the hydrogen source used by an MHHC is 99.99% pure hydrogen from the coal chemical industry, and the output hydrogen pressure is generally greater than 10 MPa, it is necessary to consider that the content of carbon monoxide (≤5 ppm) can be harmful to hydrogen compression materials. Therefore, the first-stage hydrogen compression material should have a good tolerance to impurity gases such as water vapor and oxygen, and a certain resistance to carbon monoxide. Due to the purifying effect that the first-stage hydrogen compression material has on hydrogen, the poisoning degree of the second-stage, third-stage, and fourth-stage hydrogen compression materials via impurity gases can be reduced.

MHHCs need to be used within a certain temperature range, and using water as a heat exchange medium has many advantages. Compared to oil, water has a larger specific heat capacity, which is conducive to the stable operation of the system. Water is environmentally friendly and can use industrial wastewater and waste heat. Therefore, the best working temperature for hydrogen compression materials is between room temperature (298 ± 10 K) and 373 K.

All the stages of MHHCs' output pressures should meet the high-pressure hydrogen demand for most scenarios. The first-stage hydrogen compression material, with regard to hydrogen storage and purification, should provide a hydrogen source for the second-stage hydrogen compression material, and its output pressure is about 8 MPa at a high temperature (363 K in this paper). The output pressure of the second-stage hydrogen compression material is greater than 20 MPa at a high temperature, which can fill the long tube trailer and low-pressure gas cylinder with hydrogen. The output pressure of the third-stage hydrogen compression material is greater than 40 MPa at a high temperature, which can fill the 35 MPa high-pressure hydrogen storage tank with hydrogen. The output pressure of the fourth-stage hydrogen compression material at a high temperature is about 80 MPa, which can fill the 70 MPa high-pressure hydrogen storage tank with hydrogen.

Based on the above considerations, the $LaNi_5$ hydrogen storage alloy was selected as the first-stage hydrogen compression material, which has the advantages of a strong toxicity resistance, excellent hydrogenation/dehydrogenation kinetics, and a good cyclic stability [18]. The $TiCr_2$ hydrogen storage alloy was selected as the second- and third-stage hydrogen compression materials [30]. Its advantages include a higher intrinsic plateau pressure, better kinetic performance, and higher hydrogen storage capacity than the $LaNi_5$ alloy. The fourth-stage hydrogen compression material was the $ZrFe_2$-type hydrogen storage alloy, whose advantage is that its inherent plateau pressure can be as high as 1000 MPa [24,25], which makes it easy to develop hydrogen compression materials with a high-pressure hydrogen output.

In order to obtain all the stages of the target hydrogen storage alloys between room temperature and 363 K, the compositions of the alloys were adjusted by the multi-element alloying method, and the desired effect was achieved by alloy preparation technology. The plateau pressure of the binary $LaNi_5$ alloy could be adjusted between 0 and 10 MPa. The replacement of the La by Ce could improve the plateau pressure of the alloy, but the hysteresis coefficient would increase and the hydrogen storage would decrease [22]. The Y element could also increase the alloy's plateau pressure and improve the alloy's kinetic performance. The effect of the hysteresis coefficient was lower than that of the Ce element, but it would increase the alloy's plateau slope [23]. The Ca element could improve the activation property of the alloy, and, as a light hydrogen-absorbing element, could effectively improve the hydrogen absorption capacity and reduce the plateau pressure of the alloy [19,21]. However, the alloy containing the Ca element had a high vapor pressure and was volatile, so it was difficult to control the composition during preparation. Considering the effect of the alloying elements and preparation technology, two $LaNi_5$ hydrogen storage alloys satisfying the MHHC primary hydrogen compression materials were developed:

$La_{0.6}Ce_{0.15}Y_{0.25}Ni_5$ (First-1) and $La_{0.5}Ce_{0.2}Y_{0.15}Ca_{0.15}Ni_5$ (First-2). Their compositions and main performance indexes are listed in Tables 1 and 3, respectively. The influence of the alloyed elements on the hydrogen storage properties of the $LaNi_5$ alloys is summarized in Table 4.

Table 4. Effect of alloying on hydrogen storage performance of $LaNi_5$ alloys.

Elements	Capacity	Plateau Pressure	Hysteresis	Slope	Activation Performance
La	↑	↓	↓	↓	↑
Ce	↓	↑	↑	↑	↓
Y	↑	↑	↑	↑	↑
Ca	↑	↓	-	-	↑

The plateau pressure of the $TiCr_2$ hydrogen storage alloy could be adjusted between 0 and 20 MPa at room temperature, which mainly solved the problem of a large slope factor and hysteresis coefficient [28]. Mn, Fe, V, and Cu elements were used to adjust the composition of the alloy. Cr and Mn are the two most important transition metal elements on the B side of an AB_2 hydrogen storage alloy, and the inter-substitution of Cr and Mn is a common alloy design method. The Cr element is beneficial for reducing an alloy's plateau pressure and hysteresis coefficient, but it makes the hydrogen absorption plateau of the alloy narrow and its hydrogen storage capacity decrease. The Mn element can improve the kinetic properties of an alloy, but significantly increases the alloy's hysteresis coefficient [51]. Due to its small atomic radius, Fe is mainly used to improve the plateau pressure of an alloy [31]. V is a kind of hydrogen absorption element and is an appropriate replacement for the Cr element as it can improve the hydrogen absorption amount of an alloy and reduce its slope, but its price is expensive. The Cu element can increase the plateau pressure of an alloy through the gap size effect, reduce the alloy's slope, and improve the alloy's properties. However, due to the large atomic weight of the Cu and Fe elements, the hydrogen storage capacity of the alloy will be reduced in large quantities. Based on a comprehensive consideration of these alloying elements, the $TiCr_2$ hydrogen storage alloys, $TiCr_{1.1}Mn_{0.7}V_{0.1}Cu_{0.1}$ and $TiCr_{1.4}Mn_{0.2}Fe_{0.4}$, which met the requirements of the second- and third-stage hydrogen compression materials of MHHCs, were developed, respectively. Their compositions and main performance indexes are listed in Tables 2 and 3, respectively. The influence of these alloyed elements on the hydrogen storage properties of the Ti-Cr-based alloys is summarized in Table 5.

Table 5. Effect of alloying on hydrogen storage performance of Ti-Cr-based alloys.

Elements	Capacity	Plateau Pressure	Hysteresis	Slope	Activation Performance
Cr	↓	↓	↓	-	-
Mn	-	-	↑	-	↑
V	↑	-	-	↓	-
Cu	↓	↑	-	↓	-
Fe	↓	↑	-	-	-

The hydrogen absorption/desorption pressure of the $ZrFe_2$-type hydrogen storage alloy at room temperature was ~60 MPa, so it was necessary to further reduce the plateau pressure to improve the characteristics of the plateau. The hydrogen-absorbing element Ti was used to replace the partial Zr and the appropriate plateau pressure was achieved by adjusting the Ti/Zr ratio [25]. Replacing part of the Fe element with the V element could effectively reduce the plateau pressure and the hysteresis coefficient, improving the hydrogen storage capacity of the alloy [33]. Comprehensively considering the effect of the alloying elements, the hydrogen storage alloy $Zr_{0.2}Ti_{0.8}Fe_{1.7}V_{0.3}$, a fourth-stage hydrogen compression material satisfying the MHHCs, was developed. Its compositions and main performance indexes are listed in Tables 2 and 3, respectively. The influence of the alloyed elements on the hydrogen storage properties of the Zr-Fe-based alloys is summarized in Table 6.

Table 6. Effect of alloying on hydrogen storage performance of Zr-Fe-based alloys.

Elements	Capacity	Plateau Pressure	Hysteresis	Slope	Activation Performance
Ti	↓	↓	↓	↓	↑
V	↑	↓	↓	↑	-

The above 1–4 stage alloys could realize the goals of providing high-pressure pure hydrogen with water as its heat exchange medium and promoting the rapid and good development of a hydrogen energy field through the construction of MHHCs. However, the difficulties in the performance testing of these hydrogen storage alloys with high-pressure hydrogen output characteristics and application efficiency declines caused by non-ideal plateau characteristics are still challenges to be faced (discussed in this paper in Section 2.5). In the future, it is necessary to continue to optimize the comprehensive properties of these alloys, such as their pressure plateaus and hydrogen storage capacities, and to try to improve the accuracy and repeatability of the alloy performance testing.

2.2. X-ray Diffraction

Figure 1 shows the XRD finishing patterns of the two first-stage $LaNi_5$ hydrogen storage alloys, second-stage and third-stage $TiCr_2$ hydrogen storage alloys, and fourth-stage $ZrFe_2$ hydrogen storage alloy. It can be seen that the $LaNi_5$ alloy corresponds to a hexagonal $CaCu_5$-type structure (space group P6/mmm); the $TiCr_2$ alloy mainly contains hexagonal $MgZn_2$ phase (C14) and a small amount of Ti_3Cr_3O heterophase. The $ZrFe_2$ alloy is composed of hexagonal $MgZn_2$ phase (C14). Table 7 lists the component phase abundance and lattice constants of the five alloys.

Table 7. Lattice parameters and phase abundance of 1–4 stage hydrogen storage alloys.

Alloys	Phase	Abundance/wt.%	a/Å	c/Å	V/Å³	Parameters of Fit
First-1	$CaCu_5$	100%	4.9596(1)	3.9885(1)	84.96(1)	Rw = 4.20 Rp = 2.53
First-2	$CaCu_5$	100%	4.9610(1)	3.9862(1)	84.97(1)	Rw = 4.08 Rp = 2.57
Second	C14 Laves Ti_3Cr_3O	91.85 8.15	4.8700(1) 11.3090(7)	7.9880(2) 11.3090(7)	164.07(1) 1446.46(8)	Rw = 5.23 Rp = 3.59
Third	C14 Laves Ti_3Cr_3O	93.35 6.85	4.8619(1) 11.2823(2)	7.97338(1) 11.2823(2)	163.22(1) 1436.14(8)	Rw = 4.06 Rp = 2.87
Fourth	C14 Laves	100%	4.8816(1)	7.9482(2)	164.03(1)	Rw = 3.57 Rp = 2.55

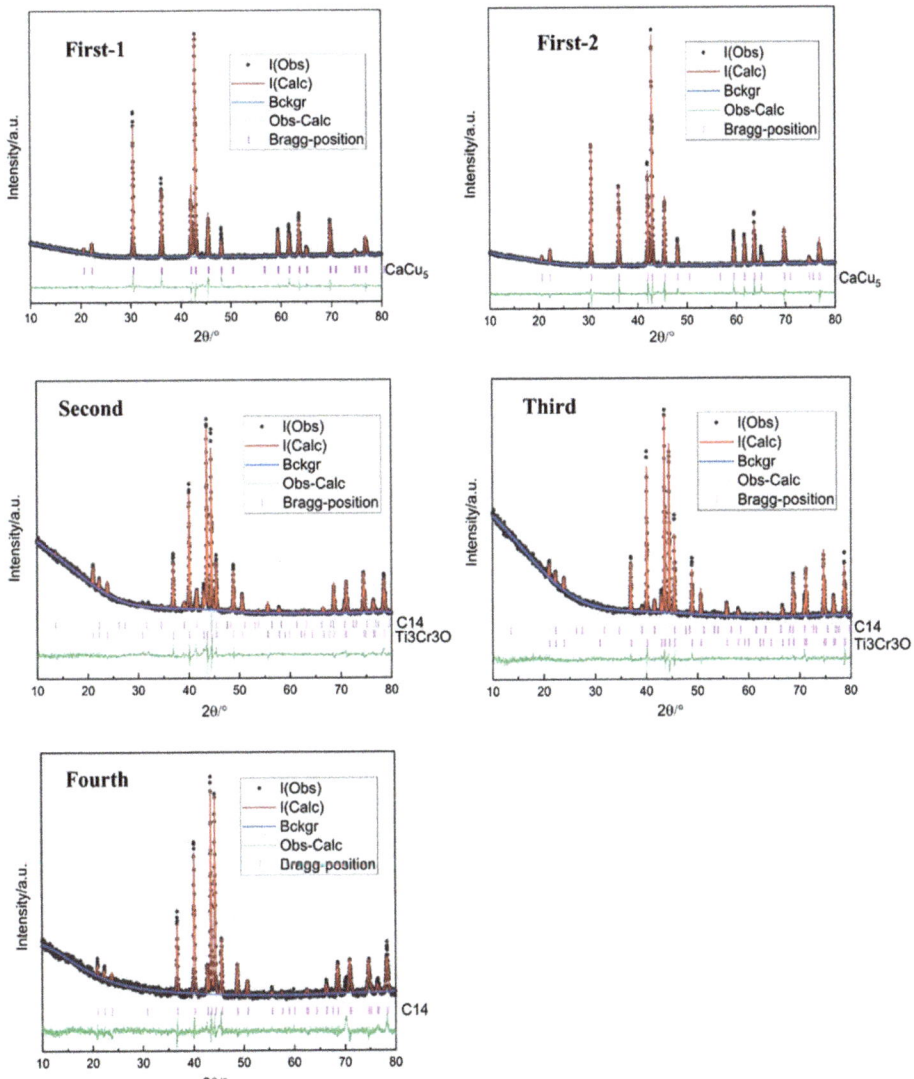

Figure 1. Results of the Rietveld analysis for 1–4 stage hydrogen storage alloys.

The lattice constant of the binary LaNi$_5$ hydrogen storage alloy is a = 5.0201 Å, c = 3.9881 Å, and cell volume V = 87.03 Å^3 [52]. The atomic radius of the La element is 1.87 Å. The La element in the quaternary La$_{0.6}$Ce$_{0.15}$Y$_{0.25}$Ni$_5$ alloy (First-1) is partially replaced by the Ce (1.82 Å), Y (1.81 Å) elements with smaller atomic radii, thus reducing the lattice constant a and cell volume V. The La element in the quinary La$_{0.5}$Ce$_{0.2}$Y$_{0.15}$Ca$_{0.15}$Ni$_5$ alloy (First-2) is partially replaced by the Ce and Y elements with smaller atomic radii, as well as the Ca (1.96 Å) element with a larger atomic radius. Nonetheless, the lattice constant a and cell volume V also decrease (slightly larger than First-1), which is related to the abnormal decrease in the lattice constant of the cerium-containing alloy that is caused by the change in the valence state of the Ce [53]. The lattice constant c of the LaNi$_5$ alloy is mainly related to the radius of the Ni atom on the B side. Since the Ni in First-1 and First-2 is not replaced by other elements, the c value is similar to that of the binary LaNi$_5$ [52].

The modified TiCr$_2$ alloy is generally still a single-phase C14 structure, and the lattice constant and cell volume vary with the atomic radii of the alloyed elements. The Cr element is usually partially replaced by Mn and Fe. When Mn partially replaces Cr, the a value of the alloy is slightly reduced, but the effects on the c value and V value may be different [31,54]. Fe partially replaces Cr and the a, c, and V values of the alloy decrease [31,54,55]. Therefore, the cell volume of the second-stage alloy, TiCr$_{1.1}$Mn$_{0.7}$V$_{0.1}$Cu$_{0.1}$, and the third-stage alloy, TiCr$_{1.4}$Mn$_{0.2}$Fe$_{0.4}$, is similar in this paper. The Ti$_3$Cr$_3$O heterophase in the alloys may be related to the high melting power (20~30 KW) and the short cooling time (8~10 min) during the alloy preparation (see Section 3).

The binary ZrFe$_2$ hydrogen storage alloy has a cubic C15 Laves phase structure [33]. In this paper, the fourth-stage Zr$_{0.2}$Ti$_{0.8}$Fe$_{1.7}$V$_{0.3}$ alloy transforms into a hexagonal C14 Laves phase structure because the Ti partially replaces the Zr and the V partially replaces the Fe. The atomic radius of the Ti (1.45 Å) is smaller than that of the Zr (1.60 Å), and the atomic radius of the V (1.31 Å) is larger than that of the Fe (1.24 Å). Compared to the lattice constants (a = 5.023 Å; b = 8.205 Å; and V = 179.35 Å^3) of the Zr$_{1.05}$Fe$_{1.8}$Mo$_{0.2}$ (Mo has an atomic radius of 1.36 Å) alloy with a C14 Laves phase structure [33], the a (4.8816 Å), c (7.9482 Å), and V (164.03 Å^3) values of the Zr$_{0.2}$Ti$_{0.8}$Fe$_{1.7}$V$_{0.3}$ alloy are significantly reduced. The results show that the lattice constants of the Zr$_{0.2}$Ti$_{0.8}$Fe$_{1.7}$V$_{0.3}$ alloy are significantly affected by the substitution of the Zr by Ti elements.

2.3. PCT Test Analysis

The reaction of the hydrogen storage alloy (M) with hydrogen (H$_2$) to form metal hydride (MH$_x$) is a reversible heat (Q)-driven process:

$$M + x/2\ H_2 \leftrightarrows MH_x + Q \tag{3}$$

The equilibrium characteristic of Reaction (3) is the relationship between the hydrogen pressure (P), hydrogen concentration in the solid phase (C), and temperature (T) (PCT diagram), and determines the thermodynamic properties of the hydrogenation of a particular hydrogen storage alloy. At a certain temperature, the hydrogen concentration C absorbed in M exceeds the concentration of the saturated solid solution (α phase) to form a hydride phase (β phase), which shows the characteristics of the first-order phase transition. With the increase in the C, the hydrogen pressure PH_2 remains unchanged, which is called plateau pressure, and the plateau width corresponds to the reversible hydrogen storage capacity of the alloy. The equilibrium of Reaction (3) in the plateau region can be described by the Van't Hoff equation:

$$\ln PH_2 = -\Delta S^0/R + \Delta H^0/RT \tag{4}$$

ΔS^0 and ΔH^0 are the standard formation entropy and standard formation enthalpy of the hydrides, respectively, and R is the gas constant [1].

Using a PCT tester to measure the PCT curves for at least three different temperatures, the hydrogenation/dehydrogenation plateau pressure (P_a and P_d) and maximum hydrogen storage capacity at a certain temperature can be obtained. Equation (2) was used to calculate the plateau slope factor S_f and Equation (1) for the plateau hysteresis coefficient H_f. According to Equation (4), Van't Hoff graphs (ln PH_2-1/T graphs) were drawn and the ΔS^0 and ΔH^0 were calculated. For the high-pressure alloy, the hydrogen emission pressure at a high temperature T_H (T_H is 363 K in this paper) can be obtained by an extrinsic method.

The multi-stage compression operation puts forward higher requirements for regulating the PCT characteristics of the hydrogen compression materials. Figure 2 shows the PCT curves of the 1–4 stage alloys at different temperatures, among which, the PCT curves of the first-stage alloys (First-1 and First-2) at a high temperature T_H (363 K) can be directly measured. The highest temperature of the second-, third-, and fourth-stage alloys is 353 K due to the limitation of the testing capability of the instrument. As can be seen from Figure 2, the pressure plateau of the first-stage alloy is relatively flat but has a large hysteresis, while the pressure plateau of the second-, third-, and fourth-stage alloys is steep but has a small hysteresis. The PCT characteristic parameters of all the alloys are listed in Table 3. Figure 3 shows the Van't Hoff diagram of the 1–4 stage hydrogen storage alloys. The inclination and hysteresis of the pressure plateau are non-idealized properties of the hydrogen storage alloy, and it is difficult to reach the ideal state at the same time.

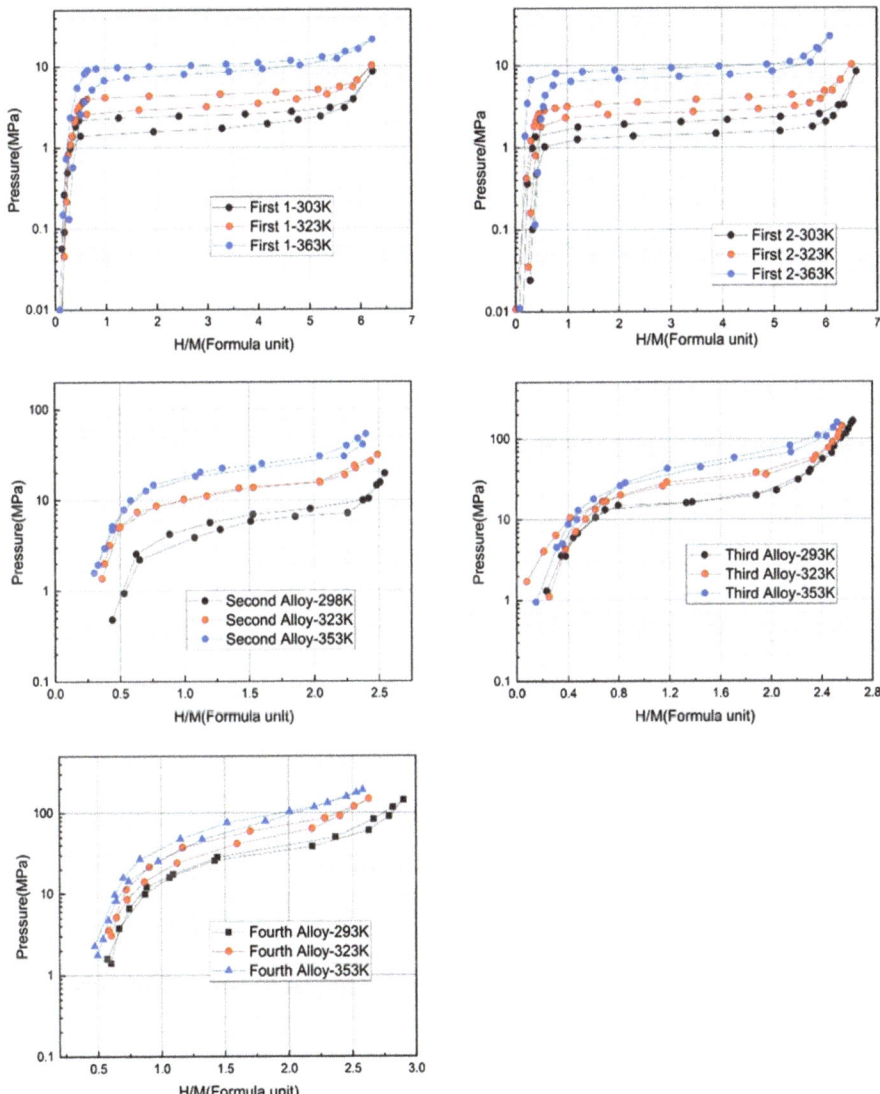

Figure 2. PCT curves of 1–4 stage hydrogen storage alloys at different temperatures.

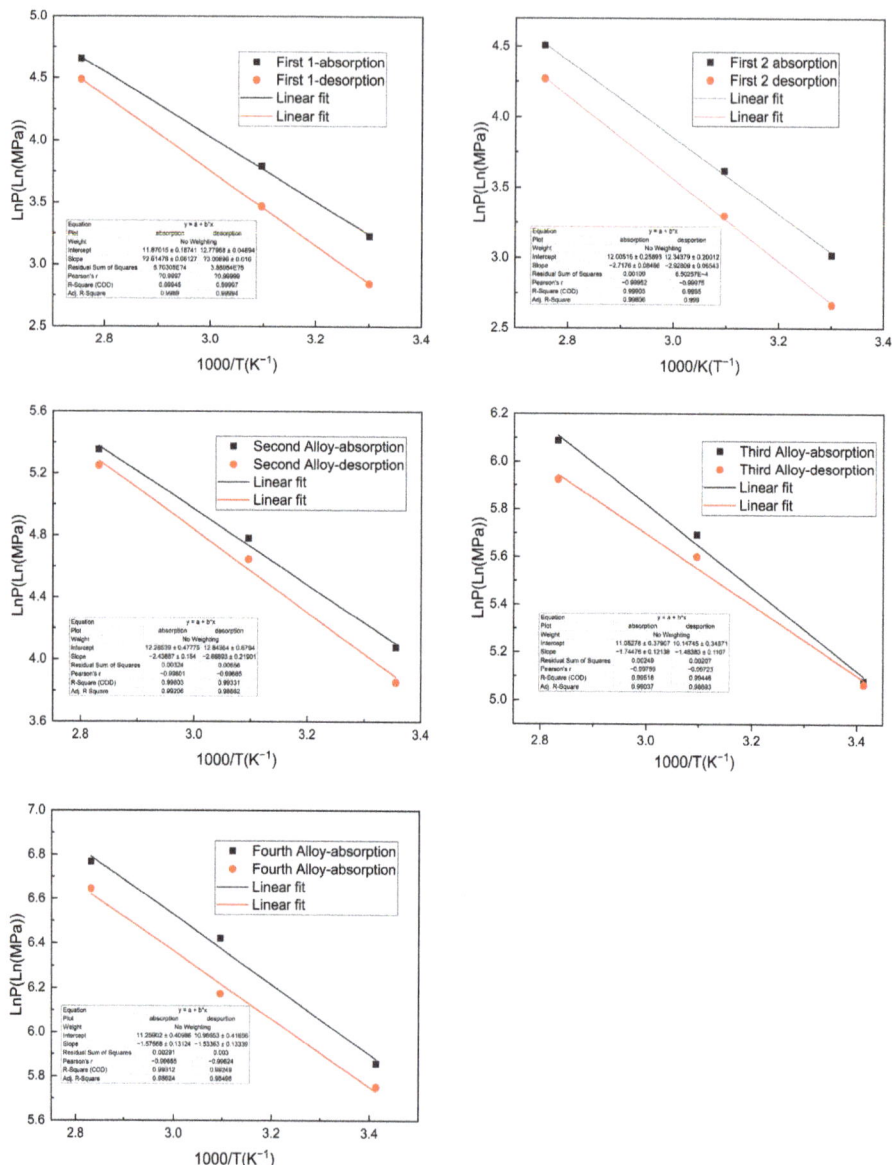

Figure 3. Van't Hoff diagram of 1–4 stage hydrogen storage alloys.

The compression ratios R_P [$=P_d (T_H)/P_a (T_L)$] of the 1–4 stage alloys are 3.53 (First-1) or 3.49 (First-2), 4.24 (second), 2.67 (third), and 2.42 (fourth), respectively. The total compression ratios of the constructed MHHC system are 96.71 or 95.61 (the product of the compression ratios at all the stages) [5]. The compression ratio mainly reflects the compression capacity of the MHHCs. The output pressures of the alloys in this paper are relatively large and, due to this and the plateau hysteresis, the compression ratio is at a moderate level and needs to be further improved.

In the MHHC system, which is constructed of hydrogen compression material, each stage of the hydrogen compression material absorbs hydrogen at a low temperature T_L

(usually room temperature) with a maximum hydrogen absorption capacity of C_A (corresponding to the right end of the pressure plateau in the PCT curve), and releases hydrogen at a high temperature T_H (363 K) with a maximum hydrogen emission of C_B (corresponding to the left end of the pressure plateau in the PCT curve). Then, the cycle yield of the MHHC system can be expressed as the effective hydrogen compression capacity, ΔC:

$$\Delta C = C_A(T_L) - C_B(T_H) \qquad (5)$$

Based on this calculation, the ΔCs of the First-1 and First-2 alloys are about 5.25 H/f.u. (1.01 wt%) and 5.90 H/f.u. (1.42 wt%), respectively. The highest temperature measured by the PCT curves of the second-, third-, and fourth stage alloys is 353 K. Calculated according to the amount of hydrogen released at 353 K, the ΔCs of the second-, third-, and fourth stage alloys are about 1.85 H/f.u. (1.18 wt%), 1.85 H/f.u. (1.19 wt%), and 1.90 H/f.u. (1.13 wt%), respectively. The ΔC values of the second-, third-, and fourth stage alloys at 363 K should be slightly lower than those calculated above due to the temperature increasing, the plateau pressure increasing, the plateau narrowing, and the C_B point shifting to the right and increasing. Considering the compression ratios and cycle yields of the hydrogen compression materials, the First-2 alloy is superior to the First-1 alloy as a primary hydrogen compression material. The cyclic yield reflects the hydrogen supply capacity of MHHCs. The effective hydrogen storage capacity of an alloy can be increased by adjusting the alloy's composition and improving the characteristics of its plateau.

The plateau pressure of the hydrogen storage alloys is related to the types, phase compositions, and cell parameters of the alloys. The two first-stage alloys in this paper are both $CaCu_5$-type single-phase structures, their lattice constants and cell volumes are very similar, and their plateau pressures are also similar. The main phase of the second-, third-, and fourth stage alloys is the C14 Laves phase, and their lattice constants and cell volumes are also very similar, but their plateau pressures vary greatly, which may be related to the composition elements of the alloys, and needs further study.

2.4. Cyclic Performance Test

MHHCs are a continuous cycle system of the hydrogen absorption and desorption of hydrogen compression materials. The toxic effect of some of the impurities in the hydrogen on hydrogen compression materials seriously affects the cyclic performance of the materials. In order to investigate this effect, the cyclic performances of the first- and second-stage alloys were tested using ordinary pure hydrogen. Table 8 lists the gas composition test results, which meet Chinese standards. Due to the purification effect of the first- and second-stage alloys on the hydrogen, the content of the impurity gases in the hydrogen entering the third- and fourth-stage alloys will be significantly reduced. Moreover, the third- and fourth-stage alloys are of the same type as the second-stage alloy, so the cyclic performances of the third- and fourth-stage alloys in ordinary pure hydrogen were not investigated.

Table 8. Composition of ordinary pure hydrogen.

Gas Composition	Standard Indicators	Measured Content
Hydrogen (H_2) (%)	≥99.99	99.99
Oxygen (O_2) (ppm)	≤5	3
Nitrogen (N_2) (ppm)	≤60	50
Carbon monoxide (CO) (ppm)	≤5	3
Carbon dioxide (CO_2) (ppm)	≤5	2
Methane (CH_4) (ppm)	≤10	5
Water vapor (H_2O) (ppm)	≤10	8

Figure 4 shows the change curves of the hydrogen storage capacities of the first-stage alloys (First-1 and First-2) and second-stage alloy for 10 cycles, respectively. It can be seen that the hydrogen storage capacity retention rates are 99.0%, 97.7%, and 98.6%, respectively, which have a good cycle stability. The cycle stability of the First-2 alloy is less than that of the First-1 alloy due to the inclusion of the Ca element.

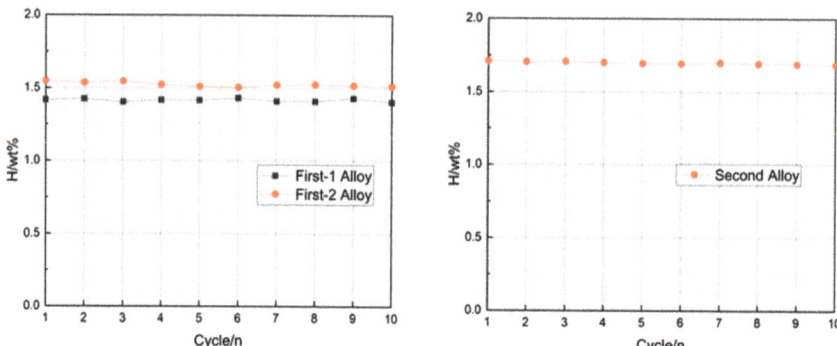

Figure 4. Hydrogenation/dehydrogenation cycle curves of 1–2 stage alloys in ordinary pure hydrogen.

2.5. Application Analysis

There are two problems to be considered in the practical application of hydrogen storage alloys as hydrogen compression materials. First is the question of: is the plateau pressure of the hydrogen storage alloy consistent with the pressure designed in the MHHC system? The second question is: can the former and latter hydrogen storage alloys work together efficiently?

The hydrogen compression materials used in MHHCs are related to the reaction characteristics at a high temperature and high pressure. It is difficult to measure the high-pressure characteristics of alloys with the current PCT test device, such as the second-, third-, and fourth stage hydrogen storage alloys in this paper. In order to solve this problem, we can use the Van't Hoff equation to draw the linear relationship between the pressure (lnP) and temperature reciprocal ($1/T$) (see Figure 3) and obtain the alloy pressure under a certain high-temperature condition by linear extrapolation. However, due to factors such as hydrogen fugacity and temperature fluctuation, the Van't Hoff diagram deviates from the linear relationship at a high temperature [17,31], and the corrected pressure value is lower than the pressure value obtained by extrapolation [56]. Therefore, the high-temperature output pressure of the hydrogen storage alloy obtained by the Van't Hoff diagram should be higher than the target output pressure, and its applicability should be further evaluated through practical application.

During the actual operation of an MHHC system, in order to ensure the matching and coupling between the alloys at all the stages, the hydrogenation pressure of the latter alloy at a low temperature must be lower than that of the former alloy at a high temperature. In an ideal state, the matching coupling between two adjacent alloys is considered by using the plateau pressure mid-value of the PCT curve (as shown in Figure 5), that is, the plateau pressure mid-value of the former alloy at a high temperature is greater than that of the latter alloy at a low temperature. However, the pressure plateau of any alloy is inclined, for example, the second-, third-, and fourth stage alloys in this paper have obvious plateau slopes. In this case, the dehydrogenation process can be completed only when the former-stage alloy dehydrogenates to the low point of the plateau pressure (left side of the plateau), and the hydrogen absorption process can be completed only when the latter-stage alloy absorbs hydrogen to the high point of the plateau pressure (right side of the plateau) (as shown in Figure 6). Comparing Figures 5 and 6, it can be seen that the plateau pressures of the second-, third-, and fourth stage alloys in this paper cannot fully meet the practical application. The minimum pressure of the former-stage alloys for high-temperature dehydrogenation needs to be increased, or the maximum hydrogen absorption pressure of the latter-stage alloys at a low temperature needs to be reduced.

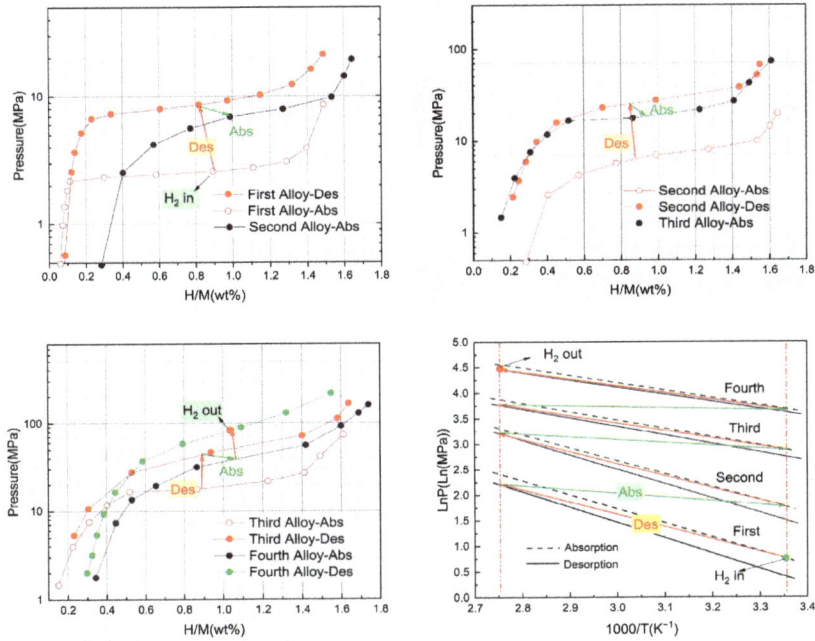

Figure 5. Matching coupling relationship of 1–4 stage hydrogen storage alloys in ideal state.

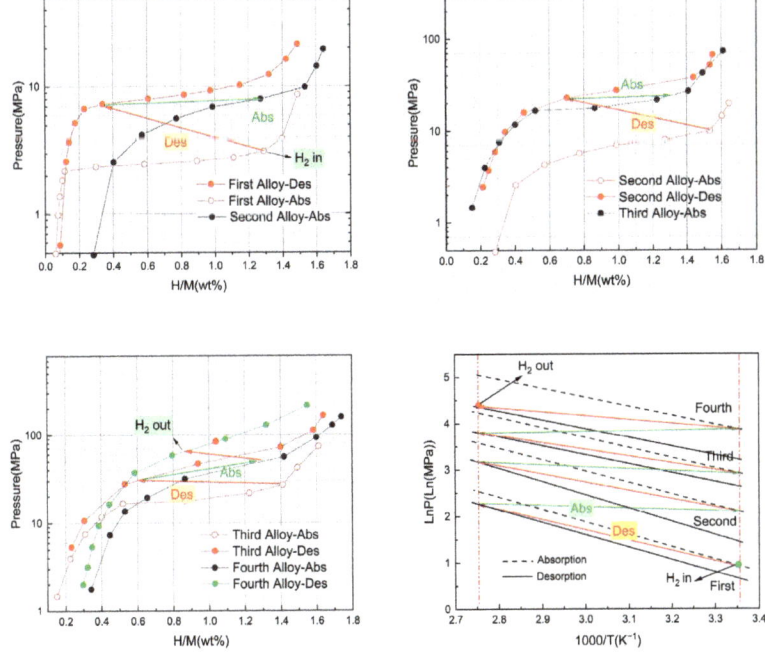

Figure 6. Matching and coupling relationship of 1–4 stage hydrogen storage alloys in actual state.

3. Materials and Methods

The metal raw materials required for the alloy preparation were calculated and weighed according to the composition formula. The purity of the metal raw materials was ≥99 wt%. Considering the burning loss of each single metal in the melting process, the La, Ce, and Y in the first-stage alloy were added by 1 wt%, and the Ca was added by 20 wt%. The Ti, Zr, Mn, and V in the second-, third-, and fourth-stage alloys were added with 3 wt%, 2 wt%, 5 wt%, and 6 wt%, respectively.

The first-stage alloy was prepared in a vacuum induction melting furnace under the protection of a 0.055 MPa argon atmosphere. The smelting process was: preheating a power of 7 KW for 3–4 min, maintaining a power of 15 KW until the alloy was completely melted, holding a power of 12 KW for 3–4 min, and then pouring into a water-cooled ingot mold with a sample weight of 1.5 kg each time.

The second-, third-, and fourth-stage alloys were melted in a magnetic levitation induction furnace under the protection of a 0.055 MPa argon atmosphere. The mass of each sample was 80 g. The smelting process was: 10–12 KW for 1 min during preheating, 20 KW for melting, 25–30 KW for 1 min so that the alloy was completely alloyed, and finally reducing the power to 0, then cooling for 8–10 min, after which, the sample was taken out. In order to ensure a uniform alloy composition and structure, each alloy sample was melted and turned over three times.

After the sample was taken out, the oxide layer on the surface was removed, and after crushing, it was passed through a 200 mesh screen. The +200 mesh powder was used for the composition test and PCT performance test. The −200 mesh powder was used for the X-ray powder diffraction test.

The composition of the alloy sample was tested by a Thermo Electron-iCAP 6300 Inductively Coupled Plasma Emission Spectrometer (ICP), and the actual composition contents of the various elements in the alloy sample were calculated according to the results.

The crystal structure of the alloy was determined by a Philips-PW 1700X powder diffractometer (XRD). The test conditions were: a Cu target, a Kα radial, a tube voltage of 40 kV, a tube current of 40 mA, a scanning range of 10°–80°, and a scanning speed of 0 01°/s. The phase structure analysis software used was Jade 6.0 and the Rietveld full-spectrum fitting and structure refinement used GSAS software [57] to obtain the phase abundance and crystal cell parameters.

The PCT performances of the hydrogen storage alloys were tested with a Sievert-type device (manufactured by the Beijing Research Institute of Engineering Technology). The PCT performances of the alloys for the third and fourth stages were tested in MSU in a high-pressure device, as described earlier in [58]. During the activation process, the +200 mesh sample ~2 g was taken and put into the sample tank. The sample tank was evacuated for 0.5 h at 423 K to ensure that the vacuum was less than 5×10^{-4} MPa. The sample was cooled to 313 K in a water bath and charged with a certain hydrogen pressure, according to the design characteristics of the alloy for the sample activation treatment. The high-temperature vacuum low-temperature hydrogen charging process was repeated to ensure that the sample was fully activated, and then the PCT curve test was carried out at different temperatures. Before testing the PCT curve, the sample tank should be evacuated at 423 K for 0.5 h to ensure the complete dehydrogenation of the sample. In order to ensure the complete balance of each measuring point during the PCT performance test, the time of each measuring point should be maintained for more than 30 min until the pressure change of ≤0.001 MPa/min. The purity of the hydrogen used in the activation and PCT curve tests was 99.999%, and the purity of the hydrogen used in the normal hydrogen cycle performance test was 99.99%.

4. Conclusions

According to the demand of metal hydride–hydrogen compressors (MHHC) for hydrogen compression materials, the first-stage LaNi$_5$, the second- and the third-stage TiCr$_2$, and the fourth-stage ZrFe$_2$ hydrogen storage alloys were developed. The output pressures of the 1–4 stage hydrogen storage alloys at 363 K were 8.90 MPa, 25.04 MPa, 42.97 MPa, and 84.73 MPa, respectively. The maximum output hydrogen pressure exceeded 80 MPa, which could meet the requirements of the 20 MPa, 35 MPa, and 70 MPa high-pressure hydrogen injections.

The $LaNi_5$ type alloy was a single-phase $CaCu_5$ type hexagonal structure, the $TiCr_2$-type alloy main phase was a hexagonal C14 Laves phase and Ti_3Cr_3O heterophase, and the $ZrFe_2$ type alloy was composed of a hexagonal C14 Laves single phase. The total compression ratio of the MHHC system constructed with the 1–4 stage alloys exceeded 95, and the hydrogen yield of each stage was more than 1 wt%. The retention rate of the hydrogen storage capacities of the first- and second-stage alloys in the 99.99 wt% ordinary pure hydrogen containing H_2O (8 ppm), O_2 (3 ppm), and CO (3 ppm), for 10 cycles of hydrogenation/dehydrogenation, exceeded 98%. The deviation between the plateau pressure and the system design output pressure, as well as the efficient and cooperative coupling of the adjacent two stage alloys, should also be considered in the practical application of the four-stage hydrogen storage alloys.

Author Contributions: S.M., E.M. and M.P. completed the preliminary statistical model design of alloy composition and the high pressure PCT tests of the third and fourth stage alloys. H.-Z.Y., X.Z., Y.-Y.Z., B.-Q.L., J.X. and W.X. completed the application design of alloy composition, alloy preparation and all other tests. All authors have read and agreed to the published version of the manuscript.

Funding: This research was funded by the National Key Research and Development Program of China (2018YFE0124400) and Major Science and Technology Projects of Inner Mongolia (2021ZD0029). Cooperative project BRICS2019-032 was supported by BRICS national annexes of Russian Federation, China and India. Part of this work was performed according to the Development program of the Interdisciplinary Scientific and Educational School of Lomonosov Moscow State University "The future of the planet and global environmental change", and was supported by the Ministry of Science and Higher Education of the Russian Federation, projects # AAAA-A16-116053110012-5 and 122012400186-9.

Data Availability Statement: Data is contained within the article.

Conflicts of Interest: The authors declare no conflict of interest.

References

1. Lototskyy, M.V.; Yartys, V.A.; Pollet, B.G.; Bowman, R.C., Jr. Metal hydride hydrogen compressors: A review. *Int. J. Hydrogen Energy* **2014**, *39*, 5818–5851. [CrossRef]
2. Hosseini, S.E.; Wahid, M.A. Hydrogen production from renewable and sustainable energy resources: Promising green energy carrier for clean development. *Renew. Sustain. Energy Rev.* **2016**, *57*, 850–866. [CrossRef]
3. Li, H.-W.; Zhu, M.; Buckley, C.; Jensen, T.R. Functional Materials Based on Metal Hydrides. *Inorganics* **2018**, *6*, 91. [CrossRef]
4. Ouyang, L.; Jiang, J.; Chen, K.; Zhu, M.; Liu, Z. Hydrogen production via hydrolysis and alcoholysis of light metal-based materials: A review. *Nano Micro Lett.* **2021**, *13*, 134. [CrossRef]
5. Peng, Z.Y.; Li, Q.; Ouyang, L.; Jiang, W.B.; Chen, K.; Wang, H.; Liu, J.W.; Li, Z.N.; Wang, S.M.; Zhu, M. Overview of hydrogen compression materials based on a three-stage metal hydride hydrogen compressor. *J. Alloy. Compd.* **2022**, *895*, 162465–162488. [CrossRef]
6. Schlapbach, L.; Züttel, A. Hydrogen-storage materials for mobile applications. *Nature* **2001**, *414*, 353–358. [CrossRef]
7. Jacobson, M.Z. Cleaning the air and improving health with hydrogen fuel-cell Vehicles. *Science* **2005**, *308*, 1901–1905. [CrossRef] [PubMed]
8. Debe, M.K. Electrocatalyst approaches and challenges for automotive fuel cells. *Nature* **2012**, *486*, 43–51. [CrossRef]
9. Sdanghi, G.; Maranzana, G.; Celzard, A.; Fierro, V. Review of the current technologies and performances of hydrogen compression for stationary and automotive applications. *Renew. Sustain. Energy Rev.* **2019**, *102*, 150–170. [CrossRef]
10. Yamashita, A.; Kondo, M.; Goto, S.; Ogami, N. Development of high-pressure hydrogen storage system for the Toyota "Mirai". In Proceedings of the SAE 2015 World Congress & Exhibition, Detroit, MI, USA, 21–23 April 2021; pp. 1169–1177.
11. Stamatakis, E.; Zoulias, E.; Tzamalis, G.; Massina, Z.; Analytis, V.; Christodoulou, C.; Stubos, A. Metal hydride hydrogen compressors: Current developments and early markets. *Renew. Energy* **2018**, *127*, 850–862. [CrossRef]
12. Sakaki, K.; Kim, H.; Enoki, H.; Yoshimura, S.-I.; Ino, S.; Nakamura, Y. Development of TiZrMn Based Hydrogen Storage Alloys for a Soft Actuator. *Mater. Trans.* **2014**, *55*, 1168–1174. [CrossRef]

13. Gkanas, E.I.; Christodoulou, C.N.; Tzamalis, G.; Stamatakis, E.; Chroneos, A.; Deligiannis, K.; Karagiorgis, G.; Stubos, A.K. Numerical investigation on the operation and energy demand of a seven-stage metal hydride hydrogen compression system for Hydrogen Refuelling Stations. *Renew. Energy* **2020**, *147*, 164–178. [CrossRef]
14. Lototskyy, M.; Davids, M.W.; Swanepoel, D.; Louw, G.; Klochko, Y.; Smith, F.; Haji, F.; Tolj, I.; Chidziva, S.; Pasupathi, S.; et al. Hydrogen refuelling station with integrated metal hydride compressor: Layout features and experience of three year operation. *Int. J. Hydrogen Energy* **2020**, *45*, 5415–5429. [CrossRef]
15. Wang, X.; Liu, H.; Li, H. A 70 MPa hydrogen-compression system using metal hydrides. *Int. J. Hydrogen Energy* **2011**, *36*, 9079–9085. [CrossRef]
16. Mitrokhin, S.; Zotov, T.; Movlaev, E.; Verbetsky, V. Hydrogen interaction with Intermetallic compounds and alloys at high pressure. *J. Alloy. Compd.* **2013**, *580*, S90–S93. [CrossRef]
17. Gray, E.M. Alloy selection for multistage metal-hydride hydrogen compressors: A thermodynamic model. *Int. J. Hydrogen Energy* **2021**, *46*, 15702–15715. [CrossRef]
18. Au, M.; Wang, Q. Rare earth-nickel alloy for hydrogen compression. *J. Alloy. Compd.* **1993**, *201*, 115–119. [CrossRef]
19. Kim, K.; Lloyd, G.; Feldman Jr., K.; Razani, A. Thermal analysis of the $Ca_{0.4}Mm_{0.6}Ni_5$ metal-hydride reactor. *Appl. Therm. Eng.* **1998**, *18*, 1325–1336. [CrossRef]
20. Wang, X.; Chen, R.; Zhang, Y.; Chen, C.; Wang, Q. Hydrogen storage properties of $(La-Ce-Ca)Ni_5$ alloys and application for hydrogen compression. *Mater. Lett.* **2007**, *61*, 1101–1104. [CrossRef]
21. Bhuiya, M.M.H.; Lee, C.Y.; Hopkins, R.; Yoon, H.; Kim, S.; Park, S.H.; Kim, K.J. A high performance dual-stage hydrogen compressor system using $Ca0.2Mm0.8Ni5$ metal hydride. In Proceedings of the ASME 2011 Conference on Smart Materials, Adaptive Structures and Intelligent Systems, SMASIS, Scottsdale, AZ, USA, 18–21 September 2011; pp. 745–751.
22. Odysseos, M.; De Rango, P.; Christodoulou, C.N.; Hlil, E.K.; Steriotis, T.; Karagiorgis, G.; Charalambopoulou, G.; Papapanagiotou, T.; Ampoumogli, A.; Psycharis, V.; et al. The effect of compositional changes on the structural and hydrogen storage properties of $(La-Ce)Ni_5$ type intermetallics towards compounds suitable for metal hydride hydrogen compression. *J. Alloy. Compd.* **2013**, *580*, S268–S270. [CrossRef]
23. Zhu, Z.; Zhu, S.; Zhao, X.; Cheng, H.; Yan, K.; Liu, J. Effects of Ce/Y on the cycle stability and anti-plateau splitting of $La_{5-x}Ce_xNi_4Co$ (x=0.4, 0.5) and $La_{5-y}Y_yNi_4Co$ (y=0.1, 0.2) hydrogen storage alloys. *Mater. Chem. Phys.* **2019**, *236*, 121725–121735. [CrossRef]
24. Filipek, S.; Jacob, I.; Paul-Boncour, V.; Percheron-Guegan, A.; Marchuk, I.; Mogilyanski, D.; Pielaszek, J. Investigation of $ZrFe_2$ and $ZrCo_2$ under very high pressure of gaseous hydrogen and deuterium. *Pol. J. Chem.* **2001**, *75*, 1921–1926.
25. Zotov, T.; Movlaev, E.; Mitrokhin, S.; Verbetsky, V. Interaction in $(Ti,Sc)Fe_2$-H_2 and $(Zr,Sc)Fe_2$-H_2 systems. *J. Alloy. Compd.* **2008**, *459*, 220–224. [CrossRef]
26. Guo, X.; Wang, S.; Liu, X.; Li, Z.; Lü, F.; Mi, J.; Hao, L.; Jiang, L. Laves phase hydrogen storage alloys for super-high-pressure metal hydride hydrogen compressors. *Rare Met.* **2011**, *30*, 227–231. [CrossRef]
27. Cao, Z.; Ouyang, L.; Wang, H.; Liu, J.; Sun, L.; Zhu, M. Composition design of Ti-CrMn-Fe alloys for hybrid high-pressure metal hydride tanks. *J. Alloy. Compd.* **2015**, *639*, 452–457. [CrossRef]
28. Cao, Z.; Ouyang, L.; Wang, H.; Liu, J.; Sun, D.; Zhang, Q.; Zhu, M. Advanced high pressure metal hydride fabricated via Ti-Cr-Mn alloys for hybrid tank. *Int. J. Hydrogen Energy* **2015**, *40*, 2717–2728. [CrossRef]
29. Sleiman, S.; Huot, J. Microstructure and Hydrogen Storage Properties of $Ti1V_{0.9}Cr_{1.1}$ Alloy with Addition of x wt %Zr (x = 0, 2, 4, 8, and 12). *Inorganics* **2017**, *5*, 86. [CrossRef]
30. Li, S.; Wang, S.; Sheng, P.; Li, Z.; Wu, Y.; Guo, X. Overview of Ti-based laves phase hydrogen storage alloys for hydrogen compressor. *Chin. J. Rare Met.* **2019**, *43*, 754–764.
31. Li, J.; Jiang, X.; Li, G.; Li, X. Development of $Ti_{1.02}Cr_{2-x-y}Fe_xMn_y$ ($0.6 \leq x \leq 0.75$, y=0.25, 0.3) alloys for high pressure metal hydride system. *Int. J. Hydrogen Energy* **2019**, *44*, 15087–15099. [CrossRef]
32. Li, J.; Jiang, X.; Li, Z.; Jiang, L.; Li, X. High-pressure hydrogen storage properties of $Ti_xCr_{1-y}Fe_yMn_{1.0}$ alloys. *Int. J. Energy Res.* **2019**, *43*, 5759–5774. [CrossRef]
33. Zhou, C.; Wang, H.; Ouyang, L.Z.; Liu, J.W.; Zhu, M. Achieving high equilibrium pressure and low hysteresis of Zr-Fe based hydrogen storage alloy by Cr/V substitution. *J. Alloy. Compd.* **2019**, *806*, 1436–1444. [CrossRef]
34. Cao, Z.; Zhou, P.; Xiao, X.; Zhan, L.; Li, Z.; Wang, S.; Chen, L. Investigation on Ti-Zr-Cr-Fe-V based alloys for metal hydride hydrogen compressor at moderate working temperatures. *Int. J. Hydrogen Energy* **2021**, *46*, 21580–21589. [CrossRef]
35. Cao, Z.; Zhou, P.; Xiao, X.; Zhan, L.; Jiang, Z.; Piao, M.; Wang, S.; Jiang, L.; Chen, L. Studies on Ti-Zr-Cr-Mn-Fe-V based alloys for hydrogen compression under mild thermal conditions of water bath. *J. Alloy. Compd.* **2022**, *892*, 162145–162452. [CrossRef]
36. Peng, Z.; Li, Q.; Sun, J.; Chen, K.; Jiang, W.; Wang, H.; Liu, J.; Ouyang, L.; Zhu, M. Ti-CrMn-Fe-based alloys optimized by orthogonal experiment for 85 MPa hydrogen compression materials. *J. Alloy. Compd.* **2022**, *891*, 161791–161816. [CrossRef]
37. Li, Q.; Peng, Z.; Jiang, W.; Ouyang, L.; Wang, H.; Liu, J.; Zhu, M. Optimization of Ti-ZrCr-Fe alloys for 45 MPa metal hydride hydrogen compressors using orthogonal analysis. *J. Alloy. Compd.* **2022**, *889*, 161629–161636. [CrossRef]
38. Li, Z.; Yan, Y.H.; Huang, H.X.; Liu, B.G.; Lv, Y.J.; Zhang, B.; Lv, W.; Yuan, J.G.; Wu, Y. Effects of the different element substitution on hydrogen storage properties of $Ti_{0.8}Zr_{0.2}Mn_{0.9}Cr_{0.6}V_{0.3}M_{0.2}$ (M=Fe, Ni, Co). *J. Alloy. Compd.* **2022**, *908*, 164605–164613. [CrossRef]
39. Challet, S.; Latroche, M.; Heurtaux, F. Hydrogenation properties and crystal structure of the single BCC $(Ti_{0.355}V_{0.645})_{100-x}M_x$ alloys with M=Mn, Fe, Co, Ni (x=7, 14 and 21). *J. Alloy. Compd.* **2007**, *439*, 294–301. [CrossRef]

40. Mitrokhin, S.V.; Tepanov, A.A.; Verbetsky, V.N. Hydrogen interaction with alloys of $NdNi_{5-x}Al_x$ system. *Int. J. Hydrogen Energy* **2017**, *42*, 22353–22357. [CrossRef]
41. Lototskyy, M.V.; Yartys, V.A.; Tarasov, B.P.; Davids, M.W.; Denys, R.V.; Tai, S. Modelling of metal hydride hydrogen compressors from thermodynamics of hydrogen-metal interactions viewpoint: Part I. Assessment of the performance of metal hydride materials. *Int. J. Hydrogen Energy* **2021**, *46*, 2330–2338. [CrossRef]
42. Flanagan, T.B.; Clewley, J.D. Hysteresis in metal hydrides. *J. Less Common Met.* **1982**, *83*, 127–141. [CrossRef]
43. Balasubramaniam, R. Hysteresis in metal-hydrogen systems. *J. Alloy. Compd.* **1997**, *253*, 203–206. [CrossRef]
44. Yao, Z.; Liu, L.; Xiao, X.; Wang, C.; Jiang, L.; Chen, L. Effect of rare earth doping on the hydrogen storage performance of $Ti_{1.02}Cr_{1.1}Mn_{0.3}Fe_{0.6}$ alloy for hybrid hydrogen storage application. *J. Alloy. Compd.* **2018**, *731*, 524–530. [CrossRef]
45. Bloch, J.; Mintz, M.H. Kinetics and mechanisms of metal hydrides formation-a review. *J. Alloy. Compd.* **1997**, *253*, 529–541. [CrossRef]
46. Fujitani, S.; Nakamura, H.; Furukawa, A.; Nasako, K.; Satoh, K.; Imoto, T.; Saito, T.; Yonezu, I. A method for numerical expressions of PC isotherms of hydrogen absorbing alloys. *Z. Phys. Chem.* **1993**, *179*, 27–33. [CrossRef]
47. Park, C.-N.; Luo, S.; Flanagan, T.B. Analysis of sloping plateaus in alloys and intermetallic hydrides. *J. Alloy. Compd.* **2004**, *384*, 203–207. [CrossRef]
48. Williams, M.; Lototsky, M.V.; Davids, M.W.; Linkov, V.; Yartys, V.A.; Solberg, J.K. Chemical surface modification for the improvement of the hydrogenation kinetics and poisoning resistance of TiFe. *J. Alloy. Compd.* **2011**, *509*, S770–S774. [CrossRef]
49. Modibane, K.D.; Williams, M.; Lototskyy, M.; Davids, M.W.; Klochko, Y.; Pollet, B.G. Poisoning-tolerant metal hydride materials and their application for hydrogen separation from CO_2/CO containing gas mixtures. *Int. J. Hydrogen Energy* **2013**, *38*, 9800–9810. [CrossRef]
50. Iosub, V.; Joubert, J.M.; Latroche, M.; Cerny, R.; Percheron-Guegan, A. Hydrogen cycling induced diffraction peak broadening in C14 and C15 Laves phases. *J. Solid State Chem.* **2005**, *178*, 1799–1806. [CrossRef]
51. Corgnale, C.; Sulic, M. High pressure thermal hydrogen compression employing $Ti_{1.1}CrMn$ metal hydride material. *J. Phys. Energy* **2019**, *2*, 014003–014015. [CrossRef]
52. Corré, S.; Bououdina, M.; Fruchart, D.; Adachi, G.-Y. Stabilisation of high dissociation pressure hydrides of formula $La_{1-x}Ce_xNi_5$ (x=0-0.3) with carbon monoxide. *J. Alloy. Compd.* **1998**, *275–277*, 99–104. [CrossRef]
53. Meyer-Liautaud, F.; Pasturel, A.; Allibert, C.H.; Colinet, C. thermodynamic study of valence state of cerium and hydrogen storage in $Ce(Ni_{1-x}Cu_x)_5$ compounds. *J. Less-Common Met.* **1985**, *110*, 119–126. [CrossRef]
54. Chen, Z.W.; Xiao, X.Z.; Chen, L.X.; Fan, X.L.; Liu, L.X.; Li, S.Q.; Ge, H.W.; Wang, Q.D. Development of Ti-Cr-Mn-Fe based alloys with high hydrogen desorption pressures for hybrid hydrogen storage vessel application. *Int. J. Hydrogen Energy* **2013**, *38*, 12803–12810. [CrossRef]
55. Li, J.G.; Xu, L.; Jiang, X.J.; Li, X.G. Study on the hydrogen storage property of $(TiZr_{0.1})_xCr_{1.7-y}Fe_yMn_{0.3}$ (1.05<x<1.2, 0.2<y<0.6) alloys. *Prog. Nat. Sci. Mater. Int.* **2018**, *28*, 470–477.
56. Charbonnier, V.; Enoki, H.; Asano, K.; Kim, H.; Sakaki, K. Tuning the hydrogenation properties of $Ti_{1+y}Cr_{2-x}Mn_x$ laves phase compounds for high pressure metal hydride compressors. *Int. J. Hydrogen Energy* **2021**, *46*, 36369–36380. [CrossRef]
57. Toby, B.H.; Von Dreele, R.B. GSAS-II: The genesis of a modern open-source all purpose crystallography software package. *J. Appl. Crystallogr.* **2013**, *46*, 544–549. [CrossRef]
58. Verbetsky, V.N.; Mitrokhin, S.V.; Zotov, T.A.; Movlaev, E.A. Intermetallic hydrides with high dissociation pressure. *Mater. Matters* **2007**, *2*, 9–12.

Disclaimer/Publisher's Note: The statements, opinions and data contained in all publications are solely those of the individual author(s) and contributor(s) and not of MDPI and/or the editor(s). MDPI and/or the editor(s) disclaim responsibility for any injury to people or property resulting from any ideas, methods, instructions or products referred to in the content.

Article

Stress Reduction of a V-Based BCC Metal Hydride Bed Using Silicone Oil as a Glidant

Xin Zheng [1,†], Hanyang Kong [2,†], Desheng Chu [1], Faping Hu [1], Yao Wang [2], Yigang Yan [2] and Chaoling Wu [2,3,*]

1. Electrical Science Institute of Yunnan Power Grid Co., Ltd., Kunming 650214, China
2. Institute of New Energy and Low-Carbon Technology, Sichuan University, Chengdu 610207, China
3. College of Materials Science and Engineering, Sichuan University, Chengdu 610065, China
* Correspondence: wuchaoling@scu.edu.cn
† These authors contributed equally to this work.

Abstract: The large volume expansion and self-locking phenomenon of metal hydride particles during hydrogen sorption often leads to a high stress concentration on the walls of a container, which may cause the collapse of the container. In present study, silicone oil was investigated as a glidant for a V-based BCC metal hydride bed to alleviate the stress concentration during hydrogen sorption. The results indicated that the addition of 5 wt% silicone oil slightly reduced the initial hydrogen storage capacity of $V_{40}Ti_{26}Cr_{26}Fe_8$ (particle size: ~325 µm) but improved the absorption reversibility, regardless of the oil viscosity. It was observed that silicone oil formed a thin oil layer of 320~460 nm in thickness on the surface of the $V_{40}Ti_{26}Cr_{26}Fe_8$ particles, which might improve the fluidity of the powder, reduce the self-locking phenomenon and alleviate the stress concentration on the container walls. Consequently, the maximum strain on the surface of the hydrogen storage container decreased by $\geq 22.5\%$ after adding 5 wt% silicone oil with a viscosity of 1000 cSt.

Keywords: hydrogen storage; metal hydride; volume expansion; stress concentration; silicone oil

Citation: Zheng, X.; Kong, H.; Chu, D.; Hu, F.; Wang, Y.; Yan, Y.; Wu, C. Stress Reduction of a V-Based BCC Metal Hydride Bed Using Silicone Oil as a Glidant. *Inorganics* **2022**, *10*, 167. https://doi.org/10.3390/inorganics10100167

Academic Editor: Maurizio Peruzzini

Received: 17 September 2022
Accepted: 29 September 2022
Published: 9 October 2022

Publisher's Note: MDPI stays neutral with regard to jurisdictional claims in published maps and institutional affiliations.

Copyright: © 2022 by the authors. Licensee MDPI, Basel, Switzerland. This article is an open access article distributed under the terms and conditions of the Creative Commons Attribution (CC BY) license (https://creativecommons.org/licenses/by/4.0/).

1. Introduction

With the growing energy demand and the concerns of environmental pollution, the development of emission-free energy carriers based on abundant resources such as hydrogen is arousing great interest [1,2]. Hydrogen can be used in internal combustion engines to produce mechanical energy or in fuel cells to convert chemical energy to electrical energy [3]. However, before the large-scale utilization of hydrogen as an energy carrier, the development of technologies to safely and efficiently store hydrogen, hydrogen storage being the bridge between its production and consumption, is essential [4,5]. To date, three main methods have been reported, i.e., hydrogen can be physically stored as a compressed gas, a cryogenic liquid, or in metal hydrides [6]. Among them, metal hydrides, which could potentially allow reversible hydrogen absorption and desorption at ambient conditions, are considered an ideal mean for hydrogen storage [7,8].

However, a large volume expansion occurs during the hydrogen absorption process due to the insertion of hydrogen atoms into the lattices [9–11]. For example, an expansion of 20% was observed for $LaNi_5$ during the absorption of hydrogen [12,13], of about 30% for Mg-based alloys [14], and of 9–23% for a Ti–Cr–Mn-based alloy [11,15]. A body-centered cubic (BCC) alloy showed a much higher expansion of more than 40% during the first hydrogen absorption, with the lattice parameters growing from 3.0327 Å for the BCC lattice to 4.2714 Å for the FCC lattice [16]. During the expansion of metal hydrides, the particles are gradually reduced in size, e.g., from 150 µm to 5 µm for the Ti–Zr–Mn-based alloy after 180 cycles [17], and meanwhile, lattice expansion leads to the concentration of stress on the walls of the hydrogen storage container, especially on the bottom of the container, due to the flow of the alloy particles under gravity and the interlock of irregular particles [9,18,19]. It is critical to ensure that the design of the hydrogen storage container allows it to withstand the expansion of metal hydrides. The stress concentration on the surface of a hydrogen storage container depends on the type, particle shape, and loading fraction of the hydrogen storage materials [12,13,20–22], as well as on how the hydrogen storage container is placed, e.g., horizontally or vertically [12,18].

To alleviate the stress concentration on the container walls, approaches focusing on the container or the metal hydride (MH) bed have been proposed. As regards the container, leaving enough free expansion space for the hydrogen alloy is a common method [23]. C. K. Lin [20] reported a structural design of the hydrogen storage container, in which the container is divided into three layers to release the stress concentration. For the MH bed, hinders or glidants are generally applied to reduce the stress concentration [24–28]. The addition of hinders, such as expanded graphite, could limit the free moving of metal hydride powders [28]. Glidants, such as silicone oil, MoS_2 nanopowders, and carbon black, can improve the fluidity of metal hydride powders and reduce the self-locking phenomenon between powders [24–27].

Vanadium-based hydrogen storage alloys have a BCC crystal structure with a hydrogen storage capacity of ~3.8 wt % and a fast hydrogen sorption kinetics at room temperature. By using a FeV80 master alloy to replace the pure vanadium metal, the cost of V-based BCC alloys has been significantly reduced [29]. However, the large volume expansion, of ~40%, during the first hydrogen absorption cycle, may cause a serious stress concentration on the container walls. In the present study, silicone oil with low cost and high temperature stability, which ranges from −50 to 200 °C during long-term use [30], was applied as a glidant for a V-based BCC alloy powder bed. The influence of silicone oil on the hydrogen storage properties of the V-based BCC alloy was systematically investigated. Furthermore, the effect of silicone oil addition on stress concentration during the first hydrogen sorption was evaluated using a swelling stress measurement system.

2. Results

2.1. Hydrogen Storage Properties

Initially, the contact angle between silicone oils with different viscosity and the $V_{40}Ti_{26}Cr_{26}Fe_8$ alloy (s1) was determined, as shown in Figure 1. The contact angle increased from 9.006° for a silicone oil of 50 cSt to 17.719° for a silicone oil of 10,000 cSt. Smaller contact angles indicate a good infiltration of the investigated silicone oil in the $V_{40}Ti_{26}Cr_{26}Fe_8$ alloy. Hence, 5 wt% silicones oil of 50 cSt (s2), 1000 cSt (s3) and 10,000 cSt (s4), respectively, were mixed with the $V_{40}Ti_{26}Cr_{26}Fe_8$ powder, and hydrogen sorption was measured.

Figure 1. Contact angles between silicone oils with different viscosity and the $V_{40}Ti_{26}Cr_{26}Fe_8$ alloy (s1).

Figure 2a compares the hydrogen absorption kinetics in the first cycle of sample s1 and samples containing 5 wt% silicone oils of 50 cSt (s2), 1000 cSt (s3) and 10,000 cSt (s4). Sample s1, the pristine $V_{40}Ti_{26}Cr_{26}Fe_8$ alloy, started to absorb hydrogen immediately without incubation at 298 K and 7 MPa H_2, with a fast absorption >3 wt% H in the first 5 min and a full capacity of 3.43 wt% H. Samples s2, s3 and s4 showed incubation periods of 0.5, 0.3 and 0.1 min, respectively. The incubation period decreased for samples containing silicone oil with higher viscosity. The hydrogen absorption rates also declined for samples s2, s3 and s4, which showed hydrogen absorption amounts of 2.29, 2.42 and 2.62 wt% H, respectively, in the first 5 min. The full capacities of samples s2, s3 and s4 decreased to 3.231, 3.362 and 3.325 wt%, respectively. In the PCT curves of the first dehydrogenation process (Figure 2b), the samples s1 to s4 showed a similar plateau pressure of ~0.4 MPa, but the hydrogen desorption amount decreased from 2.20 wt% for s1 to 1.95 wt% for s2, 2.06 wt% for s3 and 2.04 wt% for s4, within the pressure range of 0.01–7 MPa.

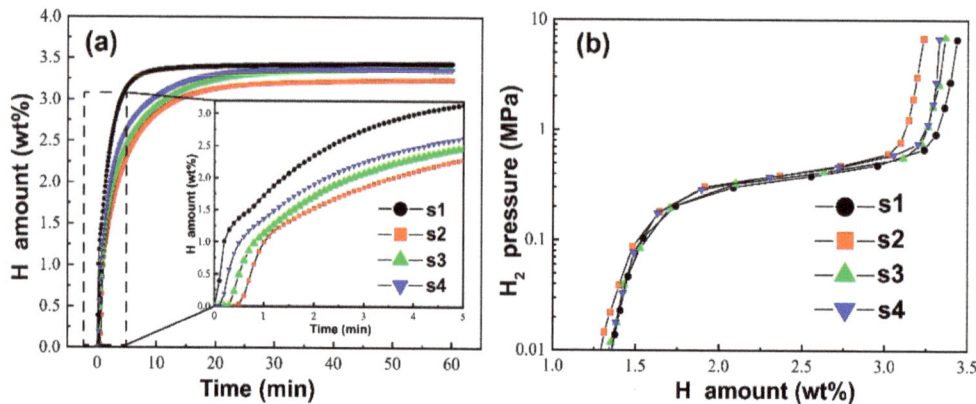

Figure 2. (a) Initial hydrogen absorption kinetics and (b) PCT plots of samples s1 to s4 during the dehydrogenation process at 298 K.

To further investigate the cycling stability, samples s1 to s4 were evacuated after dehydrogenation at RT for 30 min and were then hydrogenated under 7 MPa H_2 at RT. Figure 3a,d show the hydrogen absorption kinetics of samples s1 to s4 up from the 2nd to the 20th cycle. Starting from the second cycle, all samples could absorb hydrogen immediately without incubation and absorbed the maximum hydrogen amount within 5 min. Moreover, hydrogen absorption sped up within the first minute as the number of cycles increased for all samples (Figure 3e). For samples s2, s3 and s4, the amounts of hydrogen absorbed within the first minute were almost the same at each cycle and reached 1.7 wt% at the 20th cycle (Figure 3e), which was slightly lower than the amount of 1.87 wt% measured for sample s1.

Figure 3f compares the reversibly absorbed amounts of samples s1 to s4. The reversibly absorbed hydrogen amounts of the four samples showed a fast decay in the first 10 cycles and became relatively steady afterwards. Sample s1 showed a reversible hydrogen amount of 2.42 wt%, which decreased to 2.12 at the 10th cycle, corresponding to a capacity retention of 87.6%, and to 2.06 wt% at the 20th cycle, with a capacity retention of 85.1%. Interestingly, samples s2–s4 showed a lower initial reversible hydrogen storage capacity but a higher capacity retention of 90%, 87.3% and 87%, respectively. That is, the initial reversibly absorbed hydrogen amounts were 2.23, 2.36 and 2.32 wt% for samples s2–s4, which decreased to 2.0, 2.06 and 2.02 wt% at the 20th cycle, respectively.

Figure 4a shows the XRD pattern of the pristine $V_{40}Ti_{26}Cr_{26}Fe_8$ alloy (s1), which was composed of a BCC phase as the major phase and CeO_2 as the minor phase. After 20 ab/desorption cycles, sample s1 contained mainly a BCT phase and minor FCC and CeO_2 phases. The formation of the BCT phase, a monohydride of the V-based BCC alloy, was due to the dehydrogenation of the dihydride (FCC phase) at room temperature [31]. Note that after 20 ab/desorption cycles, the BCT phase of samples s1 to s4 showed very broad reflections (Figure 4b–e), implying the presence of large microstrains. The initial microstrain of sample s1 was around 0.36% and increased to 1.282% after 20 ab/desorption cycles. Compared to sample s1, samples s2–s4 showed very similar microstrains of around 1.3% (Figure 5f) after 20 ab/desorption cycles.

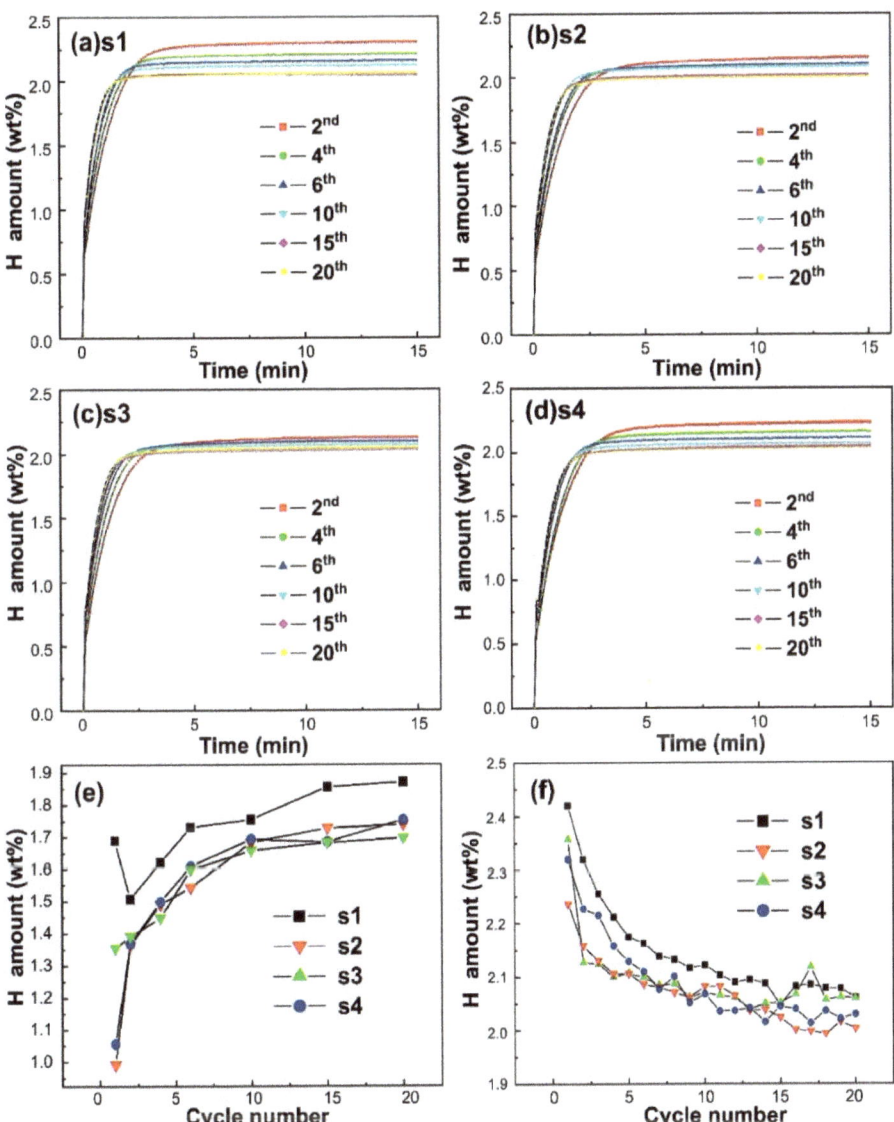

Figure 3. (a–d) Hydrogen absorption kinetics of sample s1 to s4, (e) hydrogen absorption amounts within the first minute in different cycles, and (f) reversible hydrogen absorption amount from the 1st to the 20th cycle for samples s1 to s4.

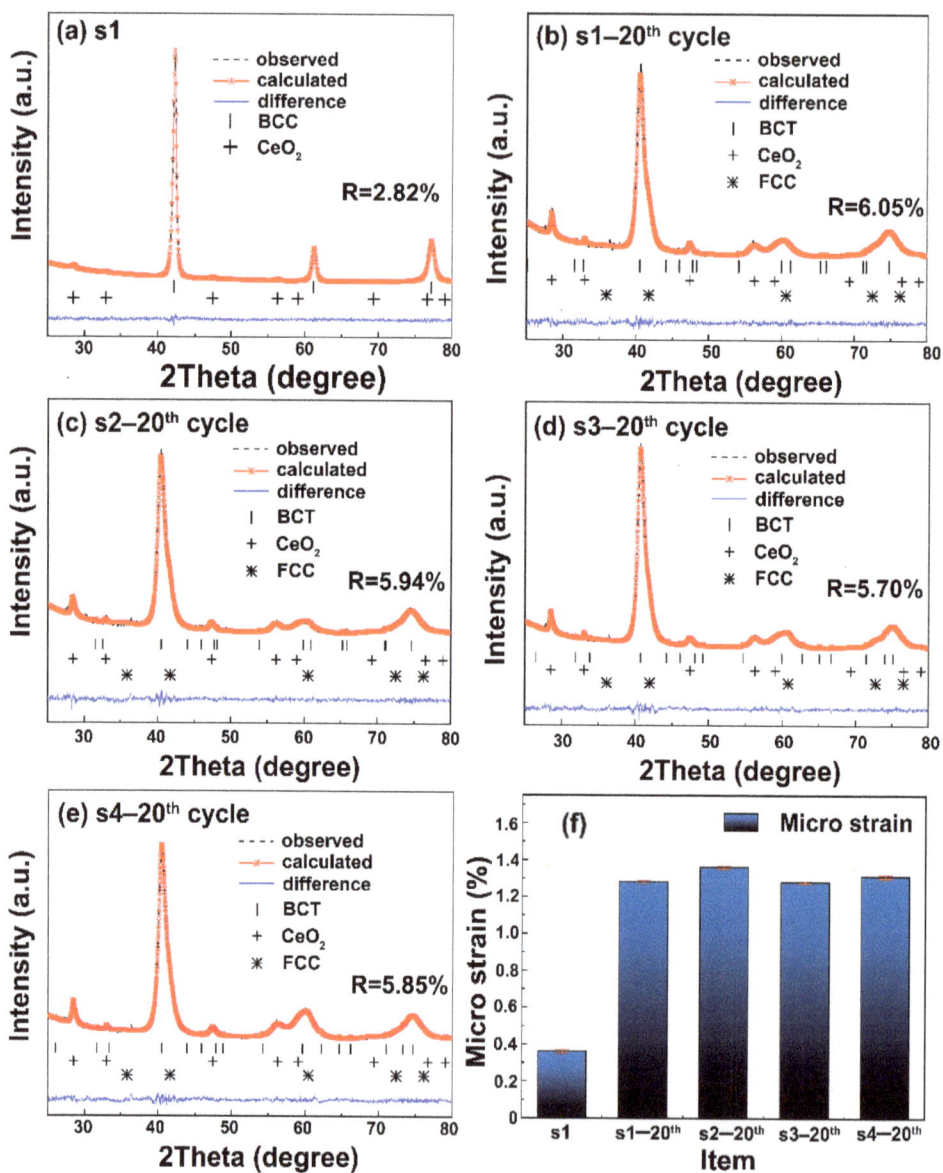

Figure 4. (**a**–**e**) Rietveld refinement of the XRD data of sample s1 and samples s1 to s4 after 20 hydrogen ab/desorption cycles and (**f**) microstrains of sample s1 and samples s1 to s4 after 20 hydrogen ab/desorption cycles.

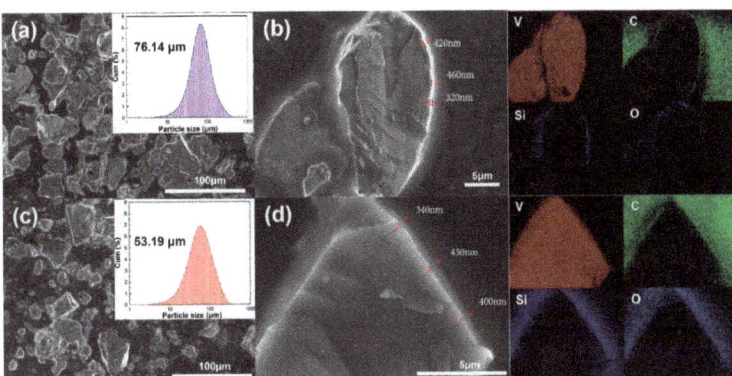

Figure 5. (**a,c**) SEM images of sample s3 and (**b,d**) EDS mapping images of selected particles after the 1st and the 20th cycle. Insets of (**a,c**) display the particle size distribution.

Sample s3 was chosen to further investigate the morphology evolution during the hydrogen ab/desorption cycles by SEM and EDS, as shown in Figure 5. The average particle size after the first ab/desorption cycle was 76.14 μm (Figure 5a) and decreased to 53.19 μm after 20 cycles (Figure 5c). With the decrease in the particle size of sample s3, a larger fresh surface was generated, improving the hydrogen absorption dynamics (Figure 3e). The presence of silicone oil on the surface of the alloy particles of sample s3 was confirmed by EDS elemental mapping, which allowed observing C, Si and O elements on the surface of the alloy particles, as shown in Figure 5b,d. Furthermore, the thickness of the silicone oil film was approximately 320–460 nm after the 1st cycle (Figure 5b) and slightly decreased to 340–450 nm after the 20th cycle (Figure 5d).

2.2. Influence of the Silicone Oil on Strain Distribution

To investigate whether the addition of silicone oil could lower the strain on the container walls during the hydrogen absorption process of the BCC alloy, three kinds of samples were loaded into a home-made MH reactor (Scheme 1), including $V_{40}Ti_{26}Cr_{26}Fe_8$ powders with particle size of 2~3 mm (s5), pre-hydrogenated $V_{40}Ti_{26}Cr_{26}Fe_8$ powder with particle size of 325 μm (s6) and a mixture of s6 and silicone oil of 1000 cSt (s7). Four strain gauges at positions 3a to 3d were used to measure the distribution of the circumferential strain on the container walls during the first hydrogen absorption by sample s5, s6 and s7.

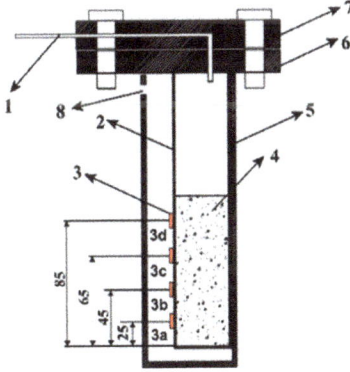

Scheme 1. A schematic image of the strain test reactor. 1, hydrogen pipeline, 2, inner container, 3, strain gauges, 4, metal hydride, 5, outer jacket, 6, lower flange, 7, upper flange, 8, wiring hole.

For sample s5 (Figure 6a), the strains at positions 3a to 3d increased slowly in the first 5 min, rose rapidly to the vertex after about 10 min and declined slowly afterwards. The strain at position 3a near the bottom of the container, with the maximum of 1750 με, was obviously higher that at other positions, indicating that stress was concentrated at the bottom of the container. This was due to the small expansion space near the bottom of the container, which allowed the fine powders to easily pass through the metal hydride bed, reaching the bottom of the container and filling the empty space [7]. As for sample s6 with a smaller particle size, the strain increased rapidly after hydrogen absorption started and reached the maximum after around 3 min. Compared to the strain at positions 3a to 3d for s5, shown in Figure 6b, the strain on the container walls for all samples were much smaller. For example, the maximum strain at position 3a decreased by about 56.89 %, from 1750 με for s5 to 750.84 με for s6. For s7 sample, the strain at the four positions was further reduced. For instance, the maximum strain at position 3a decreased to 582.13 με for s7, which is a drop by 22.5% compared to the value (750.84 με) measure for s6. Additionally, for s5 and s6, the time required to reach the maximum strain at the four positions was almost the same, i.e., 10 min for s5 and 3 min for s6. For sample s7, the time to reach the maximum strain was 7, 7, 8, and 10 min for positions 3a to 3d, respectively; the delay in reaching the maximum strain at positions 3c and 3d implied an alleviation of the stress concentration.

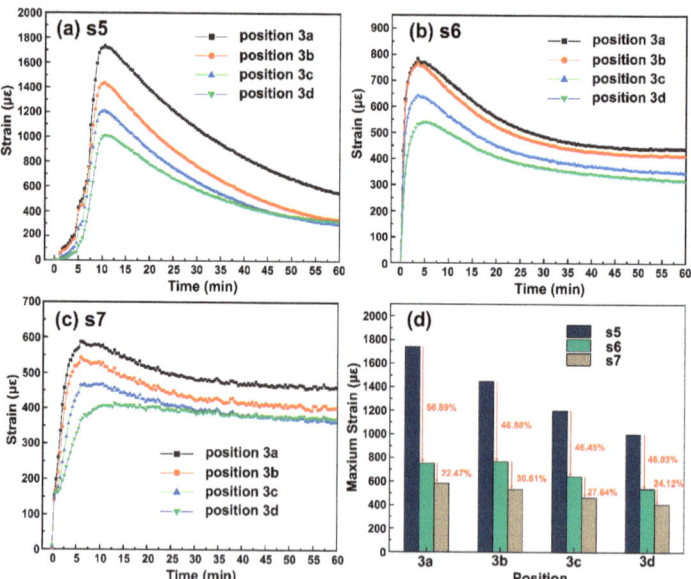

Figure 6. (a–c) Strain distribution at positions 3a to 3d on the container for sample s5, s6, and s7 and (d) comparison of the maximum strain at position 3a to 3d for sample s5, s6 and s7.

Figure 6d further compares the maximum strain at the four positions on the container walls for the three samples. For sample s7, which was composed of a fine $V_{40}Ti_{26}Cr_{26}Fe_8$ powder (325 μm) and silicone oil, the maximum strain was much lower, compared to those measured for the other two samples. This can be explained by the presence of a silicone oil film with a thickness of 320–460 nm on the surface of the metal hydride powder, which might improve the fluidity of the $V_{40}Ti_{26}Cr_{26}Fe_8$ powder, reduce the self-locking phenomenon and alleviate the stress concentration.

3. Materials and Methods

3.1. Sample Preparation

The purity of Ti, Cr and Ce was 99.5 wt%, 99.9 wt%, 99.0 wt%, respectively. A FeV80 master alloy was used as the V source; it contained 78.14 wt% V, 19.16 wt% Fe and minor amounts of Al, Si, C and O. We added 3 wt% Ce to the alloy to remove the oxygen introduced by the FeV80 master alloy and improve the activation performance [29]. The $V_{40}Ti_{26}Cr_{26}Fe_8$ alloy (denoted as s1) was prepared using a suspension melting furnace, followed by annealing at 1673 K for 30 min under vacuum (5×10^{-3} Pa) in a ZM-16 vacuum molybdenum wire furnace. The $V_{40}Ti_{26}Cr_{26}Fe_8$ alloy was crushed into a powder with particle size of 325 µm in an Argon glove box.

The silicone oils with different viscosities of 50, 1000 and 10,000 cSt were purchased from the Dow Corning company. We added 5 wt% silicone oil to the $V_{40}Ti_{26}Cr_{26}Fe_8$ powder by hand grinding for 15 min in a mortar. The mixtures of $V_{40}Ti_{26}Cr_{26}Fe_8$ powder and silicone oil of 50, 1000 and 10,000 cSt were denoted as samples s2, s3, and s4, respectively.

We placed ~2 g of the samples in a stainless-steel autoclave, which was then air-evacuated by a rotary pump at 373 K for 1 h. After the autoclave was cooled down to room temperature (RT), H_2 (purity: 99.999 wt%) at 7 MPa was introduced. The hydrogenation kinetics and pressure composition isotherm (PCT) curves during the dehydrogenation process were measured using a Sieverts-type apparatus. The hydrogen amounts of the samples were calculated without considering the mass of silicone oil.

The morphologies of the samples were observed on a scanning electron microscope (SEM, Hitachi REGULUS8230, Japan). The distribution of the particle size was analyzed by a Malvern particle analyzer (Bettersize 1900, China). The contact angles between the alloy and silicone oil were obtained by a contact angle meter (SL250, USA KINO Co., Ltd., USA). X-ray diffraction data of the samples were collected on an X-ray diffractometer (Oxford Xcalibur E diffractometer, UK) using Cu K_α radiation ($\lambda = 1.5418$ Å) at a scanning speed of 0.01°/s and 2θ of 25–80°. The microstrain of the samples was evaluated by refining the XRD patterns using the Rietveld method with the software Jade 6.0.

3.2. Strain Test

The home-made metal hydride (MH) reactor for the strain test is illustrated in Scheme 1. The reactor consists of two parts, an outer jacket and an inner MH container, made of stainless steel. The flanges were sealed by spiral wound gaskets (PTFE), and the upper flange contained a pipeline for hydrogen supply. The inner MH container with an outer diameter of 33 mm and a wall thickness of 3 mm was equipped with four strain gauges (Zemic, China). To measure the strain in the tangential (q) directions on the inner container, strain gauges were placed at positions 3a to 3d, which were 25, 45, 65 and 85 mm away from the bottom of the container, respectively. Strain is the ratio of the amount of deformation with respect to the original size and is denoted by ε. The measurement range of the strain gauges was $-11{,}000 \leq \mu\varepsilon \leq 11{,}000$. The outer jacket with an outer diameter of 45 mm and a wall thickness of 3 mm had an eccentric circular structure, leaving a 3 mm gap for attaching the strain gauges. On the top of the outer jacket, a wiring hole allowed the connection of the strain gauges with the signal acquisition device.

For each measurement, approximately 400 g of the MH powder was loaded into the MH reactor in an Ar glove box, which occupied roughly 40% of the space of the inner container. Afterward, the reactor was heated by a heating jacket to 373 K and evacuated for 3 h. Subsequently, the MH reactor was placed in a water bath for the hydrogen absorption and desorption test.

4. Conclusions

Silicone oils with viscosity from 50 to 10,000 cSt were investigated as a glidant for a V-based BCC alloy, $V_{40}Ti_{26}Cr_{26}Fe_8$ (particle size: ~325 µm). We added 5 wt% silicone oil to the $V_{40}Ti_{26}Cr_{26}Fe_8$ powder, which formed a 320~460 nm-thick layer covering the surface of the $V_{40}Ti_{26}Cr_{26}Fe_8$ particles. Regardless of viscosity, the addition of 5 wt% silicone oil slightly reduced the initial hydrogen storage capacity of $V_{40}Ti_{26}Cr_{26}Fe_8$ but improved the reversibility of hydrogen absorption. During the first 20 hydrogen sorption cycles, the reversible capacity of $V_{40}Ti_{26}Cr_{26}Fe_8$ dropped from 2.42 to 2.06 wt%, with a capacity retention of 85.1%, while the capacity retention improved to 90%, 87.3% and 87% after the addition of silicone oil with viscosity of 50, 1000, 10,000 cSt, respectively. Furthermore, for the $V_{40}Ti_{26}Cr_{26}Fe_8$ powder bed with 5 wt% silicone oil, the maximum strain on the container walls decreased by $\geq 22.5\%$. These results indicate that silicone oil is an effective glidant to alleviate the stress concentration in hydrogen storage containers with a V-based BCC alloy.

Author Contributions: Conceptualization, Y.Y.; Formal analysis, F.H. and D.C.; Funding acquisition, Y.Y. and C.W.; Investigation, X.Z. and H.K.; Resources, Y.Y.; Supervision, Y.W. and C.W.; Validation, X.Z. and H.K.; Writing—original draft, H.K. and X.Z.; Writing—review & editing, Y.Y. All authors have read and agreed to the published version of the manuscript.

Funding: This work was funded by NSAF (Grant No. U2130208), the Key R&D Program of Sichuan Province (2022YFG0116) and the Sichuan University-Panzhihua City University-Enterprise Co-project (2020CDPZH-1).

Data Availability Statement: Not applicable.

Conflicts of Interest: The authors declare no conflict of interest.

References

1. Züttel, A.; Remhof, A.; Borgschulte, A.; Friedrichs, O. Hydrogen: The future energy carrier. *Philos. Trans. R. Soc. London. Ser. A Math. Phys. Eng. Sci.* **2010**, *368*, 3329–3342. [CrossRef] [PubMed]
2. Tarhan, C.; Çil, M.A. A study on hydrogen, the clean energy of the future: Hydrogen storage methods. *J. Energy Storage* **2021**, *40*, 102676. [CrossRef]
3. Manoharan, Y.; Hosseini, S.E.; Butler, B.; Alzhahrani, H.; Senior, B.T.F.; Ashuri, T.; Krohn, J. Hydrogen Fuel Cell Vehicles; Current Status and Future Prospect. *Appl. Sci.* **2019**, *9*, 2296. [CrossRef]
4. Usman, M.R. Hydrogen storage methods: Review and current status. *Renew. Sustain. Energy Rev.* **2022**, *167*. [CrossRef]
5. Mori, D.; Hirose, K. Recent challenges of hydrogen storage technologies for fuel cell vehicles. *Int. J. Hydrogen Energy* **2009**, *34*, 4569–4574. [CrossRef]
6. Eberle, U.; Felderhoff, M.; Schüth, F. Chemical and Physical Solutions for Hydrogen Storage. *Angew. Chem. Int. Ed.* **2009**, *48*, 6608–6630. [CrossRef]
7. Sakintuna, B.; Lamari-Darkrim, F.; Hirscher, M. Metal hydride materials for solid hydrogen storage: A review. *Int. J. Hydrog. Energy* **2007**, *32*, 1121–1140. [CrossRef]
8. Pasini, J.M.; Corgnale, C.; van Hassel, B.; Motyka, T.; Kumar, S.; Simmons, K.L. Metal hydride material requirements for automotive hydrogen storage systems. *Int. J. Hydrogen Energy* **2013**, *38*, 9755–9765. [CrossRef]
9. Matsushita, M.; Monde, M.; Mitsutake, Y. Experimental formula for estimating porosity in a metal hydride packed bed. *Int. J. Hydrogen Energy* **2013**, *38*, 7056–7064. [CrossRef]
10. Matsushita, M.; Tajima, I.; Abe, M.; Tokuyama, H. Experimental study of porosity and effective thermal conductivity in packed bed of nano-structured FeTi for usage in hydrogen storage tanks. *Int. J. Hydrogen Energy* **2019**, *44*, 23239–23248. [CrossRef]
11. Smith, K.C.; Fisher, T.S. Models for metal hydride particle shape, packing, and heat transfer. *Int. J. Hydrogen Energy* **2012**, *37*, 13417–13428. [CrossRef]
12. Lin, C.-K.; Chen, Y.-C. Effects of cyclic hydriding–dehydriding reactions of LaNi5 on the thin-wall deformation of metal hydride storage vessels with various configurations. *Renew. Energy* **2012**, *48*, 404–410. [CrossRef]
13. Ao, B.; Chen, S.; Jiang, G. A study on wall stresses induced by LaNi5 alloy hydrogen absorption–desorption cycles. *J. Alloys Compd.* **2005**, *390*, 122–126. [CrossRef]
14. Tran, X.; McDonald, S.; Gu, Q.; Nogita, K. In-situ synchrotron X-ray diffraction investigation of the hydriding and dehydriding properties of a cast Mg–Ni alloy. *J. Alloys Compd.* **2015**, *636*, 249–256. [CrossRef]
15. Kojima, Y.; Kawai, Y.; Towata, S.-I.; Matsunaga, T.; Shinozawa, T.; Kimbara, M. Development of metal hydride with high dissociation pressure. *J. Alloys Compd.* **2006**, *419*, 256–261. [CrossRef]
16. Hu, H.; Ma, C.; Chen, Q. Mechanism and microstructural evolution of TiCrVFe hydrogen storage alloys upon de-/hydrogenation. *J. Alloys Compd.* **2021**, *877*, 160315. [CrossRef]
17. Saito, T.; Suwa, K.; Kawamura, T. Influence of expansion of metal hydride during hydriding–dehydriding cycles. *J. Alloys Compd.* **1997**, *253–254*, 682–685. [CrossRef]
18. Qin, F. Pulverization, expansion of La0.6Y0.4Ni4.8Mn0.2 during hydrogen absorption–desorption cycles and their influences in thin-wall reactors. *Int. J. Hydrogen Energy* **2008**, *33*, 709–717. [CrossRef]

19. Charlas, B.; Gillia, O.; Doremus, P.; Imbault, D. Experimental investigation of the swelling/shrinkage of a hydride bed in a cell during hydrogen absorption/desorption cycles. *Int. J. Hydrogen Energy* **2012**, *37*, 16031–16041. [CrossRef]
20. Lin, C.-K.; Huang, S.-M.; Jhang, Y.-H. Effects of cyclic hydriding–dehydriding reactions of Mg2Ni alloy on the expansion deformation of a metal hydride storage vessel. *J. Alloys Compd.* **2011**, *509*, 7162–7167. [CrossRef]
21. Wu, K.; Cai, B.; Fan, L.; Qin, L.; Chen, D.; Huang, Y. Stress measurement of MlNi4.5Cr0·45Mn0.05 alloy during hydrogen absorption-desorption process in a cylindrical reactor. *Int. J. Hydrogen Energy* **2020**, *45*, 28175–28182. [CrossRef]
22. Heubner, F.; Hilger, A.; Kardjilov, N.; Manke, I.; Kieback, B.; Gondek, Ł.; Banhart, J.; Röntzsch, L. In-operando stress measurement and neutron imaging of metal hydride composites for solid-state hydrogen storage. *J. Power Sources* **2018**, *397*, 262–270. [CrossRef]
23. Duan, W.; Du, J.; Wang, Z.; Niu, Y.; Huang, T.; Li, Z.; Pu, C.; Wu, Z. Strain variation on the reaction tank of high hydrogen content during hydrogen absorption-desorption cycles. *Int. J. Hydrogen Energy* **2013**, *38*, 2347–2351. [CrossRef]
24. Melnichuk, M.; Cuscueta, D.J.; Silin, N. Effect of glidants on LaNi5 powder flowability. *Int. J. Hydrogen Energy* **2018**, *43*, 6219–6228. [CrossRef]
25. Cuscueta, D.; Silin, N.; Melnichuk, M. Stress reduction in a hydride container by the addition of a glidant agent. *Int. J. Hydrogen Energy* **2020**, *45*, 27452–27456. [CrossRef]
26. Meyer, K.; Zimmermann, I. Effect of glidants in binary powder mixtures. *Powder Technol.* **2004**, *139*, 40–54. [CrossRef]
27. Wang, Q.-D.; Wu, J.; Chen, C.-P.; Li, Z.-P. An investigation of the mechanical behaviour of hydrogen storage metal beds on hydriding and dehydriding and several methods of preventing the damage of hydride containers caused by the expansion of hydrogen storage metals. *J. Less Common Met.* **1987**, *131*, 399–407. [CrossRef]
28. Heubner, F.; Pohlmann, C.; Mauermann, S.; Kieback, B.; Röntzsch, L. Mechanical stresses originating from metal hydride composites during cyclic hydrogenation. *Int. J. Hydrogen Energy* **2015**, *40*, 10123–10130. [CrossRef]
29. Yan, Y.; Chen, Y.; Wu, C.; Tao, M.; Liang, H. A low-cost BCC alloy prepared from a FeV80 alloy with a high hydrogen storage capacity. *J. Power Sources* **2007**, *164*, 799–802. [CrossRef]
30. Yu, H.; Zhai, G.; Zhang, Y.; Li, Y.; Liu, Y.; Liu, Y. Structure and thermal stability study of dimethicone (in Chinese). *Shanghai Meas. Test.* **2020**, *47*, 28–32.
31. Wu, C.; Zheng, X.; Chen, Y.; Tao, M.; Tong, G.; Zhou, J. Hydrogen storage and cyclic properties of V60Ti(21.4+x)Cr(6.6−x)Fe12 ($0 \leq x \leq 3$) alloys. *Int. J. Hydrogen Energy* **2010**, *35*, 8130–8135. [CrossRef]

Review

Metal Hydride Composite Structures for Improved Heat Transfer and Stability for Hydrogen Storage and Compression Applications

Liang Liu [1], Alexander Ilyushechkin [2], Daniel Liang [3], Ashleigh Cousins [2,*], Wendy Tian [3], Cherry Chen [3], Jon Yin [1] and Liezl Schoeman [2]

1. Mineral Resources, CSIRO, Brisbane, QLD 4069, Australia
2. Energy, CSIRO, Brisbane, QLD 4069, Australia
3. Manufacturing, CSIRO, Melbourne, VIC 3168, Australia
* Correspondence: ashleigh.cousins@csiro.au

Citation: Liu, L.; Ilyushechkin, A.; Liang, D.; Cousins, A.; Tian, W.; Chen, C.; Yin, J.; Schoeman, L. Metal Hydride Composite Structures for Improved Heat Transfer and Stability for Hydrogen Storage and Compression Applications. *Inorganics* 2023, *11*, 181. https://doi.org/10.3390/inorganics11050181

Academic Editors: Craig Buckley, Mark Paskevicius, Torben R. Jensen and Terry Humphries

Received: 17 March 2023
Revised: 11 April 2023
Accepted: 14 April 2023
Published: 24 April 2023

Copyright: © 2023 by the authors. Licensee MDPI, Basel, Switzerland. This article is an open access article distributed under the terms and conditions of the Creative Commons Attribution (CC BY) license (https://creativecommons.org/licenses/by/4.0/).

Abstract: Metal alloys and intermetallic compounds offer an attractive method for safely storing hydrogen (H_2). The metal alloys absorb H_2 into their structure, often swelling and fracturing as a result of phase transformation during hydride formation/decomposition cycles. The absorption of H_2 is an exothermic process, requiring the effective and efficient removal of heat. This can be challenging as heat transfer to/from powdered beds is notoriously difficult, and often limited by poor thermal conductivity. Hence, the observed reaction kinetics for absorption and desorption of H_2 is dominated by heat flow. The most common method for improving the thermal conductivity of the alloy powders is to prepare them into composite structures with other high thermal conductivity materials, such as carbons and expanded natural graphite. Such composite structures, some also combined with polymers/resins, can also mitigate safety issues related to swelling and improve cyclic durability. This paper reviews the methods that have been used to prepare such composite structures and evaluates the observed impact on thermal conductivity.

Keywords: metal hydride; composite; thermal conductivity; expanded natural graphite

1. Introduction

Metal hydrides (MH) are attractive for hydrogen (H_2) storage and compression applications, particularly where waste heat is available. Metal hydride-based compressors have the advantages of no moving parts, simplicity in design, compactness, and safety and reliability [1]. Metal hydrides have higher volumetric hydrogen storage density (e.g., 150 kgH_2/m^3 in Mg_2FeH_6) than compressed hydrogen gas (<40 kgH_2/m^3 at 800 bar) or liquid hydrogen (70.8 kgH_2/m^3 at $-252\ °C$ and 1 bar) [2].

Metal hydrides are formed via the reversible reaction of a hydride-forming metal/alloy or intermetallic compound with H_2 gas:

$$M(s) + \frac{x}{2}H_2(g) \leftrightarrow MH_x(s) + Q \qquad (1)$$

where M is a metal/alloy, (s) and (g) relate to the solid and gas phases, respectively, and Q is the energy released [1]. This leads to the potential for solid-state storage of H_2 under moderate temperatures and pressures. H_2 absorption is an exothermic process, as outlined in Equation (1). Heat released during hydride formation must be continuously removed to prevent the hydride/alloy system from heating up. If the temperature is allowed to increase, the corresponding equilibrium pressure will also increase, which will result in no further H_2 sorption taking place. Conversely, when H_2 is recovered, heat must be added to release hydrogen.

To improve kinetics, metal alloys used for H_2 storage/compression are often in powdered form. These alloy powders swell as they absorb H_2 as the crystalline cell volume increases with H_2 entering the interstitial positions in the crystalline structure of the metal alloy. This, in turn, creates internal stresses which are sufficient to fracture the metal hydride with continued cycling, resulting in the creation of submicron particles [3]. A smaller particle size increases the surface area of the powders, promoting a faster diffusion of hydrogen into the alloy powders. For example, it has been demonstrated that Mg nanoparticles with smaller size (i.e., 25 nm) had much faster H_2 sorption kinetics [4]. Mg colloids with a particle size of 5 nm began to dehydrogenate at 85 °C, which was much lower than that of commercial micrometer-size MgH_2 (400 °C) [5]. Moreover, morphologies can also play a vital role in the materials' performance [6–9]. Although nanostructured metal hydrides show improved performance in sorption/desorption kinetics, they tend to aggregate, resulting in decreased sorption capacity with multiple cycles. Other challenges faced when using powdered metal hydride materials include effective removal of the surface oxide layer of the powders (needed to allow the hydrogen absorption reaction to proceed), higher heat flow (due to increased reaction kinetics), and limited heat transfer to/from powdered beds. The heat transfer is limited as the thermal conductivity of MH powdered beds is poor, owing to their porosity and inter-particle contact resistance. For example, thermal conductivities of packed $LaNi_5$ powders are very low, in the range of 0.1–1 W/m·K, compared with a bulk thermal conductivity at 12.5 W/m·K [10,11]. As a result of the poor thermal conductivity, the observed charging/discharging rate for absorption and desorption of hydrogen is dominated by heat flow [12].

For a commercial system, the practicalities of filling and long-term operation of MH-filled vessels needs to be improved, particularly at the scale required for H_2 fuel cell electric vehicles (FCEV). In order to fill a MH vessel quickly, heat must be removed efficiently and effectively from the system (and reversely quickly added for discharge of H_2). The poor heat transfer capability of a powdered metal hydride bed is a considerable restriction on the design and construction of hydrogen storage and compression systems [13]. Various modifications to MH vessels such as introducing metal fins, cooling tubes, and phase change materials (PCMs) have been suggested to improve heat transfer characteristics [14]. These modifications, however, significantly increase the cost and complexity of MH storage vessels. Potentially more economical alternatives involve mixing the MH with conductive materials and forming composite structures. These thermally conductive additions could be metals (e.g., Al powder) or carbon-based materials such as expanded natural graphite (ENG) [15–18], activated carbon [19], carbon nanotube [20], and carbon fiber [21]. With the addition of 20 wt% carbon fiber, the thermal conductivity of $TiFe_{0.9}Ni_{0.1}$ composite was increased from about 0.5 to 2 W/m·K using aramid pulp as a binder [21]. A solid compact combining $LaNi_5$ powder with 20 wt% Al powder was found to dramatically increase the thermal conductivity of the $LaNi_5$ hydride from around 1 W/m·K to 32.5 W/m·K [13].

To improve the handling capability of the hydride, contain its expansion, and improve durability, composites with polymers/resins have also been suggested [22,23]. Polymers can also be used as a binder to bind fine MH powders onto surfaces of microporous carbon scaffolds or substrates [24]. Tokiwa et al. [22] measured the distortion imparted on an MH storage vessel as a result of H_2 sorption. While the addition of the resin to the composite was effective in reducing the distortion of the storage vessel, the performance varied depending on the polymer used.

To solve the stability issue of the nanostructured metal hydrides, nanoconfinement provides a feasible solution to prevent particle aggregation by limiting its mobility in a nanoporous support [25]. Combining MH powders in structures with components of high thermal conductivity can also help overcome the challenges associated with heat transfer. As a result, MH composite materials have attracted great interest [26,27]. This paper provides a review of methods that have been explored to generate MH composite structures for improved heat transfer for H_2 storage and compression applications. For a practical reactor, it is desirable to maximize its gravimetric and volumetric sorption capacities. Therefore, it is best to minimize the amount of support material to achieve a high hydrogen storage capacity and avoid the need for excessive heating/cooling of inert material. The MH mass fraction in the composite is expected to be as high as possible. This review focuses on MH composite materials with a relatively high MH mass fraction.

2. Composite Materials

The "ideal" composite material for hydrogen storage and compression should satisfy several requirements:

- High hydrogen permeability.
- High thermal conductivity.
- Certain degree of elasticity to accommodate metal particles expansion/contraction during hydride formation/decomposition.
- Certain degree of mechanical strength to be resistant to mechanical degradation though charging cycles.
- Thermal stability to work within desired temperature interval.
- Absence of chemical interactions with metal/alloy.

To accommodate the expansion during hydrogenation, MH in pellet/granule form could be compacted with other flexible materials. In earlier studies, several different metal matrix materials and fabrication techniques were tested for practical fabrication of the composites containing the hydrogenated metal alloy and aluminium, nickel, and copper as matrix [3,28]. Using an aluminium matrix was abandoned due to the difficulty of sintering aluminium [3]. The compacts with nickel powder had a very low green strength, could not form stable pellets under cold compaction, and had to be vacuum sintered at temperatures from 1200 °C to 1300 °C, while the compacts made with copper powders demonstrated relatively high green strengths and the unsintered pellets could survive 138 sorption/desorption cycles with minor degradation on the edges [3]. Moreover, compacts with metal additions such as nickel and copper have a substantially increased reactor bed parasitic thermal mass due to their relatively high densities (~8.9 g/cm^3) and require complex chemical processes (e.g., solution coating, sintering under 300 bar H_2) [15]. More recently, porous graphite was suggested as a matrix material which provides a simple, economic, and effective way to increase the thermal conductivity of MH compacts. For example, it has been reported that compacts with 2.28 wt% ENG had a higher effective thermal conductivity of about 19.45 W/m·K compared with 8 W/m·K for compacts with 6.14 wt% Al-foams [17].

Compacts with polymers as a binder between the metal alloy powders have also been explored to accommodate metal hydride expansion and/or provide a support for the powdered alloy material. Although a polymer has relatively low thermal conductivity, it can prevent the metal alloy from forming more porous structures between the metal powders during sorption/desorption cycles, which can limit the reduction in thermal conductivity. When the amount of polymer addition is controlled at a low level, the polymer is only present on limited areas of the powder surfaces. It acts as a binder to hold the powder together, improving the quality of the direct contact between the powder particles, which prevents excessive loss of thermal conductivity during thermal cycles. Additives such as ENG can also be incorporated to further improve the thermal conductivity of MH composites. The key factor that influences the thermal conductivity of the composite is the quality of the direct contact between the metal hydride powders during the sorption and desorption cycles.

Mixtures of polymer/resin with hydrides and conductivity enhancers have also been suggested [22]. In polymer-bonded composites, hydrogen transport to or from the MH particles occurs through pre-existing cracks and porosity and/or by molecular diffusion through the polymer binders. Factors that can increase the diffusion coefficient of hydrogen through the polymer binders include chemistries of polymers free of Cl, O, N, and S that reduce polarization of the polymer, a larger free volume due to molecular-scale gaps or pores between the chains of polymers, lower-density molecular structures, and complexities of the side-branch groups. For semi-crystalline polymers, there are crystalline and amorphous regions in the polymer, and the crystalline region has less permeability due to the lack of molecular-scale gaps or pores. Consequently, a higher fraction of crystalline regions could hinder hydrogen transport [29–34].

Polyethylene (PE) is the most common and cost-effective nonpolar semicrystalline polymer used including low-density PE (LDPE) and high-density PE (HDPE). These polymers have different densities varying between 0.88–0.96 g/cm^3 and melting temperatures in the range of 115–135 °C [34]. LDPE has demonstrated a hydrogen permeability of ~2.0–3.0 × 10^{-15} mol·m/(m^2·s·Pa), which is 2–3 times higher than HDPE ($p < 5$ bar and 50 °C, [35]), due to its lower crystallinity and greater free volume. Temperatures above 50 °C will increase the free volume in polymeric materials [36], leading to a greater hydrogen permeability in either LDPE or HDPE. Hydrogen diffusion through the PE polymers at this permeability is considered faster than the reaction kinetics of MHs, and thus not the factor limiting hydrogen transport in a polymer matrix MH composite [37]. Since the melting point of PE is low, low-temperature metal hydrides (e.g., LaNi$_5$ with a reaction temperature < 100 °C) are commonly selected for manufacturing PE-based metal hydride composites [38].

Polypropylene (PP) is another nonpolar semicrystalline polymer utilized as either matrix or binder for low-temperature metal hydride composites. The melting point of the PP is between 130 °C and 171 °C, with densities in the range of 0.855 (amorphous) to 0.946 (crystalline) g/cm^3. Due to its similar characteristics to PE polymers, H$_2$ permeability is almost identical to the PE polymer [39]. The use of the PP in polymer MH composites, therefore, is exchangeable with PE [40–44], in terms of permeability, working temperature, and weight.

A higher working temperature polymer becomes essential when high-temperature metal hydrides are involved, such as MgH$_2$ (working temperature > 300 °C). One example of a high-temperature polymer is high-temperature-vulcanized polysiloxanes, especially polydimethylsiloxanes (PDMS). The glass transition temperature of PDMS is -149 °C [45] and it can operate under extreme environments and temperature from -55 to 300 °C [46]. The hydrogen permeability of PDMS is measured as ~3 × 10^{-13} mol·m/(m^2·s·Pa) at 35 °C and up to 16 bar [47,48], which is significantly higher than ~3 × 10^{-15} mol·m/(m^2·s·Pa) determined for LDPE [49]. The PDMS polymer is often deployed to form a porous scaffold system, rather than a continuous matrix, in the metal hydride composites that contain either high-temperature non-Pd hydrides (e.g., MgH$_2$) or volatile/liquid complex metal hydrides [50].

Polymethylpentene (PMP) is a semicrystalline polymer with the chemical formula (C$_6$H$_{12}$)$_n$ and is commonly called TPX^{TM}. It has a side branch at the molecular level, which makes it a promising candidate for gas transport. It has a melting temperature of around 230 °C, suitable for most of H$_2$ and MH reactions. PMP is also used to form the scaffold system for polymer metal hydride composites [37,40,51,52].

As high-temperature polymers typically have a poor hydrogen permeability, the role of the polymers in the high-temperature metal hydride composites has been changed to hold MH particles onto a scaffold or substrate that are based on carbon or ceramic materials or bind the particles together [53,54]. The polymers are often coated onto the surfaces of the powders, rather than as a continuous matrix. While the polymer coating on the MH particles is thin, it is still desirable for the polymer to provide surface protection for the MH powders from oxidation and other contaminates [44,55]. A wide range of other high-temperature polymers has been explored as the scaffold system for MH composites. These polymers include polyvinyl chloride (PVC, (C$_2$H$_3$Cl)$_n$, melting point < 260 °C) [56], Phenol formaldehyde resins (PF) that decompose at >220 °C [53,54], thermoplastic polyesters, e.g., Nylon^{TM}, melting point of heating 210–260 °C and of cooling 180–210 °C [57], and thermoplastic fluoropolymers (e.g., polyvinylidene difluoride (PVDF), melting point of 177 °C) [41,44].

2.1. Composites with Expanded Natural Graphite (ENG)

MH composite compacts with ENG as the binder were suggested as an alternative and efficient solution for MH reaction beds instead of commercially available Al-foams [17]. The effective thermal conductivity is reduced by increasing the MH to ENG mass ratio. It was concluded that the MH to ENG ratio should be higher than 20 and the effective thermal conductivity should be >8 W/m·K in order to be competitive to Al-foams, which can be achieved in composites with relatively low porosities [58].

Various materials compacted at different pressures with different proportions of ENG (in range of 2.5–34 wt%) were prepared and tested [16–18,58–60]. Melt-spun magnesium alloy flakes and ENG compacts prepared as cylindrical pellets have demonstrated good effective thermal conductivity [18]. Later, this research group demonstrated that Hydralloy C5 (AB_2 type metal hydride)/ENG composite pellets can be used for hydrogen storage applications [59]. Hydrogenation performance and the pellet evolution (composites with 5 wt% ENG compacted at 75 MPa were tested) during the observed hydrogenation cycles was found to be promising for use in tubular storage tanks.

Another composite of a mixture of $Mg(NH_2)_2$, LiH, and KOH (molar ratio of 1:2:0.07) with 9 wt% ENG addition were prepared at a compaction pressure of 156 MPa [60]. The increase in the compaction pressure led to the reduction in the desorption kinetics for the 1st sorption cycle but had a minor effect for the following cycles. The composite of $La_{1-x}Ce_xNi_5$ and ENG was fabricated by compaction at 30 MPa, and the thermo-physical properties of the composite were measured [16]. The thermal conductivity was increased from 2 to 8 W/m·K.

Figure 1 shows the influence of ENG content on the thermal conductivity of $Mg_{90}Ni_{10}$-ENG compacts and their approximate hydrogenation states (i.e., MgH_2-ENG compacts) [18]. The thermal conductivity of all composites showed a high degree of anisotropy. It was much higher in the radial direction, which is perpendicular to the compaction direction. The compaction pressure had a significant impact on the radial thermal conductivity for $Mg_{90}Ni_{10}$-ENG compacts. On the other hand, it had little impact for MgH_2-ENG compacts. For lower ENG content compacts, there was a decrease in thermal conductivity when it was in a hydrogenated state. It is suggested that the heat transfer characteristics can be tuned in a wide range depending on ENG content and compaction pressure.

2.2. Composites with Polymers

The Japan Steel Works have developed composites of $MmNi_{4.4}Mn_{0.1}Co_{0.5}$ powder with two resins (i.e., LASTOSIL® M4648 and WACKER SilGel® 612). Pressure-composition-temperature (PCT) isotherms of the raw alloy and resin composites were found to be similar, however, the H_2 reaction rate was noted to be slower for the composite compared to the alloy powder [61]. Strain measurements performed on the MH containers confirmed lower strain recorded for the containers containing the resin composites. This lower strain was thought to be due to the immobilization and uniform distribution of the MH powder [61]. A 1000 Nm^3-class hydrogen storage system was fabricated, filled with 7.2 tons of MH composite materials. This system could work at maximum hydrogen charging/discharging rates of 70 Nm^3/h at a medium flow rate of 30 NL/min at temperatures in the range of 25–35 °C [61]. Resin composites prepared by Watanabe et al. [62] similarly showed no signs of damage after five absorption/desorption cycles. Even though disintegration of the MH powders took place, they were found to be firmly fixed in the resin matrix.

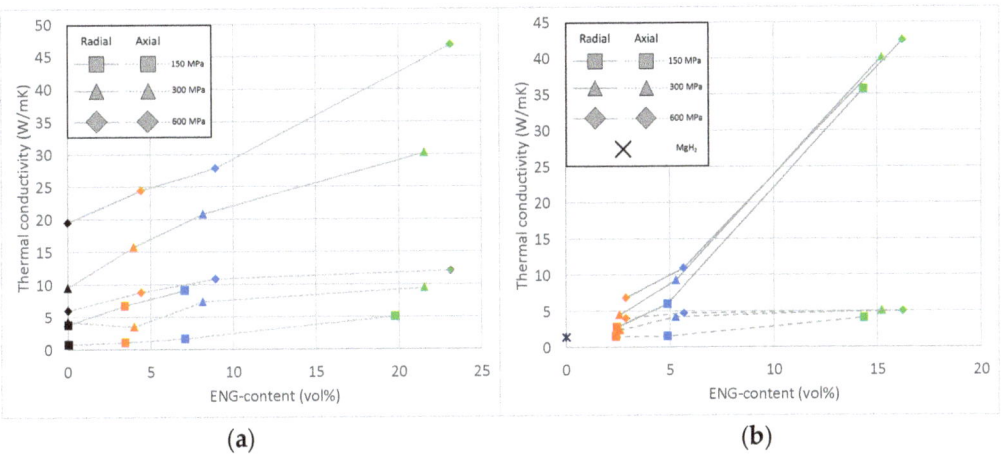

Figure 1. (a) Radial (solid line) and axial (dashed line) thermal conductivities of the Mg$_{90}$Ni$_{10}$-ENG compacts at different compaction pressures as a function of ENG content. (b) Radial (solid line) and axial (dashed line) thermal conductivities of the MgH$_2$-ENG compacts at different compaction pressures as a function of ENG content (0 wt% black, 5 wt% orange, 10 wt% blue, 25.5 wt% green). Note: MgH$_2$ is treated as the approximate hydrogenated state of Mg$_{90}$Ni$_{10}$. Figure adapted from Pohlmann et al., International Journal of Hydrogen Energy; published by Elsevier, 2010 [18].

Checcetto et al. [38] prepared composites of LaNi$_5$ powders distributed in a H$_2$ permeable elastomer (polysiloxane), with MH contents of 50 and 83 wt%. The composite with 50 wt% alloy exhibited a low H$_2$ storage capacity, thought to be due to the MH particles being separated and the chemistry at the MH–polymer interface, limiting H$_2$ absorption by the metal particles. When the MH content was increased to 83 wt%, H$_2$ sorption was noted to increase; however, this was still lower than anticipated. It was suggested that H$_2$ disassociation needed to occur at a metal surface, which was hindered in the polymer composites.

In further work, Checcetto et al. [43] showed that surface chemistries in metal–polymer composites were important and can impede the activation of the metal alloy surface. It was speculated that this is the underlying mechanism as to why some polymer–metal composites showed reduced H$_2$ storage capacity. LaNi$_5$ particles (<30 μm) dispersed in polysiloxane showed negligible H$_2$ storage capacity while LaNi$_5$ particles dispersed in polyethylene were completely hydrogenated. Conversely, Pd powders (average size 1 μm) dispersed in polysiloxane were found to completely hydrogenate. For complete hydrogenation of a metal–polymer composite, weak chemical interactions between the polymeric chain and the metal surface are required (i.e., nonpolar polymers). H$_2$ desorption rates were noted to be slower for all composite materials, thought to be due to slow H$_2$ diffusion into the polymeric part of the composite.

This work shows that combining polymers with metal alloys can be effective in accommodating the expansion that occurs as the hydride phase is formed. However, care needs to be taken in the selection of the polymer to ensure that it not only maintains suitable properties at the operating temperatures of H$_2$ charging and discharging, but also does not impede the activation and subsequent sorption of H$_2$ by the composite. There is also potential to explore the possibility that permeative polymers could play a role in protecting the metal powders during prolonged operation.

3. Liquid-State Fabrication Methods to Produce Metal Hydride Composites

3.1. High-Temperature Melt Infiltration to Form Supported Composites

For some metallic-bonded metal hydrides, melt infiltration can load their metals into certain supports by melting the metal alloys under appropriate conditions [63]. Loading is determined by the wettability of the melt on the support [64]. If the contact angle is smaller than 90°, the pores of the support can be filled by capillary force. On the other hand, if the contact angle is larger than 90° (non-wetting), external pressure is needed to fill the pores of the support [65]. This method can achieve relatively high loadings in a single step. However, this is only applicable to metal hydrides with melting temperatures where the support materials are stable. Therefore, the maximum temperature that can be applied is also limited by the nature of the support materials. Magnesium nanoparticles were prepared by infiltration of nanoporous carbon with molten magnesium [66]. The size of the Mg crystallites was affected by the pore size of the carbon and can be less than 2 nm. The maximum loading without bulk magnesium aggregation was up to 13 wt%.

3.2. Low-Temperature Solution Impregnation to Form Supported Composites

For the ionic- or covalent-bond metal hydrides that can form complex metal hydrides in a solution, the supported composites can be fabricated at low temperature using a solution method. Solution impregnation is widely used for catalyst preparation. A solution of the MH is allowed to infiltrate the porous support materials, and then the solvent can be removed by a drying process. This method relies on the solubility of the hydride in an organic solvent. The most common solvents are ethers, diethyl ether, and tetrahydrofuran (THF). A main disadvantage is that high loadings can only be obtained by repeated processes as the metal hydride solubility is often low. Cahen et al. [67] incorporated $LiBH_4$ into a mesoporous carbon by room temperature impregnation using 0.029 wt% $LiBH_4$ in methyl tert-butyl ether (MTBE). The composite with a 33:67 weight ratio ($LiBH_4$/carbon) showed excellent desorption kinetics with a hydrogen release of 3.4 wt% in 90 min at 300 °C, whereas the decomposition of neat $LiBH_4$ was not significant at the same temperature.

3.3. Low-Temperature In-Situ Solution Synthesis to Form Supported Composites

In-situ solution synthesis uses MH precursors instead of as-prepared metal or MH. The MH nanoparticles are formed in the pores of the support. It can achieve relatively high loadings in a single step. However, it highly relies on the nature of the MH and the properties of the support. For example, $Mg(Bu)_2$ has been widely investigated as a precursor to synthesis of MgH_2 [68]. MgH_2 nanoparticles with a size less than 3 nm were formed inside the pores of a carbon scaffold and a significant reduction in reaction enthalpy and entropy was found for the composite. A MgH_2/graphene composite with MgH_2 loading of 75 wt% (MHGH-75) was synthesized via the hydrogenation of $Mg(Bu)_2$ in cyclohexane [69]. Nickel was further introduced as a catalyst to enhance hydrogen storage performance of MgH_2. The MHGH-75 composite showed H_2 sorption capacity of 4.3 wt% in 60 min at 200 °C under 30 bar H_2 pressure. Moreover, the sorption capacity of the composite with nickel catalyst (Ni-MHGH-75) was 5.4 wt%. Conversely, there was no sorption observed from ball-milled MgH_2 and MgH_2/graphene under the same conditions. Figure 2 shows the cyclic performance and the thermal conductivity of the composites [69]. Overall, the Ni-MHGH-75 composite demonstrated stable cycling performance up to 100 cycles. Compared with MgH_2 nanoparticles (NPs) and ball-milled MgH_2/graphene composite, the thermal conductivity of MHGH-75 composite was increased by ≈93.5-fold and 5.6-fold, respectively, at both room temperature and 200 °C. It is suggested that graphene acts not only as a structural support, but also as a space barrier to prevent the growth of MgH_2 nanoparticles and as a thermally conductive pathway. Similar results have been reported for a $NaBH_4$/graphene composite [70].

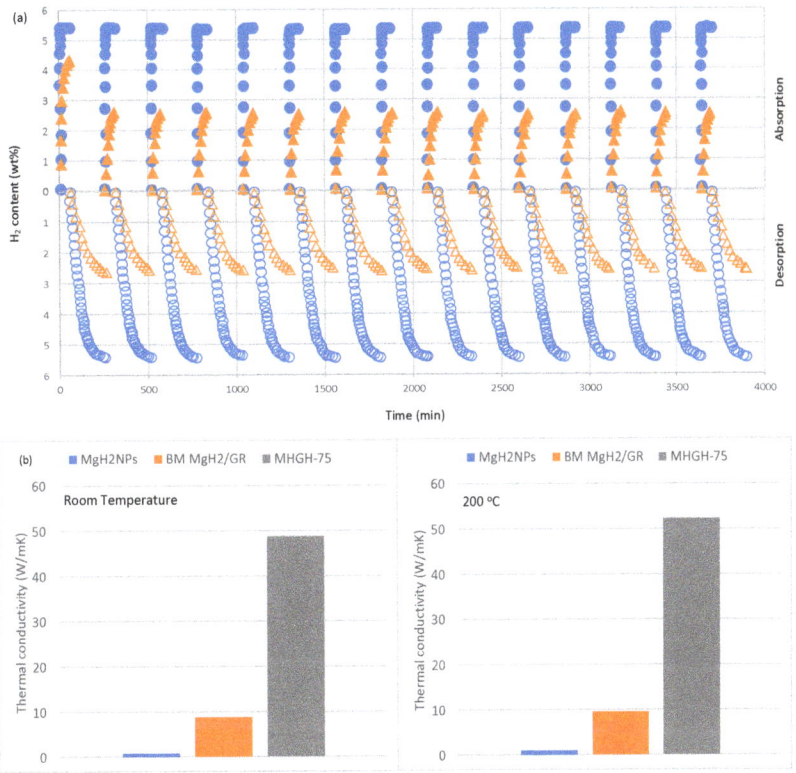

Figure 2. (**a**) Reversible H_2 absorption (under 30 atm hydrogen pressure) and desorption (under 0.01 atm hydrogen pressure) of Ni-MHGH-75 (circles) and MHGH-75 (triangles) at 200 °C. (**b**) Thermal conductivity of MHGH-75 in comparison with MgH_2 NPs and ball-milled MgH_2 containing 25 wt% graphene (BM MgH_2/GR) at room temperature and 200 °C, respectively. Figure adapted from Xia et al., Advanced Materials; published by Wiley, 2015 [69].

4. Powder-Based Fabrication Methods

Most literature reports investigated MH/composite in the powder form. For the metallic metal hydrides, e.g., based on Mg, V, Ti, La, and Al metals, combing the powders of these alloys with support materials without melting the metal is a popular method as they are simple, low cost, and consume less energy in comparison with the liquid fabrication route.

However, the powder composite has some disadvantages, especially for large-scale systems. First, the breakdown of metal particles into fine powders after hydrogenation cycles is well-known. This will limit the loading of the material in the reactor and can introduce internal stress to the reactor wall [71]. Second, it will reduce the thermal conductivity of the material, increasing heat transfer resistance, which may adversely impact sorption/desorption kinetics. Fine MH powders are found to have a typical thermal conductivity of ~0.1 W/m·K [15]. As a result, additional heat management measures such as fins, Al-foams, and phase change materials must be considered [14], which will increase the system cost and further reduce the packing density of the reactor. Although most pilot and industrial hydride tanks are using MH/composite powder, it is important to explore properties of composite materials in other forms such as pellet/disc and film/sheet. These composites can be synthesized by applying additional molding steps. It is expected that they have a much higher thermal conductivity and better structural integrity, which is beneficial to the overall performance of the reactor.

4.1. Powder Preparation by Ball Milling

Ball milling is one of the most-common techniques to reduce the particle size of MH [72] as either standalone (unsupported) hydrogen storage materials or as the constituents to be incorporated into a supported composite. It can readily reduce the particle size of MH down to a few nanometers and introduce many defects that can enhance hydrogen diffusion processes. Carbon material in this case is used as an additive rather than a support. Hydrogen or solvents can be introduced during the milling process [73,74]. It is a convenient method for the activation and formation of MH. Moreover, it can prevent or minimize the formation of a metal oxide layer when an inert atmosphere is applied. It has been reported that the hydrogen sorption rate increased significantly as a result of reactive ball milling of magnesium with different carbon allotropes (e.g., graphite, ultrafine diamonds, carbon nanotubes, and amorphous carbon powder) [75]. Figure 3 shows the impact of activated carbon (AC) on the hydrogen sorption capacity of ball-milled MgH_2-AC composites. Overall, the addition of AC can improve H_2 sorption kinetics, particularly at low temperature conditions, compared with ball-milled MgH_2 nanoparticles. The MgH_2-5 wt% AC composite shows H_2 sorption capacities of 6.5, 6.7, and 6 wt% in 10, 120, and 600 min at 300, 200, and 150 °C, respectively.

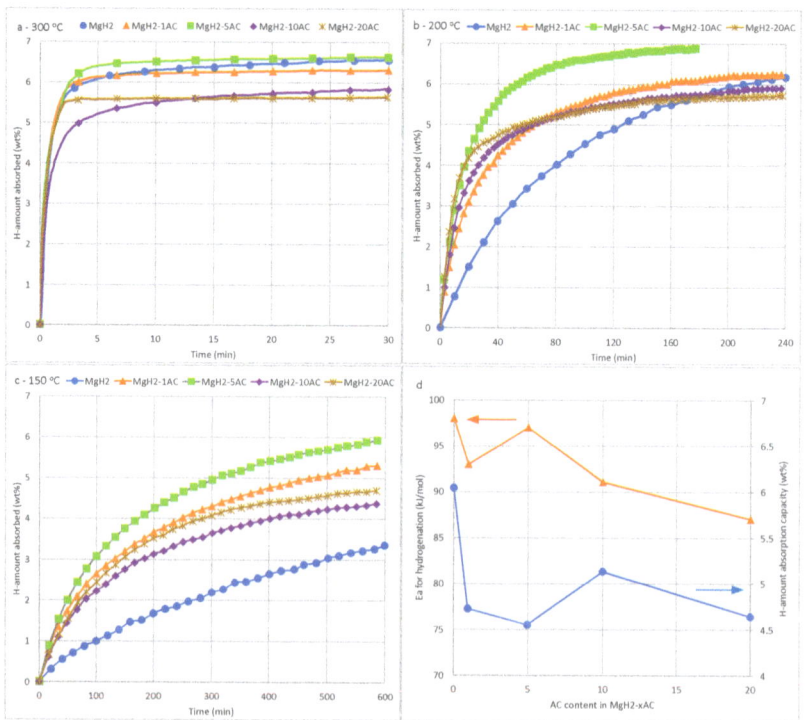

Figure 3. Hydrogen sorption isotherms of the as-prepared MgH_2-xAC composites (x represents the weight percentage of AC) at (**a**) 300 °C, (**b**) 200 °C, and (**c**) 150 °C with an initial hydrogen pressure of 2 MPa, and (**d**) the comparison of the capacities of hydrogenation within 2 h at 300 °C and relevant activation energies for hydrogenation. Figure adapted from Jia et al., International Journal of Hydrogen Energy; published by Elsevier, 2012 [76].

The hydrogenation process can also be accelerated by the addition of catalysts [72,77]. Meng et al. [78] reported an improvement in the hydrogen storage properties of MgH_2 by introducing electrospun carbon fiber-encapsulated nickel catalyst. For instance, the MgH_2-10 wt% Ni/Carbon fiber composite demonstrated dehydrogenation capacities of 5.79 wt% and 6.12 wt% at 280 °C and 300 °C, respectively, whereas the as-milled MgH_2 hardly decomposed at the same temperature.

Ball milling has also been used to coat alloy particles, e.g., with polymer materials. Rafatnejad et al. [55] used ball milling to coat MgH_2 powders with poly(methyl methacrylate) (PMMA). The aim here was to apply a protective coating to the alloy particles to protect them from oxidation.

If the operating temperature is moderate but above the polymer melting point, the solid-state metal powders can be utilized to form the composites without melting.

4.2. Polymer Solution Forming (Casting)

A composite of polyethylene (PE) and $LaNi_5$ alloy was prepared by dissolving pellets of low-density PE in boiling heptane [43]. A fine powder of $LaNi_5$ was added to the solution, and heptane was removed via evaporation. A final homogenous composite was prepared by melting and then shaping the resulting material into disks by hot-pressing. Checchetto et al. [43] also achieved a dispersion of Pd powder into poly(N-vinyl pyrrolidone) by dissolving the polymer into a small amount of ethanol at room temperature. The Pd powder was added to the resulting viscous mixture, and ultrasound was used to increase homogeneity. The ethanol was then evaporated under vacuum over three days.

Japan Steel Works have applied this method to produce composites with $MmNi_{4.4}Mn_{0.1}Co_{0.5}$ powder and resins [61]. The powder and resins were mixed sufficiently to uniformly distribute the alloy particles (1 mm), before being poured into a low-strength aluminium container. Microstructural analysis confirmed the alloy powder to be uniformly distributed in the composite. Watanabe et al. [62] developed cylindrical composites by solidifying a mixture of polyvinyl alcohol, starch, and $CaNi_5$ powder in a cylindrical vessel at 120–180 °C. The composite was then washed to remove starch after solidification. This resulted in a porous composite with the $CaNi_5$ bound by the resin. A thin film composite was also prepared by pouring the mixture onto a plate and solidifying at 100 °C. They noted a slight decrease in the H_2 absorption capacity of the composite materials compared to the raw alloy, but little damage to the resin was observed.

To avoid settling and/or segregation of alloy particles as the polymer cures, the polymer/alloy composite mixtures are often quite viscous. This can make filling tanks difficult, particularly through narrow openings. Japan steel works have a patented process using vibrations to assist filling MH storage tanks with a resin/alloy/carbon fiber composite that has a pre-cured consistency similar to wet sand [23]. Vibrating the tank during filling is noted to result in more even distribution of the composite within the MH tank.

4.3. Solid State Compacting

Compaction is commonly used to prepare MH composites in a pellet form. Cold compaction was found as the most practical method of forming the composite. Acceptable microstructures were fabricated using uniaxial pressing, while isostatic pressing could be used if more uniform matrix density is required or if it is able to decrease residual matrix stresses [3].

The properties of the composite such as density, porosity, and thermal conductivity [79] can be affected by the applied compaction pressure [60]. In general, the porosity will decrease when the compaction pressure increases, but it could limit the mass transfer of hydrogen [80]. There are currently commercially available MHs developed for storage applications that are supplied in the form of compacted pellets (e.g., GKN [81]).

4.4. Solid-State Cold Rolling

Cold rolling is another method for MH composite preparation and molding [82–84]. Mg-Mg$_2$Ni-carbon soot/graphite composites were prepared by cold rolling and their sorption/desorption kinetics are shown in Figure 4 [85]. The graphite composite delivered the fastest absorption kinetics (4.5 wt% in 80 s), while it was only 0.7 wt% for the Mg-Mg$_2$Ni alloy.

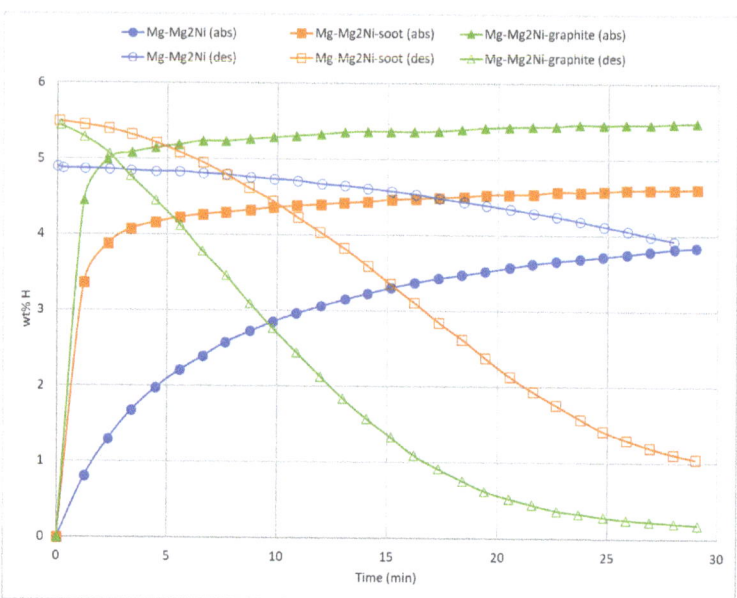

Figure 4. Hydrogenation (20 bar H$_2$ pressure, filled markers) and dehydrogenation (0.1 bar vacuum, empty markers) capacity as a function time for Mg–Mg$_2$Ni and Mg–Mg$_2$Ni–soot/graphite hybrids at 300 °C. Figure adapted from Gupta et al., International Journal of Alloys and Compounds; published by Elsevier, 2015 [85].

4.5. Hot Pressing

Hot pressing uses high pressure and temperature to form powder compacts. Pentimalli et al. [86] used a ball mill to develop a MH–polymer composite resulting in the metal particles being coated in the polymer material. The composite was consolidated by hot pressing into pellets (6 kN force for 30 min, 150–175 °C). The composite materials showed no losses in loading capacity or kinetic properties compared to the raw alloy powder. Micro cracks and channels were noted to form in the polymer matrix after cycling, allowing the easy passage of H$_2$ to the embedded alloy particles.

Selection of hot-pressing process parameters, such as temperature and pressure, is influenced by the melting or glass transition temperatures of the polymer and the desired porosity level in the formed composites. One hot-forming example conducted at 100 MPa and 165 °C [41] aimed to fabricate composites consisting of LaNi$_5$, polymer, and graphitic flakes with about 25–30% porosity, 55–60% LaNi$_5$, and 10–15% polymers and graphite.

4.6. Extrusion

Extrusion is a deformation-based manufacturing process and can produce MH composites of various forms. Watanabe et al. [62] developed a CaNi$_5$ composite with resin by extrusion. A mixture of a fine granular phenol resin, a liquid-state phenol resin, and CaNi$_5$ was extruded into a cylindrical form and solidified at 147–157 °C. It was confirmed that the metal hydride powders were well fixed in the resin matrix although the breakdown of the powders was observed after a few cycles.

Pentimalli et al. [87] combined high-energy ball milling with extrusion as a low-cost method to generate polymeric composites with LaNi$_5$-type MH. Acrylonitrile butadiene styrene (ABS) copolymer was selected due to its rheological and thermal properties that make it suitable for extrusion, its wide availability, and low cost. It does not exhibit significant gas barrier behavior and the presence of a rubber component is able to compensate for the volumetric expansion/contraction resulting from hydrogen absorption/desorption by the hydride particles. The polymer was supplied as spherical pellets and reduced to powder by centrifugal milling after embrittlement in liquid nitrogen. Graphite powder was also added to the final composite to increase thermal conductivity. The three components were blended using high-energy ball milling. The blends obtained were then extruded using a bench top compounder (110–220 °C) and cut to form pellets. This resulted in a composite with a homogeneous porous microstructure. Thermal conductivity increased from 0.2 W/m·K for the raw polymer to 2 W/m·K for the composite. PCT isotherms showed little difference between the composite and raw alloy.

5. Cyclic Stability and Composite Evolution

Cyclic stability is an important parameter for composite materials. It is imperative that the composite maintains its structural integrity during cycling. Otherwise, its properties can change significantly, and the composite may eventually break down into fine powders.

It was reported that the unsintered composites of LaNi$_{4.25}$Al$_{0.75}$ with copper could be handled and survived a 138-cycle absorption/desorption test with only minor degradation on the edges [3]. Additional vacuum sintering treatment was suggested for strengthening the copper matrix without affecting the ability of the composite to absorb hydrogen.

Figure 5 shows the hydrogen storage capacity and mass decrease in two MH sheets (70 wt% MH, 10 wt% aramid pulp, 20 wt% carbon fiber) as a function of cycle number [21]. There were prepared by wet papermaking method. For the La$_{0.6}$Y$_{0.4}$Ni$_{4.9}$Al$_{0.1}$ (LYNA) composite, the hydrogen uptake increased slowly in the first 20 cycles and a decreasing trend was observed up to 100 cycles, while the TiFe$_{0.9}$Ni$_{0.1}$ (TFN) composite maintained its sorption capacity. The mass losses were less than 1 wt% after 100 cycles for both samples, indicating the composites had a good mechanical stability.

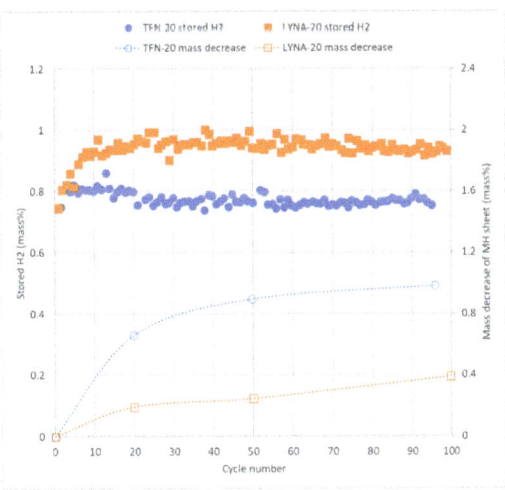

Figure 5. Hydrogen storage capacity and mass decrease in MH sheets (TFN: TiFe$_{0.9}$Ni$_{0.1}$; LYNA: La$_{0.6}$Y$_{0.4}$Ni$_{4.9}$Al$_{0.1}$) during hydrogenation cycle test at 3.8 MPa of hydrogen and was vacuumed for 1 h. Figure adapted from Yasuda et al., International Journal of Hydrogen Energy; published by Elsevier, 2013 [21].

Regarding MH/ENG composites, it was observed that lower activation temperatures require more cycles to reach full performance [58]. It was also found that the grain size of Hydralloy C5 used in the composites decreased during cyclic hydrogenation (from 108 nm to 15 nm), which is believed to be beneficial for the sorption kinetics [59].

Figure 6 shows the radial thermal conductivity of Hydralloy C5-10 wt% ENG composite throughout 1000 cycles [88]. The composite had a diameter of 13.5 mm and thickness of 6.7 mm and was compacted at 75 bar. The thermal conductivity of the composite decreased from 40 W/m·K to 12.7 W/m·K after 250 cycles. Then, it was in the range of 10–15 W/m·K up to 1000 cycles. This is much higher than that of loose powder beds [89]. The good thermal conductivity of the composite was supported by the temperature profile during the sorption process. The bed temperature increased from 50 °C to nearly 78 °C when sorption occurred, before dropping down to 50 °C in 3 min.

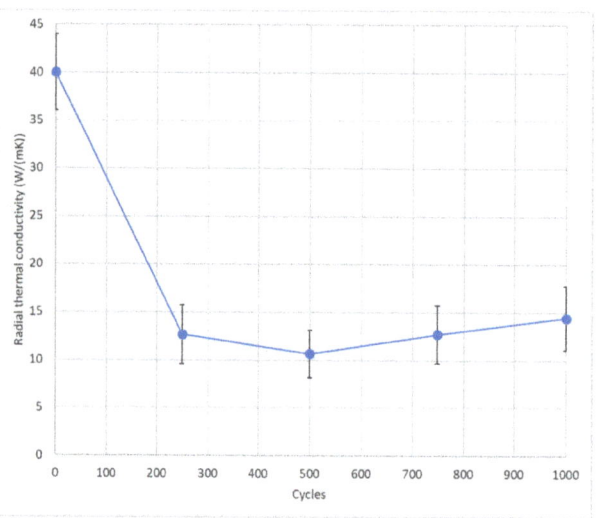

Figure 6. The radial thermal conductivity of Hydralloy C5-10 wt% ENG composite throughout 1000 cycles. Cycling temperature: 50 °C; absorption pressure: 40 bar; desorption pressure: 1 bar. Figure adapted from Dieterich et al., International Journal of Hydrogen Energy; published by Elsevier, 2015 [88].

The Japan Steel Works reported the cycling performance of $MmNi_{4.4}Mn_{0.1}Co_{0.5}$ composites with two resin materials [61]. No shape collapse of the composite materials was observed after 100 absorption and desorption cycles. As the MH powder was immobilized in high elasticity resin, cracks were also not observed in the composites after cycling.

6. Summary and Concluding Remarks

Metal alloys and intermetallic compounds are a promising option for the safe storage of H_2. Challenges relating to cyclic stability and heat transfer to/from particle beds can be improved by forming composites of the alloy powders with high thermal conductivity materials, such as ENG. Such composites can be prepared by a number of methods, often dependent on the properties of the storage alloy. It is worth noting that high MH content is favorable for the composite as additives can cause the loss of H_2 storage capacity. In general, liquid-based methods can achieve better dispersion of metal alloy, which is beneficial to the reaction kinetics, while its loading can be limited. On the other hand, powder-based methods are versatile and can be easily scaled up. However, the dispersion of metal alloy may not be as good as those produced from liquid-based methods.

Challenges related to the expansion of metal particles as they absorb H_2 and form a hydride phase can be mitigated by incorporating the metal particles into composite structures with polymers. Care needs to be taken in the selection of the polymer, however, to ensure polymer–metal surface chemistries do not impede performance. For example, if the polymer can form strong bonds with the metal surface, this bonding will decrease the contact of the hydrogen with the active metal surfaces, leading to a decrease in hydrogen absorption. With ENG additive, the applied compaction pressure should be optimized. A higher thermal conductivity can be achieved by applying a higher pressure (i.e., reduced porosity). Conversely, a denser structure may hinder H_2 Transport.

With suitable selection of conductivity enhancers and polymers, composite structures can be prepared with improved heat transfer, cycling performance, and durability.

Author Contributions: Writing—original draft preparation, L.L., A.I. and D.L.; writing – review and editing, A.C., W.T., C.C. and J.Y.; conceptualization, L.L., A.I., D.L., A.C. and L.S. All authors have read and agreed to the published version of the manuscript.

Funding: This research received no external funding.

Data Availability Statement: No new data were created or analyzed in this study. Data sharing is not applicable to this article.

Acknowledgments: Funding for this work was provided through the Hydrogen Energy Systems Future Science Platform at CSIRO.

Conflicts of Interest: The authors declare no conflict of interest.

References

1. Lototkyy, M.V.; Yartys, V.A.; Pollet, B.G.; Bowman, R.C. Metal hydride hydrogen compressors: A review. *Int. J. Hydrogen Energy* **2014**, *39*, 5818–5851. [CrossRef]
2. Züttel, A. Materials for Hydrogen Storage. *Mater. Today* **2003**, *6*, 24–33. [CrossRef]
3. Congdon, J.W. Development of Metal Hydride Composites, DOI Report WSRC-TR-92-594. 1992. Available online: https://www.osti.gov/servlets/purl/144787 (accessed on 5 February 2022).
4. Norberg, N.S.; Arthur, T.S.; Fredrick, S.J.; Prieto, A.L. Size-Dependent Hydrogen Storage Properties of Mg Nanocrystals Prepared from Solution. *J. Am. Chem. Soc.* **2011**, *133*, 10679–10681. [CrossRef]
5. Aguey-Zinsou, K.-F.; Ares-Fernández, J.-R. Synthesis of Colloidal Magnesium: A Near Room Temperature Store for Hydrogen. *Chem. Mater.* **2008**, *20*, 376–378. [CrossRef]
6. Li, W.; Li, C.; Ma, H.; Chen, J. Magnesium Nanowires: Enhanced Kinetics for Hydrogen Absorption and Desorption. *J. Am. Chem. Soc.* **2007**, *129*, 6710–6711. [CrossRef]
7. Saita, I.; Toshima, T.; Tanda, S.; Akiyama, T. Hydrogen Storage Property of MgH_2 Synthesized by Hydriding Chemical Vapor Deposition. *J. Alloys Compd.* **2007**, *446–447*, 80–83. [CrossRef]
8. Zhu, C.; Hosokai, S.; Matsumoto, I.; Akiyama, T. Shape-Controlled Growth of MgH_2/Mg Nano/Microstructures Via Hydriding Chemical Vapor Deposition. *Cryst. Growth Des.* **2010**, *10*, 5123–5128. [CrossRef]
9. Baldi, A.; Dam, B. Thin Film Metal Hydrides for Hydrogen Storage Applications. *J. Mater. Chem.* **2011**, *21*, 4021–4026. [CrossRef]
10. Kim, D.; Kim, J.B.; Lee, J.; Lee, B.J. Measurement of Effective Thermal Conductivity of $LaNi_5$ Powder Packed Bed. *Int. J. Heat Mass Transf.* **2021**, *165*, 120735. [CrossRef]
11. Hahne, E.; Kallweit, J. Thermal Conductivity of Metal Hydride Materials for Storage of Hydrogen: Experimental Investigation. *Int. J. Hydrogen Energy* **1998**, *23*, 107–114. [CrossRef]
12. Abdin, Z.; Webb, C.J.; Gray, E.M. One-Dimensional Metal-Hydride Tank Model and Simulation in Matlab–Simulink. *Int. J. Hydrogen Energy* **2018**, *43*, 5048–5067. [CrossRef]
13. Ron, M.; Gruen, D.; Mendelsohn, M.; Sheet, I. Preparation and Properties of Porous Metal Hydride Compacts. *J. Less Common Met.* **1980**, *74*, 445–448. [CrossRef]
14. Afzal, M.; Mane, R.; Sharma, P. Heat Transfer Techniques in Metal Hydride Hydrogen Storage: A Review. *Int. J. Hydrogen Energy* **2017**, *42*, 30661–30682. [CrossRef]
15. Kim, K.J.; Montoya, B.; Razani, A.; Lee, K.H. Metal Hydride Compacts of Improved Thermal Conductivity. *Int. J. Hydrogen Energy* **2001**, *26*, 609–613. [CrossRef]
16. Park, C.S.; Jung, K.; Jeong, S.U.; Kang, K.S.; Lee, Y.H.; Park, Y.S.; Park, B.H. Development of Hydrogen Storage Reactor Using Composite of Metal Hydride Materials with ENG. *Int. J. Hydrogen Energy* **2020**, *45*, 27434–27442. [CrossRef]
17. Sánchez, A.R.; Klein, H.P.; Groll, M.; Pohlmann, C.; Röntzsch, L.; Kalinichenka, S.; Hutsch, T.; Kieback, B.; Park, C.S.; Jung, K.; et al. Expanded Graphite as Heat Transfer Matrix in Metal Hydride Beds. *Int. J. Hydrogen Energy* **2003**, *28*, 515–527.

18. Pohlmann, C.; Röntzsch, L.; Kalinichenka, S.; Hutsch, T.; Kieback, B. Magnesium Alloy-Graphite Composites with Tailored Heat Conduction Properties for Hydrogen Storage Applications. *Int. J. Hydrogen Energy* **2010**, *35*, 12829–12836. [CrossRef]
19. Wu, C.; Cheng, H.-M. Effects of Carbon on Hydrogen Storage Performances of Hydrides. *J. Mater. Chem.* **2010**, *20*, 5390–5400. [CrossRef]
20. Inoue, S.; Iba, Y.; Matsumura, Y. Drastic Enhancement of Effective Thermal Conductivity of a Metal Hydride Packed Bed by Direct Synthesis of Single-Walled Carbon Nanotubes. *Int. J. Hydrogen Energy* **2012**, *37*, 1836–1841. [CrossRef]
21. Yasuda, N.; Tsuchiya, T.; Okinaka, N.; Akiyama, T. Thermal Conductivity and Cycle Characteristic of Metal Hydride Sheet Formed Using Aramid Pulp and Carbon Fiber. *Int. J. Hydrogen Energy* **2013**, *38*, 1657–1661. [CrossRef]
22. Tokiwa, T.; Iwamoto, T.; Kawaharazaki, Y.; Fujita, Y. Composition Comprising Hydrogen-Absorbing Alloy and Resin. European Patent EP 2110411 A1, 21 December 2016.
23. Kawaharazaki, Y.; Hayashi, Y. Method for Filling Hydrogen Storage Alloy. U.S. Patent US 2017/0016578 A1, 19 January 2017.
24. Kim, J.S.; Kim, D.W.; Kim, D.O.; Ahn, G.R.; JeaSung, P.; Jeon, H.J.; Ihm, J.; Cha, M.-H. Scaffold Materials-Transition Metal Hydride Complexes, Intermediates Therefor and Method for Preparing the Same. WO Patent WO2010085108 A2, 29 July 2010.
25. Gross, A.F.; Vajo, J.J.; Van Atta, S.L.; Olson, G.L. Enhanced Hydrogen Storage Kinetics of LiBH$_4$ in Nanoporous Carbon Scaffolds. *J. Phys. Chem. C* **2008**, *112*, 5651–5657. [CrossRef]
26. Yu, X.; Tang, Z.; Sun, D.; Ouyang, L.; Zhu, M. Recent Advances and Remaining Challenges of Nanostructured Materials for Hydrogen Storage Applications. *Prog. Mater. Sci.* **2017**, *88*, 1–48. [CrossRef]
27. Schneemann, A.; White, J.L.; Kang, S.; Jeong, S.; Wan, L.F.; Cho, E.S.; Heo, T.W.; Prendergast, D.; Urban, J.J.; Wood, B.C.; et al. Nanostructured Metal Hydrides for Hydrogen Storage. *Chem. Rev.* **2018**, *118*, 10775–10839. [CrossRef]
28. Ron, M.; Bershadsky, E.; Josephy, Y. Thermal Conductivity of PMH Compacts, Measurements and Evaluation. *Int. J. Hydrogen Energy* **1992**, *17*, 623–630. [CrossRef]
29. Kane, M. *Permeability, Solubility, and Interaction of Hydrogen in Polymers—An Assessment of Materials for Hydrogen Transport*; No. WSRC-STI-2008-00009; Washington Savannah River Company: Washington, DC, USA, 2008.
30. Swapna, V.P.; Abhisha, V.S.; Stephen, R. Polymer/Polyhedral Oligomeric Silsesquioxane Nanocomposite Membranes for Pervaporation. In *Polymer Nanocomposite Membranes for Pervaporation*; Elsevier: Amsterdam, The Netherlands, 2020. [CrossRef]
31. Sweed, M. Free Volume Properties of Semi-Crystalline Polymers. PhD. Thesis, University of Stellenbosch, Stellenbosch, South Africa, 2011.
32. Mostafa, N.; Ali, E.H.; Mohsen, M. Dynamic Study of Free Volume Properties in Polyethylene/Styrene Butadiene Rubber Blends by Positron Annihilation Lifetime Method. *J. Appl. Polym. Sci.* **2009**, *113*, 3228–3235. [CrossRef]
33. Petzetakis, N.; Doherty, C.M.; Thornton, A.W.; Chen, X.; Cotanda, P.; Hill, A.J.; Balsara, N.P. Membranes with Artificial Free-Volume for Biofuel Production. *Nat. Commun.* **2015**, *6*, 7529. [CrossRef] [PubMed]
34. Li, D.; Zhou, L.; Wang, X.; He, L.; Yang, X. Effect of Crystallinity of Polyethylene with Different Densities on Breakdown Strength and Conductance Property. *Materials* **2019**, *12*, 1746. [CrossRef] [PubMed]
35. Fujiwara, H.; Ono, H.; Ohyama, K.; Kasai, M.; Kaneko, F.; Nishimura, S. Hydrogen Permeation under High Pressure Conditions and the Destruction of Exposed Polyethylene-Property of Polymeric Materials for High-Pressure Hydrogen Devices (2)-. *Int. J. Hydrogen Energy* **2021**, *46*, 11832–11848. [CrossRef]
36. Mascia, L.; Kouparitsas, Y.; Nocita, D.; Bao, X. Antiplasticization of Polymer Materials: Structural Aspects and Effects on Mechanical and Diffusion-Controlled Properties. *Polymers* **2020**, *12*, 769. [CrossRef]
37. Weigelt, F.; Georgopanos, P.; Shishatskiy, S.; Filiz, V.; Brinkmann, T.; Abetz, V. Development and Characterization of Defect-Free Matrimid® Mixed-Matrix Membranes Containing Activated Carbon Particles for Gas Separation. *Polymers* **2018**, *10*, 51. [CrossRef]
38. Checchetto, R.; Carotenuto, G.; Bazzanella, N.; Miotello, A. Synthesis and Characterization of Polymer Embedded LaNi$_5$ Composite Material for Hydrogen Storage. *J. Phys. D Appl. Phys.* **2007**, *40*, 4043. [CrossRef]
39. Gahleitner, M.; Paulik, C. *Polypropylene, Ullmann's Encyclopedia of Industrial Chemistry*; Wiley-VCH: Weinheim, Germany, 2014; pp. 1–44. [CrossRef]
40. Cao, R.H.; Wang, T.; Cao, Z.Z.; Cao, J.L.; Cao, J.X. Microenvironment Hydrogen Supplement Breathable Layer and External Application Bag. China Patent CN111481425 A, 4 August 2020.
41. Zimmermann, J. Composite Hydrogen Storage Material and Methods Related Thereto. Europe Patent EP1874679 A1, 6 June 2013.
42. Tsaur, G.; Wang, T.H.; Tsaur, F.; Tsaur, N.; Tsaur, E. Microenvironment Hydrogen-Supplying Breathable Layer and Applications Thereof. U.S. Patent US1096959 B2, 30 July 2020.
43. Checchetto, R.; Bazzanella, N.; Miotello, A.; Carotenuto, G.; Nicolais, L. Hydrogen Sorption in Metal-Polymer Composites: The Role of Interfaces. *J. Appl. Phys.* **2009**, *105*, 83513. [CrossRef]
44. Bennington, S.; McGrady, S.; Royse, D.; Lovell, A.; Mooring, L.; Headen, T. A Hydrogen Storage Pellet. Europe Patent EP2935092 A1, 21 December 2012.
45. Polymer Database. Available online: https://www.polymerdatabase.com/ (accessed on 10 December 2022).
46. Allen, G.; Bevington, J.C. (Eds.) *Comprehensive Polymer Science and Supplements*; Elsevier: Amsterdam, The Netherlands, 1996; pp. 175–189. ISBN 978-0-08-096701-1.
47. Merkel, T.C.; Bondar, V.I.; Nagai, K.; Freeman, B.D.; Pinnau, I. Gas Sorption, Diffusion, and Permeation in Poly (Dimethylsiloxane). *J. Polym. Sci. Part B Polym. Phys.* **2000**, *38*, 415–434. [CrossRef]

48. Komatsuka, T.; Nagai, K. Temperature Dependence on Gas Permeability and Permselectivity of Poly (Lactic Acid) Blend Membranes. *Polym. J.* **2009**, *41*, 455–458. [CrossRef]
49. Kanesugi, H.; Ohyama, K.; Fujiwara, H.; Nishimura, S. High-Pressure Hydrogen Permeability Model for Crystalline Polymers. *Int. J. Hydrogen Energy* **2023**, *48*, 723–739. [CrossRef]
50. Sivasubramanian, P.; Pariyadath, N. Medium for the Stabilisation and Utility of Volatile or Liquid Hydrides of Boron. U.S. Patent US 8802769 B2, 12 August 2014.
51. Cao, H.; Georgopanos, P.; Capurso, G.; Pistidda, C.; Weigelt, F.; Chaudhary, A.-L.; Filiz, V.; Tseng, J.-C.; Wharmby, M.T.; Dornheim, M.; et al. Air-stable metal hydride-polymer composites of $Mg(NH_2)_2$—LiH and TPX™. *Mater. Today Energy* **2018**, *10*, 98–107. [CrossRef]
52. Le, T.T.; Pistidda, C.; Abetz, C.; Georgopanos, P.; Garroni, S.; Capurso, G.; Milanese, C.; Puszkiel, J.; Dornheim, M.; Abetz, V. Enhanced Stability of Li-RHC Embedded in an Adaptive TPX™ Polymer Scaffold. *Materials* **2020**, *13*, 991. [CrossRef]
53. Hearley, A.K.; Redmond, S.D. Solid-State Hydrogen Storage Systems. U.S. Patent US20040213998 A1, 28 October 2004.
54. Kim, K.C.; Dai, B.; Johnson, J.K.; Sholl, D.S. Assessing Nanoparticle Size Effects on Metal Hydride Thermodynamics Using the Wulff Construction. *Nanotechnology* **2009**, *20*, 204001. [CrossRef]
55. Rafatnejad, M.; Raygan, S.; Sefidmooy Azar, M. Investigation of Dehydrogenation Performance and Air Stability of MgH_2–PMMA Nanostructured Composite Prepared by Direct High-Energy Ball-Milling. *Mater. Renew. Sustain. Energy* **2020**, *9*, 14. [CrossRef]
56. Wilkes, C.E.; James, J.W.; Daniel, C.A.; Berard, M.T. (Eds.) *PVC Handbook*; Hanser Verlag: Munich, Germany, 2005; ISBN 978-1-56990-379-7.
57. Matsuda, T.; Shimomura, T.; Hirami, M. Crystallization and Melting of Binary Mixtures of Nylon 6 and Nylon 66. A Study by DSC. *Polym. J.* **1999**, *31*, 795–800. [CrossRef]
58. Klein, H.-P.; Groll, M. Heat Transfer Characteristics of Expanded Graphite Matrices in Metal Hydride Beds. *Int. J. Hydrogen Energy* **2004**, *29*, 1503–1511. [CrossRef]
59. Pohlmann, C.; Röntzsch, L.; Heubner, F.; Weißgärber, T.; Kieback, B. Solid-State Hydrogen Storage in Hydralloy-Graphite Composites. *J. Power Sources* **2013**, *231*, 97–105. [CrossRef]
60. Yan, M.Y.; Sun, F.; Liu, X.P.; Ye, J.H. Effects of Compaction Pressure and Graphite Content on Hydrogen Storage Properties of $Mg(NH_2)_2$-2LiH Hydride. *Int. J. Hydrogen Energy* **2014**, *39*, 19656–19661. [CrossRef]
61. Kubo, K.; Kawaharazaki, Y.; Itoh, H. Development of Large MH Tank System for Renewable Energy Storage. *Int. J. Hydrogen Energy* **2017**, *42*, 22475–22479. [CrossRef]
62. Watanabe, F.; Ikai, C.; Hasatanti, M.; Marumo, C. Hydration Characteristics of Metal Hydride Fixed in Resin Form. *J. Chem. Eng. Japan* **1992**, *25*, 1–5. [CrossRef]
63. Sitthiwet, C.; Thiangviriya, S.; Thaweelap, N.; Meethom, S.; Kaewsuwan, D.; Chanlek, N.; Utke, R. Hydrogen Sorption and Permeability of Compacted $LiBH_4$ Nanoconfined into Activated Carbon Nanofibers Impregnated with TiO_2. *J. Phys. Chem. Solids* **2017**, *110*, 344–353. [CrossRef]
64. De Jongh, P.E.; Eggenhuisen, T.M. Melt Infiltration: An Emerging Technique for the Preparation of Novel Functional Nanostructured Materials. *Adv. Mater.* **2013**, *25*, 6672–6690. [CrossRef]
65. Adelhelm, P.; De Jongh, P.E. The Impact of Carbon Materials on the Hydrogen Storage Properties of Light Metal Hydrides. *J. Mater. Chem.* **2011**, *21*, 2417–2427. [CrossRef]
66. De Jongh, P.E.; Wagemans, R.W.P.; Eggenhuisen, T.M.; Dauvillier, B.S.; Radstake, P.B.; Meeldijk, J.D.; Geus, J.W.; de Jong, K.P. The Preparation of Carbon-Supported Magnesium Nanoparticles Using Melt Infiltration. *Chem. Mater.* **2007**, *19*, 6052–6057. [CrossRef]
67. Cahen, S.; Eymery, J.-B.; Janot, R.; Tarascon, J.-M. Improvement of the $LiBH_4$ Hydrogen Desorption by Inclusion into Mesoporous Carbons. *J. Power Sources* **2009**, *189*, 902–908. [CrossRef]
68. Zhao-Karger, Z.; Hu, J.; Roth, A.; Wang, D.; Kübel, C.; Lohstroh, W.; Fichtner, M. Altered Thermodynamic and Kinetic Properties of MgH_2 Infiltrated in Microporous Scaffold. *Chem. Commun.* **2010**, *46*, 8353–8355. [CrossRef] [PubMed]
69. Xia, G.; Tan, Y.; Chen, X.; Sun, D.; Guo, Z.; Liu, H.; Ouyang, L.; Zhu, M.; Yu, X. Monodisperse Magnesium Hydride Nanoparticles Uniformly Self-Assembled on Graphene. *Adv. Mater.* **2015**, *27*, 5981–5988. [CrossRef] [PubMed]
70. Chong, L.; Zeng, X.; Ding, W.; Liu, D.J.; Zou, J. $NaBH_4$ in "Graphene Wrapper:" Significantly Enhanced Hydrogen Storage Capacity and Regenerability through Nanoencapsulation. *Adv. Mater.* **2015**, *27*, 5070–5074. [CrossRef]
71. Borzenko, V.I.; Romanov, I.A.; Dunikov, D.O.; Kazakov, A.N. Hydrogen Sorption Properties of Metal Hydride Beds: Effect of Internal Stresses Caused by Reactor Geometry. *Int. J. Hydrogen Energy* **2019**, *44*, 6086–6092. [CrossRef]
72. Sakintuna, B.; Lamari-Darkrim, F.; Hirscher, M. Metal Hydride Materials for Solid Hydrogen Storage: A Review. *Int. J. Hydrogen Energy* **2007**, *32*, 1121–1140. [CrossRef]
73. Guo, F.; Namba, K.; Miyaoka, H.; Jain, A.; Ichikawa, T. Hydrogen Storage Behavior of TiFe Alloy Activated by Different Methods. *Mater. Lett. X* **2021**, *9*, 100061. [CrossRef]
74. Alsabawi, K.; Webb, T.A.; Gray, E.M.; Webb, C.J. The Effect of C60 Additive on Magnesium Hydride for Hydrogen Storage. *Int. J. Hydrogen Energy* **2015**, *40*, 10508–10515. [CrossRef]
75. Rud, A.D.; Lakhnik, A.M. Effect of Carbon Allotropes on the Structure and Hydrogen Sorption during Reactive Ball-Milling of Mg-C Powder Mixtures. *Int. J. Hydrogen Energy* **2011**, *37*, 4179–4187. [CrossRef]
76. Jia, Y.; Guo, Y.; Zou, J.; Yao, X. Hydrogenation/Dehydrogenation in MgH_2-Activated Carbon Composites Prepared by Ball Milling. *Int. J. Hydrogen Energy* **2012**, *37*, 7579–7585. [CrossRef]

77. Jung, K.S.; Lee, E.Y.; Lee, K.S. Catalytic Effects of Metal Oxide on Hydrogen Absorption of Magnesium Metal Hydride. *J. Alloys Compd.* **2006**, *421*, 179–184. [CrossRef]
78. Meng, Q.; Huang, Y.; Ye, J.; Xia, G.; Wang, G.; Dong, L.; Yang, Z.; Yu, X. Electrospun Carbon Nanofibers with In-Situ Encapsulated Ni Nanoparticles as Catalyst for Enhanced Hydrogen Storage of MgH$_2$. *J. Alloys Compd.* **2021**, *851*, 156874. [CrossRef]
79. Zhao, W.; Yang, Y.; Bao, Z.; Yan, D.; Zhu, Z. Methods for Measuring the Effective Thermal Conductivity of Metal Hydride Beds: A Review. *Int. J. Hydrogen Energy* **2020**, *45*, 6680–6700. [CrossRef]
80. Yan, M.; Sun, F.; Liu, X.; Ye, J.; Wang, S.; Jiang, L. Effects of Graphite Content and Compaction Pressure on Hydrogen Desorption Properties of Mg(NH$_2$)$_2$–2LiH Based Tank. *J. Alloys Compd.* **2015**, *628*, 63–67. [CrossRef]
81. Simanullang, M.; Prost, L. Nanomaterials for on-board solid-state hydrogen storage applications. *Int. J. Hydrogen Energy* **2022**, *47*, 29808–29846. [CrossRef]
82. De Lima Andreani, G.F.; Triques, M.R.M.; Kiminami, C.S.; Botta, W.J.; Roche, V.; Jorge, A.M. Characterization of Hydrogen Storage Properties of Mg-Fe-CNT Composites Prepared by Ball Milling, Hot-Extrusion and Severe Plastic Deformation Methods. *Int. J. Hydrogen Energy* **2016**, *41*, 23092–23098. [CrossRef]
83. Huot, J.; Tousignant, M. Effect of Cold Rolling on Metal Hydrides. *Mater. Trans.* **2019**, *60*, 1571–1576. [CrossRef]
84. Lang, J.; Huot, J. A New Approach to the Processing of Metal Hydrides. *J. Alloys Compd.* **2011**, *509*, 18–22. [CrossRef]
85. Gupta, A.; Shervani, S.; Faisal, M.; Balani, K.; Subramaniam, A. Hydrogen Storage in Mg-Mg$_2$Ni-Carbon Hybrids. *J. Alloys Compd.* **2015**, *645*, S397–S399. [CrossRef]
86. Pentimalli, M.; Padella, F.; La Barbera, A.; Pilloni, L.; Imperi, E. A Metal Hydride–Polymer Composite for Hydrogen Storage Applications. *Energy Convers. Manag.* **2009**, *50*, 3140–3146. [CrossRef]
87. Pentimalli, M.; Imperi, E.; Zaccagnini, A.; Padella, F. Nanostructured Metal Hydride–Polymer Composite as Fixed Bed for Sorption Technologies. Advantages of an Innovative Combined Approach by High-Energy Ball Milling and Extrusion Techniques. *Renew. Energy* **2017**, *110*, 69–78. [CrossRef]
88. Dieterich, M.; Pohlmann, C.; Bürger, I.; Linder, M.; Röntzsch, L. Long-Term Cycle Stability of Metal Hydride-Graphite Composites. *Int. J. Hydrogen Energy* **2015**, *40*, 16375–16382. [CrossRef]
89. Matsushita, M.; Tajima, I.; Abe, M.; Tokuyama, H. Experimental Study of Porosity and Effective Thermal Conductivity in Packed Bed of Nano-Structured FeTi for Usage in Hydrogen Storage Tanks. *Int. J. Hydrogen Energy* **2019**, *44*, 23239–23248. [CrossRef]

Disclaimer/Publisher's Note: The statements, opinions and data contained in all publications are solely those of the individual author(s) and contributor(s) and not of MDPI and/or the editor(s). MDPI and/or the editor(s) disclaim responsibility for any injury to people or property resulting from any ideas, methods, instructions or products referred to in the content.

Article

The Role of Bulk Stiffening in Reducing the Critical Temperature of the Metal-to-Hydride Phase Transition and the Hydride Stability: The Case of $Zr(Mo_xFe_{1-x})_2$-H_2

Isaac Jacob [1,*], Dotan Babai [2,†], Matvey Bereznitsky [1,2] and Roni Z. Shneck [2]

[1] Unit of Nuclear Engineering, Ben Gurion University of the Negev, Beer Sheva 84105, Israel; matveyb@bgu.ac.il
[2] Department of Materials Engineering, Ben Gurion University of the Negev, Beer Sheva 84105, Israel; babaidot@gmail.com (D.B.); roni@bgu.ac.il (R.Z.S.)
* Correspondence: izi@bgu.ac.il
† Current address: Faculty of Engineering and Bar-Ilan Institute for Nanotechnology and Advanced Materials, Bar-Ilan University, Ramat-Gan 52900, Israel.

Abstract: This study aims to shed light on the unusual trend in the stabilities of $Zr(Mo_xFe_{1-x})_2$, $0 \leq x \leq 1$, hydrides. Both the rule of reversed stability and the crystal volume criterion correlate with the increased hydride stabilities from x = 0 to x = 0.5, but are in contrast with the destabilization of the end member $ZrMo_2$ hydride. The pressure-composition isotherms of $ZrMo_2$-H_2 exhibit very wide solid solubility regions, which may be associated with diminished H–H elastic interaction, u_{elas}. In order to discern this possibility, we measured the elastic moduli of $Zr(Mo_xFe_{1-x})_2$, x = 0, 0.5, 1. The shear modulus, G, shows a moderate variation in this composition range, while the bulk modulus, B, increases significantly and monotonically from 148.2 GPa in $ZrFe_2$ to 200.4 GPa in $ZrMo_2$. The H–H elastic interaction is proportional to B and therefore its increase cannot directly account for a decrease in u_{elas}. Therefore, we turn our attention to the volume of the hydrogen atom, v_H, which usually varies in a limited range. Two coexisting phases, a Laves cubic (a = 7.826 Å) and a tetragonal (a = 5.603 Å, c = 8.081 Å) hydride phase are identified in $ZrMo_2H_{3.5}$, obtained by cooling to liquid nitrogen temperature at about 50 atm. The volume of the hydrogen atom in these two hydrides is estimated to be 2.2 Å^3/(H atom). Some very low v_H values, have been reported by other investigators. The low v_H values, as well as the one derived in this work, significantly reduce u_{elas} for $ZrMo_2$-H_2, and thus reduce the corresponding critical temperature for the metal-to-hydride phase transition, and the heat of hydride formation. We suggest that the bulk stiffening in $ZrMo_2$ confines the corresponding hydride expansion and thus reduces the H–H elastic interaction.

Keywords: intermetallic hydrides; hydrogen atomic volume; $Zr(Mo_xFe_{1-x})_2$ pseudobinary intermetallics; critical temperature of metal to hydride phase transition; hydrogen–hydrogen elastic interaction; elastic moduli

Citation: Jacob, I.; Babai, D.; Bereznitsky, M.; Shneck, R.Z. The Role of Bulk Stiffening in Reducing the Critical Temperature of the Metal-to-Hydride Phase Transition and the Hydride Stability: The Case of $Zr(Mo_xFe_{1-x})_2$-H_2. *Inorganics* **2023**, *11*, 228. https://doi.org/10.3390/inorganics11060228

Academic Editor: Duncan H. Gregory

Received: 17 April 2023
Revised: 20 May 2023
Accepted: 23 May 2023
Published: 25 May 2023

Copyright: © 2023 by the authors. Licensee MDPI, Basel, Switzerland. This article is an open access article distributed under the terms and conditions of the Creative Commons Attribution (CC BY) license (https://creativecommons.org/licenses/by/4.0/).

1. Introduction

The present study demonstrates a significant impact of the bulk modulus on the hydride formation of the intermetallic compounds $ZrMo_2$ and TaV_2. It would be instructive to consider at the outset some general factors determining the hydrogenation behavior of metal alloys. It is a well-known empirical rule that hydrogen-absorbing alloys or intermetallic compounds must contain at least one elementary metal forming a reversible hydride at nearly ambient conditions, e.g., Ref. [1]. The ability of the compounds to absorb hydrogen is thus determined by the metal–hydrogen interaction. In addition, the properties of the parent intermetallics provide some guidelines regarding the stability of the corresponding hydrides: (i) the rule of reversed stability—the more stable the original intermetallic, the less stable the resulting hydride [2]; (ii) the crystal volume criterion—the hydride stability increases with the increase in the original crystal volume [3]; (iii) the crystal structure—sometimes alloys of identical or nearly identical chemical compositions crystallize in a completely different lattice structure. The one, presenting interstitial sites more abundant in hydride-forming elements, forms more stable hydrides. For example, the pseudobinary $U(Al_xNi_{1-x})_2$ system crystallizes in two different structures, a Laves phase structure in the compositional ranges $0 \leq x \leq 0.2$, $0.9 \leq x \leq 1$, and a ZrNiAl-type (Fe_2P-type) structure at $0.4 \leq x \leq 0.5$. The Laves phase compounds do not absorb hydrogen up to pressures of 100 atm, while the hydrides $UNiAlH_{2.5}$ and $UNi_{1.2}Al_{0.8}H_{2.35}$ are formed at that pressure range. This difference of hydrogen absorption is attributed to the presence of uranium-richer tetrahedral sites (3U + 1(Ni,Al)) in $U(Al_xNi_{1-x})_2$, $x = 0.4, 0.5$, as compared to the (2U + 1(Ni,Al)) sites in the Laves-phase compounds, where uranium is the hydride-forming element [4,5]; and (iv) we will consider in some detail the influence of the bulk elastic modulus on the critical temperature, T_c.

The metal to hydride phase transition in most hydrogen-absorbing intermetallic compounds is characterized by a critical temperature, T_c. A discontinuous transformation of the crystal lattice occurs below T_c, while above it the final hydride structure may be reached in a gradual, continuous way. This behavior imposes identical or very similar crystal structures of the original intermetallic compound and its corresponding hydride, besides the usual increase in the lattice volume upon hydrogenation. The metal to hydride phase transformation is driven by attractive hydrogen–hydrogen interaction, $-u$, which is composed of elastic and electronic parts [6].

$$u = u_{elas} + u_{elec}$$

The critical temperature, T_c, is related to the attractive H–H interaction:

$$T_c = u \times r/4k \tag{1}$$

r is the number of interstitial sites per metal atom available for hydrogen absorption and k is Boltzmann's constant (1.38×10^{-23} J K^{-1}). The corresponding units of u are J.

The form of the elastic part, u_{elas}, has been derived by several investigators, e.g., Refs. [6–8]:

$$u_{elas} = \left(1 - \frac{B}{B + \frac{4}{3}G}\right) B \frac{v_H^2}{V} \tag{2}$$

B and G are the bulk and shear moduli in Pa (N/m^2), respectively, v_H is the excess volume occupied by one hydrogen atom in the metal matrix in m^3, and V is the average volume per metal atom in m^3. u_{elas} is then obtained in J. u_{elas} is usually presented in eV by multiplying the value in J by the conversion factor 1 J = 6.242×10^{18} eV. The volume, v_H, of the hydrogen atom in the metal matrix and the term in the brackets usually do not change very much. Typical values of v_H stay in the range of 2.5–3 Å^3/H atom, e.g., Ref. [6]. The bulk modulus, B, is then a leading term in u_{elas}. The physical reasoning for the increase in the long-range u_{elas} interaction with B is that larger B induces larger strain fields in the crystal lattice around a hydrogen atom and provides better linkage between the hydrogen atoms. The attractive elastic interaction is considered to play a main role in the metal-to-hydride phase transition. Fen Li et al. have proposed a repulsive screened Coulomb interaction between the hydrogen atoms in $TiCr_2H_x$ [9]. It was recently demonstrated that the electronic interaction may be repulsive, negligible, or attractive by considering the corresponding critical temperatures of the $ZrCr_2$-H_2 (<300 K), $ZrMn_2$-H_2 (564 K), and Pd-H_2 (565 K) systems in relation to their relevant elastic properties, Equations (1) and (2) [10].

This study aims at elucidating a puzzling behavior of the hydride stabilities in the $Zr(Mo_xFe_{1-x})_2$ system, as presented at MH2018 and subsequently published [11]. Figure 1 [11] exhibits pressure-composition isotherms, demonstrating increased monotonic stability of the $Zr(Mo_xFe_{1-x})_2$ hydrides for $x = 0, 0.2, 0.5$, but then this trend is reversed by a stability decrease for $x = 1$. Investigating these stability trends of the $Zr(Mo_xFe_{1-x})_2$ hydrides, in view of the guidelines listed above, requires a knowledge of relevant data, namely, the heats of formation of the original intermetallic compounds, their crystal structures, lattice parameters, bulk and shear moduli, and the degree of expansion upon hydrogenation which determines v_H. Most of these data appear in the literature. In order to complete the required set, elastic properties have been determined in this work. In addition, we hydrogenated $ZrMo_2$ to the greatest possible extent under the available experimental conditions in our lab, attempting to settle some differing crystallographic results, associated with the volume of the hydrogen atom in $ZrMo_2$ hydride(s).

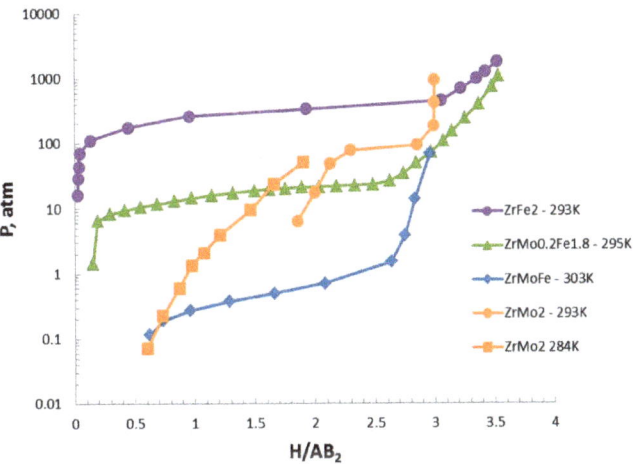

Figure 1. Pressure-composition isotherms of $Zr(Mo_xFe_{1-x})_2$-H_2 systems, x = 0, 0.1. 0.5, 1 at the indicated temperatures. (Reprinted with permission from Ref. [11] under license 5531441401625 by Elsevier).

2. Experimental Details

$Zr(Mo_xFe_{1-x})_2$, x = 0, 0.5, 1, samples were prepared by arc melting high-purity (~99.9%) elements on a water-cooled copper hearth. The pellets were homogenized by melting them at least three times, turning them over after each melting. The as-cast alloys were then sealed in argon-filled quartz ampules and annealed for 72 h at 1373 K. The crystal structure of the prepared compounds was determined and verified by X-ray diffraction (XRD) in a PW 1050/70 diffractometer from Philips, utilizing Cu K_α radiation of 1.5418 Å. Parallel-faced intermetallic slices of about 2.5 mm thickness were cut using a diamond sawing disk and subsequently polished to achieve a parallelism within the range of 3×10^{-6} m. The resulting parallel-faced samples were then subjected to sound velocity measurements using the pulse echo (PE) method. The frequency of the PE piezoelectric transducers for both the longitudinal and transverse modes was 5 MHz. The sound wave velocities were determined by measuring the time needed for the acoustic waves to travel forth and back between the opposite parallel faces of the sample. The densities of the samples were analyzed using the measured structural parameters. The obtained results were then compared to the measured Archimedes densities (both wet and dry) in order to determine the porosity ratio of the samples.

$ZrMo_2$ compound was hydrogenated at about 60 atm hydrogen pressure and room temperature in a home-built Sieverts system, equipped with 3.5, 60, and 100 atm pressure transducers. The amount of absorbed hydrogen in $ZrMo_2$ was calculated by considering the non-ideality of the hydrogen gas. For example, the difference between the real and the ideal hydrogen pressures at 100 atm and room temperature is about 6%. The reactor chamber, with the hydrogenated sample in it, was then very slowly cooled from room to liquid nitrogen temperature in order to enhance the hydrogen absorption. The equilibrium hydrogenation pressure decreases in an exponential way vs. the decreased temperature. Thus, decreasing the sample temperature at approximately constant pressure, as most of the system is kept at ambient conditions, corresponds to significantly increasing the pressure at room temperature. Hydrogen absorption is expected to be practically none at very low temperatures, e.g., 78 K, because of extremely reduced kinetics. The very slow cooling is then intended to enable hydrogen absorption at intermediate temperatures, between 295 K and 78 K. Finally, a poisoning procedure was applied. (The purpose of the poisoning procedure is a contamination of the surface of the sample, for example by water vapor, without changing its bulk content. This contamination keeps the hydride intact and prevents its decomposition for long enough time in order to enable some experimental measurements at ambient conditions, e.g., XRD.) The hydrogen was evacuated from the reaction chamber, kept at liquid nitrogen temperature, and then the sample was exposed to the ambient atmosphere. The X-ray diffraction pattern of the air-exposed sample was immediately recorded without any additional precautions. The poisoning is considered successful provided two conditions are fulfilled: (i) the XRD pattern of the hydrogenated sample is significantly different from the pattern of the original compound; and (ii) the XRD pattern of the hydrogenated sample does not change or changes very slowly with time. These two conditions were satisfied in the present case of $ZrMo_2$—see Section 3.1 below.

3. Results and Discussion

3.1. Structural and Elastic Properties of $Zr(Mo_xFe_{1-x})_2$, $x = 0, 0.5, 1$

$ZrFe_2$ and $Zr(Mo_{0.5}Fe_{0.5})_2$ exhibited single Laves phases in their XRD patterns, while 95% Laves phase was obtained for $ZrMo_2$ (Figure 2). $ZrFe_2$ and $ZrMo_2$ crystallized in the cubic C15 phase with lattice constants 7.074 Å and 7.589 Å, respectively. These values are in fair agreement with the published data for $ZrFe_2$ (e.g., 7.074 Å [12]) and $ZrMo_2$ (e.g., 7.59 Å [13]). About 5% of Mo, with lattice constant a = 3.185 Å, was present along the $ZrMo_2$ pattern (Figure 2). The cubic crystal parameter of pure Mo is 3.147 Å [14]. The increased a value indicates the dissolution of some Zr into the molybdenum matrix. $Zr(Mo_{0.5}Fe_{0.5})_2$ presents a C14 hexagonal phase with lattice constants a = 5.173 Å and c = 8.461 Å, comparable to a = 5.172 Å and c = 8.463 Å [15] or a = 5.130 Å and c = 8.440 Å [16]. The longitudinal, v_L, and transverse, v_T, ultrasonic velocities, measured by the pulse-echo method, are presented in Table 1. They were used to calculate two elastic constants according to the relations $C_{11} = v_L^2 \rho$ and $C_{44} = v_T^2 \rho$, assumed to hold for polycrystalline isotropic samples (ρ is the sample density). Bulk, B, and shear, G, moduli were derived from the relations $B = C_{11} - 4 C_{44}/3$ and $G = C_{44}$ (Table 1). Minor corrections of B and G were made in view of the small porosities of the samples [17,18]. The shear modulus changes moderately with the composition, x, along the $Zr(Mo_xFe_{1-x})_2$ series, while the bulk modulus increases significantly and monotonically as a function of x. The variations of B and G are exhibited in Figure 3 and Table 1.

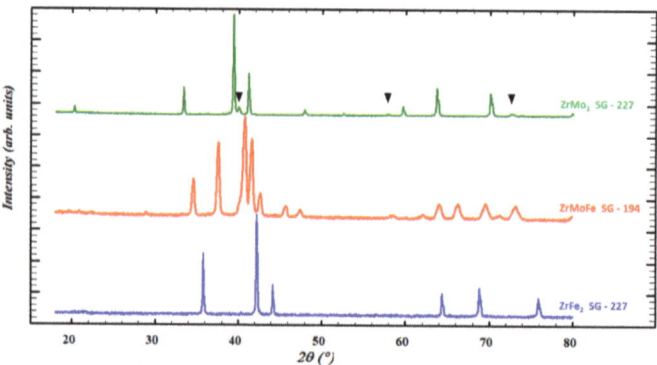

Figure 2. XRD patterns of $Zr(Mo_xFe_{1-x})_2$, $x = 0, 0.5, 1$. Single-phased cubic C15 and hexagonal C14 Laves structures were found for $ZrFe_2$ (blue pattern) and ZrMoFe (red pattern), respectively. For $ZrMo_2$, 95% C15 structure was found with 5% Mo (▼) (green pattern).

Table 1. Lattice constants, theoretical densities, porosities, longitudinal, v_L, and transverse, v_T, ultrasonic velocities, shear, G, and bulk, B, elastic moduli in the Laves phase $Zr(Mo_xFe_{1-x})_2$ system. Experimental B and G values of $Zr(Al_{0.04}Fe_{0.96})_2$ and theoretical B and G values of $ZrMo_2$ are also presented. The numbers in the parentheses present the estimated errors of the last significant digits.

x	a [Å]	c [Å]	Theoretical Density [kg/m³]	Porosity [%]	v_L [m/s]	v_T [m/s]	G [GPa]	B [GPa]
0 * [19]							75.8	148
0	7.074		7614	0.9	5674 (9)	3142 (4)	75.6 (3)	148.2 (9)
0.5	5.173	8.461	8297	0.8	5337 (2)	2608 (1)	56.3 (2)	164.2 (4)
1	7.589		8605	0.7	5733 (16)	2762 (14)	66.1 (8)	200.4 (2)
1 [20]	7.594						57.3	196.5

* $Zr(Al_{0.04}Fe_{0.96})_2$.

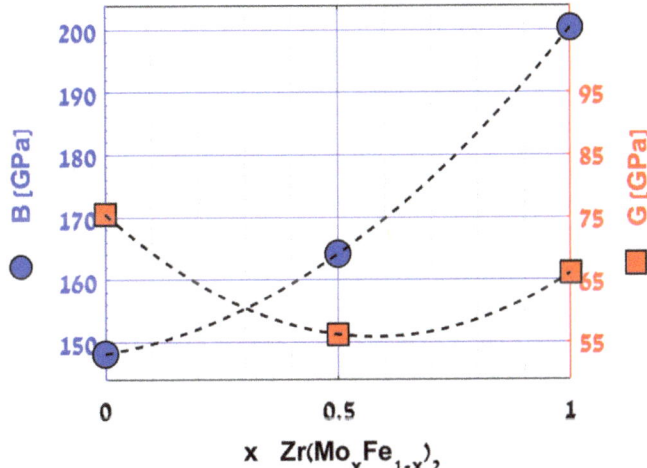

Figure 3. Bulk, B, and shear, G, moduli in $Zr(Mo_xFe_{1-x})_2$ intermetallic compounds.

$ZrMo_2$ absorbed about 2.1 H atoms per formula unit at 60 atm and room temperature, and about 3.5 H atoms per f.u. upon slow cooling to 78 K at 50 atm pressure. It should be born in mind that the hydrogenations of $ZrMo_2$ were carried out in order to estimate the corresponding v_H values. PCT (pressure-composition-temperature) isotherms of $ZrMo_2$-H_2 may be found elsewhere, e.g., Ref. [11]. The XRD pattern, recorded immediately after removing the sample from the reactor, revealed two hydride phases, cubic Laves structure with lattice constant a = 7.826 Å (26%) and tetragonal structure, space group 88, with cell parameters a = 5.603 Å, c = 8.081 Å (68%) (Figure 4). These lattice constants slowly decreased during one week of exposure to the ambient atmosphere. The crystal parameter of Mo with dissolved Zr almost did not change (a = 3.187 Å) upon hydrogenation, indicating no hydrogen absorption in it. By imposing somewhat arbitrarily identical v_H for the two existing phases in $ZrMo_2H_{3.5}$, we obtain for the composition of these two phases $ZrMo_2H_{2.39}$ ($MgCu_2$-type, a = 7.826 Å) and $ZrMo_2H_{3.98}$ (tetragonal, SG88, a = 5.603 Å, c = 8.081 Å). Their common v_H is 2.2 Å³/(H atom). Additional v_H estimations, performed in a similar way for $ZrMo_2H_{1.6}$ and $ZrMo_2H_{2.8}$, yielded values between 2.1 Å³/(H atom) and 2.2 Å³/(H atom).

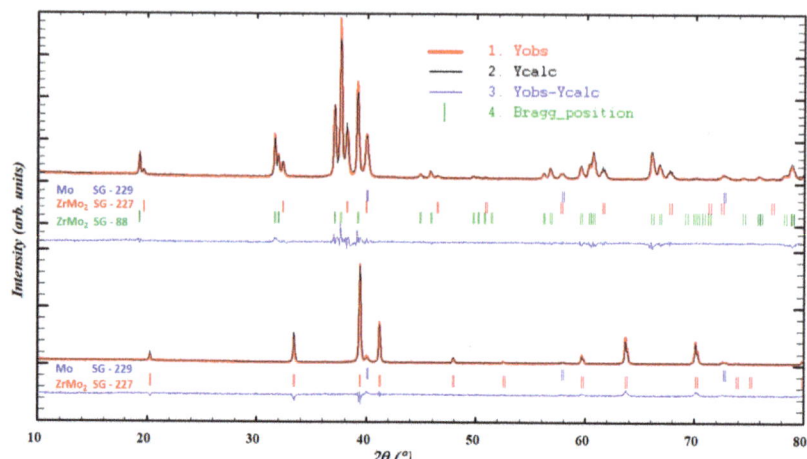

Figure 4. XRD pattern of ZrMo$_2$H$_{3.5}$ (upper plot) reveals the existence of two hydride phases (see also text). The pattern of ZrMo$_2$ before hydrogenation is shown in the lower plot for comparison.

3.2. Resolving the Peculiarity in the Stability Trends of the Zr(Mo$_x$Fe$_{1-x}$)$_2$, x = 0, 0.1, 0.5, 1, Hydrides

As cited in the Introduction, the formation pressures of the Zr(Mo$_x$Fe$_{1-x}$)$_2$ hydrides decrease between x = 0 and x = 0.5, indicating increasing stability in this compositional range. Then, this trend is unexpectedly reversed and ZrMo$_2$ hydride is formed at higher pressures than ZrMoFe and Zr(Mo$_{0.1}$Fe$_{0.9}$)$_2$ hydrides. We consider next the possible factors influencing the stability of these hydrides.

3.2.1. The Substitution of Fe by Mo

Both Fe and Mo form hydrides at very high hydrogen pressures. The equilibrium formation pressure of MoH$_{1.1}$ and FeH are about 4.3 GPa at 600 °C and 3.5 GPa at 300 °C (decomposition pressure 2.2 GPa), respectively [21,22]. It is thus not expected that the hydrogen affinity properties of Mo, as compared to Fe, play a dominant role in the variation of the hydride stabilities in Zr(Mo$_x$Fe$_{1-x}$)$_2$.

3.2.2. Trends of the Heats of Formation and the Crystal Volumes in the Zr(Mo$_x$Fe$_{1-x}$)$_2$ System

Table 2 presents the crystal volumes per formula unit (f.u.), V, and theoretically derived heats of formation, ΔH_f, of the original, non-hydrogenated Zr(Mo$_x$Fe$_{1-x}$)$_2$ intermetallics, x = 0, 0.1, 0.5, 1. The theoretically determined ΔH_f of 0.281 eV/(metal atom) for ZrFe$_2$ is in good agreement with a corresponding experimental value of 0.26 eV/(metal atom), e.g., Ref. [23]. ΔH_f values of Zr(Mo$_x$Fe$_{1-x}$)$_2$, x = 0.1, 0.5 are obtained by interpolation.

Table 2. Unit cell volumes, V, heats of intermetallic formation, ΔH_f, hydrogen contents, hydrogen atomic volumes, v_H, and elastic interactions, u_{elas}, for Zr(Mo$_x$Fe$_{1-x}$)$_2$ and the corresponding hydrides.

x	V [Å^3/f.u.]	ΔH_f [eV/Atom]	H Content [H Atoms/f.u.]	v_H [Å^3/H Atom]	u_{elas} [eV]
0 [24,25]	44.058	0.281 [26]	3.54	3.02	0.23
0.1 [25]	44.805	0.267	3.54	2.99	
0.5	48.089 [16] 49.013 [15]	0.21	2.5 2.6	2.90 2.74	0.17 0.15
1	54.677 [27] 54.829 [16] 54.634 this work	0.139 [28]	4 1.4 3.5	1.41 1.57 2.2	0.042 0.052 0.102

The heats of formation of the intermetallic compounds decrease with x from 0.281 eV/(metal atom) for ZrFe$_2$ to 0.139 eV/(metal atom) for ZrMo$_2$, while the corresponding crystal volumes increase from 44.058 Å^3/(f.u.) to 54.634 Å^3/(f.u.) (see Table 2). Thus, according to both the rule of reverse stability and the crystal volume criterion (see (i) and (ii) in the Introduction), the hydride stabilities should increase with x. These trends comply with the behavior for $0 \leq x \leq 0.5$, but disagree with the decreased stability of ZrMo$_2$ hydride.

3.2.3. Elastic Properties and H–H Elastic Interaction in Zr(Mo$_x$Fe$_{1-x}$)$_2$-H$_2$

As already noted, the H–H interaction influences the critical temperature, T_c, according to Equation (1). It also contributes to the heat of hydrogen solution or hydride formation [6]. We find two indications for the reduction in the attractive, elastically mediated H–H interaction.

(a) From inspection of Figure 1, the very wide solubility region of ZrMo$_2$-H$_2$ indicates a decrease in T_c with respect to the rest of the hydrogenated Zr(Mo$_x$Fe$_{1-x}$)$_2$ compounds, x = 0, 0.1, 0.5. This occurs in spite of the monotonic increase in the bulk modulus, B, from 148.2 GPa for ZrFe$_2$ to 200.4 GPa for ZrMo$_2$, expected in turn to increase T_c.

(b) The increased hydrogenation pressures in ZrMo$_2$-H$_2$ with regard to Zr(Mo$_x$Fe$_{1-x}$)$_2$-H$_2$, x = 0.1, 0.5, indicate a corresponding destabilization of ZrMo$_2$H$_x$. Indeed, the absolute enthalpy of ZrMo$_2$ hydride formation, 22 kJ/(mole H$_2$), is significantly lower than those of Zr(Mo$_x$Fe$_{1-x}$)$_2$ hydrides, 29.5 kJ/(mole H$_2$) for x = 0.5, and 26 kJ/(mole H$_2$) for x = 0.1. It even approaches the 21.3 kJ/(mole H$_2$) for ZrFe$_2$ hydride [11]. This destabilization behavior suggests that the attractive (negative) contribution of the H–H elastic interaction to the enthalpy of hydride formation or to the enthalpy of hydrogen solubility in ZrMo$_2$ weakens with the increase in hydrogen content (see the similar case of TaV$_2$-H$_2$ below).

As the variation in B of the Zr(Mo$_x$Fe$_{1-x}$)$_2$ compounds cannot directly account for the observed trend in the critical temperatures of the metal-to-hydride phase transitions and the heats of hydride formation at x > 0.5, we inspect another term that appears in a squared form in Equation (2), namely, v_H. A priori, v_H is not supposed to significantly affect u_{elas} according to Equation (2), as v_H usually acquires values in quite a narrow range, e.g., Ref. [6]. We find hereafter surprising experimentally derived v_H values. Lushnikov et al. report that a Laves ZrMo$_2$ compound with lattice constant a = 7.591 Å undergoes a tetragonal distortion upon deuteration to ZrMo$_2$D$_4$ with lattice constants a = 5.496 Å and c = 7.986 Å [27]. The authors note that such a tetragonal distortion is a known phenomenon in hydrogenated Laves phase compounds [29]. It is also worth noting that the composition of the interstitial sites does not change upon the tetragonal distortion. The deuterium atoms occupy (2Zr + 2Mo) tetrahedral interstitial sites [27], which are the preferable (2A + 2B) sites for hydrogen occupation in AB$_2$ Laves phase compounds [30]. A straightforward evaluation of the volume of the hydrogen atom in ZrMo$_2$D$_4$ (4 f.u./unit cell), using also the above data for ZrMo$_2$ (8 f.u./unit cell), yields v_H = 1.4 Å^3/(H atom). This is the lowest reported v_H value in intermetallic and even in binary hydrides [6] (pp. 105–109). Supporting evidence for this hydrogen atomic volume may be found in the reported cell parameter a = 7.698 Å of ZrMo$_2$H$_{1.4}$ [16]. Utilizing a = 7.598 Å for ZrMo$_2$ by the same authors [31], v_H = 1.6 Å^3/(H atom) is evaluated. It should be noted that a cell parameter a = 7.548 Å is reported for ZrMo$_2$ in [16], which is very different from other reported values [13,27], as well as from the value, reported later, by the same authors [31]. Nonetheless, v_H of 2.33 Å^3/(H atom) is obtained for ZrMo$_2$H$_{1.4}$ by employing a = 7.548 Å for ZrMo$_2$, although the stated v_H in [16] is 2.6 Å^3/(H atom). Additional v_H values, reported for various hydrogen compositions in ZrMo$_2$, span between 3.2 and 4.5 Å^3/(H atom) [31]. We recommend adopting v_H = 2.2 Å^3/(H atom) for the ZrMo$_2$ hydrides following the experimental estimations of this work. It is worthwhile noting that we found larger lattice constants of the tetragonal hydride phase (see Section 3.1) than those cited above. The latter lattice constants were obtained in the course of a seemingly very comprehensive neutron diffraction experiment, under much more extreme conditions of 2500 atm hydrogen pressure and a temperature of 78 K [27]. Table 2 presents the relevant information for the Zr(Mo$_x$Fe$_{1-x}$)$_2$ compounds and their hydrides. The small values of v_H affect u_{elas} very significantly. Table 2 lists the calculated values of u_{elas} according to Equation (2), utilizing the measured bulk and shear moduli (Table 1), and the V and v_H volumes (Table 2). The $ZrMo_2$-H_2 system clearly exhibits the smallest v_H and u_{elas}. This may explain the decreased T_c in this system according to Equation (1). The hydride stability would also be decreased as u_{elas} provides a smaller negative contribution to the heat of hydride formation of $ZrMo_2$ with regard to $ZrFe_2$ and ZrMoFe. A most probable reason for the decrease in v_H, and hence in u_{elas}, in the $Zr(Mo_xFe_{1-x})_2$ system is the large bulk modulus of $ZrMo_2$. We suggest that the large bulk modulus of $ZrMo_2$ confines the expansion of the metal lattice upon hydrogenation by restricting the H atoms in smaller interstitial volumes, v_H. The variation of v_H as a function of B, plotted in Figure 5, supports this suggestion. The depressed expansion of the crystal lattice mitigates the strain fields around a hydrogen atom and thus reduces the elastically mediated interaction between the hydrogen atoms.

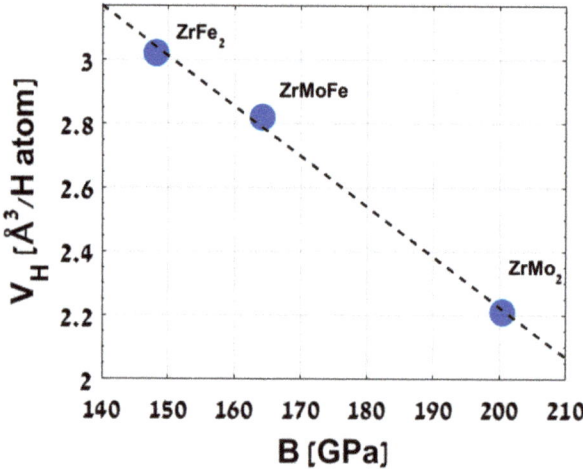

Figure 5. Variation of the hydrogen atomic volume, v_H, in the hydrogenated $Zr(Mo_xFe_{1-x})_2$, x = 0, 0.5, 1, compounds as a function of the bulk moduli, B, of the corresponding original intermetallics.

Elastic moduli rarely play such a significant role in the hydrogenation properties of intermetallic compounds. We note that a very similar behavior to that of $ZrMo_2$-H_2 is demonstrated by the C15 compound TaV_2. The experimentally derived bulk modulus of C15 TaV_2 varies between 200 GPa at low temperatures and 194 GPa at room temperature [32]. These B values of TaV_2 are very close to the 200.4 GPa bulk modulus of $ZrMo_2$, obtained in this work (see Table 1 and Figure 5). The volume of the hydrogen atom in hydrogenated TaV_2 is 2.09 Å^3/(H atom) [33] in the low range of v_H values, and in close proximity to v_H of 2.2 Å^3/(H atom), estimated here for the hydrogenated $ZrMo_2$ (see Table 2 and Figure 5). Accordingly, TaV_2-H_2 exhibits a broad solid hydrogen solubility region without a discontinuous metal-to-hydride phase transition even at temperatures as low as 195 K and pressures of 1000 atm [33]. In addition, the heat of hydrogen solution decreases with the increase in H content, i.e., becomes less exothermic. This may be regarded as surmounting the attractive elastic H–H interaction by some repulsive, probably electronic H–H interaction. It is worthwhile noting that the heats of hydrogen solution in metal–hydrogen systems usually become more exothermic with the increase in H content, e.g., Refs. [6,34]. We suggest that, in resemblance to $ZrMo_2$-H_2, the large bulk modulus of TaV_2 confines the hydrogen atoms into small interstitial volumes and consequently reduces the H–H elastic interaction, as indicated by the broad solubility region and the lack of metal to hydride phase transition in TaV_2-H_2. Another example of a significant role of elastic moduli is the shear stiffening in $Zr(Al_xM_{1-x})_2$, M=Fe, Co [19,35]. It prevents hydrogen absorption in $ZrAl_2$ even at extremely high H_2 pressures of 40 GPa [36], although

Al substitution drastically stabilizes the hydrogen absorption of $ZrFe_2$ and $ZrCo_2$ for small x values [37,38].

4. Summary and Conclusions

The elastic moduli of $Zr(Mo_xFe_{1-x})_2$, x = 0, 0.5, 1, as well as hydrogen absorption in $ZrMo_2$, were measured in an attempt to shed light on the unusual trend of hydride stabilities in this system. It seems that the bulk modulus plays a dual role in that context. Initially, B participates in generating attractive elastic interaction between the hydrogen atoms. Then, the significant increase in B in $ZrMo_2$ confines the expansion of the metal matrix and restrains the hydrogen atoms in smaller interstitial volumes. Consequently, the attractive H–H interaction is reduced, demonstrated by a sharp decrease in the critical temperature for metal-to-hydride phase transition and the hydride stability in the $ZrMo_2$-H_2 system. We also suggest a similar explanation for the lack of a discontinuous metal-to-hydride phase transition in the Laves phase TaV_2-H_2 system. It would be of interest to find a criterion that determines at which point B converts from a stabilizing to destabilizing agent during hydrogen absorption in intermetallic systems.

Author Contributions: Conceptualization, I.J.; methodology, D.B. and M.B.; validation, D.B. and M.B.; formal analysis, D.B.; investigation, D.B. and M.B.; writing—original draft preparation, I.J.; writing—review and editing, D.B., R.Z.S. and I.J.; supervision, I.J. and R.Z.S.; funding acquisition, I.J. and R.Z.S. All authors have read and agreed to the published version of the manuscript.

Funding: This research was partly funded by the Israel Science Foundation grant number 745/15.

Acknowledgments: This research was partly supported by the Israel Science Foundation (grant No. 745/15). One of the authors (IJ) is grateful to S. Mitrokhin for granting access to his presentation at MH2018, as well as for turning attention to the unusual stability trends of the $Zr(Mo_xFe_{1-x})_2$ hydrides.

Conflicts of Interest: The authors declare no conflict of interest.

References

1. Reilly, J.J. Chemistry of intermetallic hydrides, BNL report 46778, presented at the Symposium for Hydrogen Storage Materials, Battery and Electrochemistry. In Proceedings of the 180th Meeting of the Electrochemical Society, Phoenix, AZ, USA, 13–18 October 1991.
2. van Mal, H.H.; Buschow, K.H.J.; Miedema, A.R. Hydrogen absorption in LaNi$_5$ and related compounds: Experimental observations and their explanation. *J. Less-Common Met.* **1974**, *35*, 65–76. [CrossRef]
3. Lartigue, C. Etude Structurale et Thermodynamique du Systeme LaNi$_{5-x}$Mn$_x$—Hydrogene. Ph.D. Thesis, Universite Paris VI, Paris, France, 1984.
4. Jacob, I.; Hadari, Z.; Reilly, J.J. Hydrogen absorption in ANiAl (A = Zr, Y, U). *J. Less-Common Met.* **1984**, *103*, 123–127. [CrossRef]
5. Biderman, S.; Jacob, I.; Mintz, M.H.; Hadari, Z. Analysis of the hydrogen absorption in the U(Al$_x$Ni$_{1-x}$)2 system. *Trans. Nucl. Soc. Isr.* **1982**, *10*, 129–132.
6. Fukai, Y. *The Metal-Hydrogen System: Basic Bulk Properties*, 2nd ed.; Springer: Berlin/Heidelberg, Germany, 2005.
7. Alefeld, G. Phase transitions of hydrogen in metals due to elastic interaction. *Ber. Bunsenges. Phys. Chem.* **1972**, *76*, 746–755.
8. Wagner, H. Elastic Interaction and phase transitionin in coherent metal-hydrogen systems. In *Hydrogen in Metals I*; Alefeld, G., Völkl, J., Eds.; Springer: Berlin/Heidelberg, Germany, 1978; pp. 5–51.
9. Li, F.; Zhao, J.; Tian, D.; Zhang, H.; Ke, X.; Johansson, B. Hydrogen storage behavior in C15 Laves compound TiCr$_2$ by first principles. *J. Appl. Phys.* **2009**, *105*, 043707. [CrossRef]
10. Babai, D.; Bereznitsky, M.; Shneck, R.Z.; Jacob, I. The effect of Pd on hydride formation in $Zr(Pd_xM_{1-x})_2$ intermetallics where M is a 3d element. *J. Alloys Compd.* **2021**, *889*, 161503. [CrossRef]
11. Mitrokhin, S.; Verbetsky, V. Peculiarities of Hydrogen Interaction with Alloys of $ZrFe_2$-$ZrMo_2$ System. *Int. J. Hydrogen Energy* **2019**, *44*, 29166–29169. [CrossRef]
12. Muraoka, Y.; Shiga, M.; Nakamura, Y. Magnetic properties and Mössbauer effect of A(Fe$_{1-x}$B$_x$)$_2$ (A = Y or Zr, B = Al or Ni) Laves phase intermetallic compounds. *Phys. Status Solidi* **1977**, *42*, 369–374. [CrossRef]
13. Domagala, R.F.; McPherson, D.J.; Hansen, M. Systems Zirconium-Molybdenum and Zirconium-Wolfram. *JOM* **1953**, *5*, 73–79. [CrossRef]
14. Straumanis, M.E.; Shodhan, R.P. Lattice Constants, Thermal Expansion Coefficients and Densities of Molybdenum and the Solubility of Sulphur, Selenium and Tellurium in it at 1100 °C. *Z. Metallkd.* **1968**, *59*, 492–495. [CrossRef]
15. Yartys, V.A.; Burnasheva, V.V.; Fadeeva, N.V.; Solov'ev, S.P.; Semenenko, K.N. The crystal structure of the deuteride ZrMoFeD2.6. *Sov. Phys. Crystallogr. (Transl. Krist.)* **1982**, *27*, 540–543.

16. Semenenko, K.N.; Verbetskii, V.N.; Mitrokhin, S.V.; Burnasheva, V.V. Investigation of the interaction with hydrogen of Zirconium intermetallic compounds crystallised in Laves phase structure types. *Russ. J. Inorg. Chem.* **1980**, *25*, 961–964, Translated from *Zhurnal Neorgamcheskoi Khimil* **1980**, *25*, 1731–1736.
17. Mackenzie, J.K. The elastic constants of a solid containing spherical holes. *Proc. Phys. Soc. B* **1950**, *63*, 2–11. [CrossRef]
18. Masi, L.; Borchi, E.; de Gennaro, S. Porosity behavior of ultrasonic velocities in polycrystalline Y-B-C-O. *J. Phys. D Appl. Phys.* **1996**, *29*, 2015–2019. [CrossRef]
19. Willis, F.; Leisure, R.G.; Jacob, I. Elastic moduli of the Laves-phase pseudobinary compounds Zr(Al$_x$Fe$_{1-x}$)$_2$ as determined by ultrasonic measurements. *Phys. Rev. B* **1994**, *50*, 13792–13794. [CrossRef]
20. Turkdal, N.; Deligoz, E.; Ozisik, H.; Ozisik, H.B. First-principles studies of the structural, elastic, and lattice dynamical properties of ZrMo$_2$ and HfMo$_2$. *Phase Transit.* **2017**, *90*, 598–609. [CrossRef]
21. Abramov, S.N.; Antonov, V.E.; Bulychev, B.M.; Fedotov, V.K.; Kulakov, V.I.; Matveev, D.V.; Sholin, I.A.; Tkacz, M. T-P phase diagrams of the Mo-H system revisited. *J. Alloys Compd.* **2016**, *672*, 623–629. [CrossRef]
22. Tkacz, M. Thermodynamic properties of iron hydride. *J. Alloys Compd.* **2002**, *330–332*, 25–28. [CrossRef]
23. Klein, R.; Jacob, I.; O'Hare, P.A.G.; Goldberg, R.N. Solution-calorimetric determination of the standard molar enthalpies of formation of the pseudobinary compounds Zr(AlxFe1-x)2 at the temperature 298.15 K. *J. Chem. Thermodyn.* **1994**, *26*, 599–608. [CrossRef]
24. Zotov, T.; Movlaev, E.; Mitrokhin, S.; Verbetsky, V. Interaction in (Ti,Sc)Fe$_2$-H$_2$ (Zr,Sc)Fe$_2$-H$_2$ systems. *J. Alloys Compd.* **2008**, *459*, 220–224. [CrossRef]
25. Zotov, T.A.; Sivov, R.B.; Movlaev, E.A.; Mitrokhin, S.V.; Verbetsky, V.N. IMC hydrides with high hydrogen dissociation pressure. *J. Alloys Compd.* **2011**, *509*, S839–S843. [CrossRef]
26. The Materials Project. Available online: https://materialsproject.org/materials/mp-1718/ (accessed on 9 April 2023).
27. Lushnikov, S.A.; Movlaev, E.A.; Verbetsky, V.N.; Somenlov, V.A.; Agafonov, S.S. Interaction of ZrMo$_2$ with hydrogen at high pressure. *Int. J. Hydrogen Energy* **2017**, *42*, 29166–29169. [CrossRef]
28. The Materials Project. Available online: https://materialsproject.org/materials/mp-2049/ (accessed on 9 April 2023).
29. Irodova, A.V.; Glazkov, V.P.; Somenkov, V.A.; Shilstein, S.S. Hydrogen ordering in the cubic Laves phase HfV$_2$. *J. Less-Common Met.* **1981**, *77*, 89–98. [CrossRef]
30. Jacob, I.; Bloch, J.M.; Shaltiel, D.; Davidov, D. On the occupation of interstitial sites by hydrogen atoms in intermetallic hydrides: A quantitative model. *Solid State Commun.* **1980**, *35*, 155. [CrossRef]
31. Semenenko, K.N.; Verbetskii, V.N.; Pilchenko, V.A. Interaction of ZrMo$_2$ with hydrogen at low temperatures. *Mosc. Univ. Chem. Bull. (Transl. Vestn. Mosk. Univ. Ser. 2)* **1986**, *41*, 131–133.
32. Foster, K.; Hightower, J.E.; Leisure, R.G.; Skripov, A.V. Elastic moduli of the C15 Laves-phase materials TaV$_2$, TaV$_2$H(D)$_x$ and ZrCr$_2$. *Philos. Mag. B* **2000**, *80*, 1667–1679. [CrossRef]
33. Lynch, J.F. The solution of hydrogen in TaV$_2$. *J. Phys. Chem. Solids* **1981**, *42*, 411–419. [CrossRef]
34. Bereznitsky, M.; Mogilyanski, D.; Jacob, I. Destabilizing effect of Al substitution on hydrogen absorption in Zr(Al$_x$V$_{1-x}$)$_2$. *J. Alloys Compd.* **2012**, *542*, 213–217. [CrossRef]
35. Jacob, I.; Bereznitsky, M.; Yeheskel, O.; Leisure, R.G. Role of shear stiffening in reducing hydrogenation in intermetallic compounds. *Appl. Phys. Lett.* **2006**, *89*, 201909. [CrossRef]
36. Machida, A. Unpublished Results.
37. Bereznitsky, M.; Jacob, I.; Bloch, J.; Mintz, M.H. Thermodynamic and structural aspects of hydrogen absorption in the Zr(Al$_x$Co$_{1-x}$)$_2$ system. *J. Alloys Compd.* **2002**, *346*, 217. [CrossRef]
38. Bereznitsky, M.; Jacob, I.; Bloch, J.; Mintz, M.H. Thermodynamic and structural aspects of hydrogen absorption in the Zr(Al$_x$Fe$_{1-x}$)$_2$ system. *J. Alloys Compd.* **2003**, *351*, 180. [CrossRef]

Disclaimer/Publisher's Note: The statements, opinions and data contained in all publications are solely those of the individual author(s) and contributor(s) and not of MDPI and/or the editor(s). MDPI and/or the editor(s) disclaim responsibility for any injury to people or property resulting from any ideas, methods, instructions or products referred to in the content.

Article

Exploring Proton Pair Motion Away from the Global Proton–Tuple Energy Minimum in Yttrium-Doped Barium Zirconate

Yiqing Pan [1], Minh Tam Hoang [1,2], Sanaa Mansoor [1,3] and Maria Alexandra Gomez [1,*]

[1] Department of Chemistry, Mount Holyoke College, South Hadley, MA 01075, USA
[2] Department of Chemical Engineering and Materials Science, University of Minnesota Twin Cities, Minneapolis, MN 55455, USA
[3] Molecular Engineering Graduate Program, University of Washington, Washington, DC 98195, USA
* Correspondence: magomez@mtholyoke.edu

Abstract: Yttrium-doped barium zirconate is one of the fastest solid-state proton conductors. While previous studies suggest that proton–tuples move as pairs in yttrium-doped barium zirconate, a systematic catalog of possible close proton–tuple moves is missing. Such a catalog is essential to simulating dual proton conduction effects. Density functional theory with the Perdew–Burke–Ernzerhof functional is utilized to obtain the total electronic energy for each proton–tuple. The conjugate gradient and nudged elastic band methods are used to find the minima and transition states for proton–tuple motion. In the lowest-energy configuration, protons are in close proximity to each other and the dopant, significantly affecting the backbone structure. The map of moves away from the global minimum proton–tuple shows that the most critical move for long-range proton conduction is a rotation with a barrier range of 0.31–0.41 eV when the two protons are in close proximity.

Keywords: proton conduction; barium zirconate; proton pair; limiting barrier; proton–tuple

1. Introduction

Yttrium-doped barium zirconate shows great promise as one of the fastest solid-state proton conductors [1,2]. High proton conduction benefits both from high proton concentration and low long-range conduction limiting barriers. Increasing the dopant concentration in barium zirconate increases the maximum proton load, while at the same time introducing proton trap regions near the lower valency dopant which can increase the limiting barrier [3–7]. The conduction mechanism has long been understood to occur by a series of transfers and rotations away from one dopant trap to another [8–19]. While the earliest studies focused on proton conduction in idealized barium zirconate systems, the most recent studies considered the effect of multiple defects including additional protons, dopant ions, and oxygen vacancies with varying distributions as well as a greater diversity of sites [9,11,20–23]. For example, Hu et al. [8] find that the limiting barrier for a path of a rotation and transfer immediately adjacent to the dopant is a transfer at low dopant concentrations and a rotation at higher dopant concentrations. Draper et al. [9,11] find that with increasing the dopant concentration and considering a wide variety of dopant relative locations, nanoscale percolation paths along the dopant allow rapid proton conduction increasing conductivity. Vera et al. [1] and Hossain et al. [2] provide excellent reviews of both experiments and calculations of proton conduction in barium zirconate.

This contribution focuses on how two protonic defects influence each other. One might expect that repulsion would lead two protonic defects to be very far from each other. However, several studies challenge this notion. The proton–proton radial distribution function found in our earlier work [23] suggests that the most probable proton–proton distance is 4.0 Å with some protons as close as 2.0 Å from each other and the global energy

tuple minimum shows two protons in the same face of a perovskite unit cell within a larger system. More recently, Figure 2c in Draper et al. [11] shows a global ab initio minimum for a structure with proton-proton distance slightly above 2.0 Å. Quasi-elastic neutron scattering studies interpreted with a Holstein-type polaron model find that the proton–polaron effective mass is twice the proton mass, suggesting that protons travel in pairs [20,24]. Kinetic Monte Carlo (kMC) simulations [23] show an increase in long-range proton conduction barriers with increasing the number of protons while maintaining close proton–proton distances. Ab initio calculations [20] show that one proton facilitates distortions, which benefits the other nearby proton. Assistance of the lattice in maintaining high proton concentration by favoring nearby protons at the cost of slightly increased long-range barriers may be key to high conductivity. Lattice dynamics has been shown to be important to proton transport in both simulation and experiment [22,25,26]. However, most studies have focused on single proton transfer and rotation calculations. A systematic catalog of possible close proton–tuple moves is missing. Such a catalog would be helpful to understand the full mechanism of long-range proton conduction in yttrium-doped barium zirconate.

While a full catalog of ab initio energy barriers for all possible proton–tuple moves in a large system is computationally prohibitive, a systematic catalog of moves away from the global proton–tuple energy minimum is within the scope of current calculations. This contribution builds on our earlier work, which found the global energy tuple minimum in 12.5% yttrium-doped barium zirconate in $2 \times 2 \times 2$ [23] and $2 \times 2 \times 4$ [20] unit cell systems and considers all motions away from that lowest energy proton–tuple. A larger system of $4 \times 4 \times 4$ unit cells is considered to avoid spurious long-range octahedral tilting transformations when two defects are in close proximity. The climbing nudged elastic band method (cNEB) [27] is used to calculate barriers away from the global energy tuple minimum with the Vienna ab-initio simulation package (VASP) [28–32]. The resulting proton pair motion is analyzed in the context of the typical limiting barrier moves for long-range proton conduction.

2. Results

2.1. Lowest Energy Proton–Tuple and Single Proton Moves Away From It

Figure 1 shows the lowest energy proton-tuple superimposed on the minima in which one proton has moved one step away from the global minimum structure. Zirconium and yttrium ions are in green and teal, respectively. Oxygen ion types and proton site types are labeled using the same notation as in our earlier work [10,21]. Proton site types are labeled based on bonded oxygen ion type. Type I oxygen ions are the first nearest neighbors to the closest yttrium dopant ions. Type II oxygen ions are the second nearest neighbors to the closest yttrium ions. While earlier work [10,21] also highlights the third nearest neighbor oxygen ions, this work focuses on small moves away from the global proton–tuple minimum, not including moves to the third nearest neighbor oxygen ions. The lowest energy proton–tuple has the protons in the two circled sites, namely H_I^{Far} and H_{II}^{Far} sites. The subscripts refer to the type of oxygen ion the site is on, specifically whether the oxygen ion is a first nearest neighbor (I) or second nearest neighbor (II) to the closest yttrium ion. The superscript *Close* or *Far* refers to whether the proton-bonded oxygen ion and the opposite oxygen ion on the same face plane as the OH are the closer or the farther two opposite oxygen ions on that unit face.

From the lowest-energy tuple structure (H_I^{Far}, H_{II}^{Far}), each proton can rotate to two possible locations and transfer to two possible locations. As seen in Figure 1, the proton in the H_I^{Far} site can rotate forward to another H_I^{Far} site through a 0.11 eV barrier or backward to a H_I^{Close} site through a 0.26 eV barrier. This proton can also transfer to the right to a H_I^{Close} site through a 0.24 eV barrier or to the left to a H_{II}^{Close} site through a 0.11 eV barrier. The second proton in the global minimum tuple can also rotate and transfer. The proton in the H_{II}^{Far} site can rotate forward to a H_{IIND}^{Close} site through a 0.31 eV barrier or backward to a H_{IIND}^{Far} site through a 0.31 eV barrier. The addition of *ND* to the subscript is made to indicate that these two sites are on a plane without the dopant ions. The H_{II}^{Far} proton can also transfer to the right to a H_I^{Close} site through a 0.12 eV barrier or to the left to a H_{II}^{Close} site through a 0.21 eV barrier. The highest barriers in this group are the bottom two rotational barriers, which are critical to moving from near one yttrium dopant trap to another yttrium dopant trap.

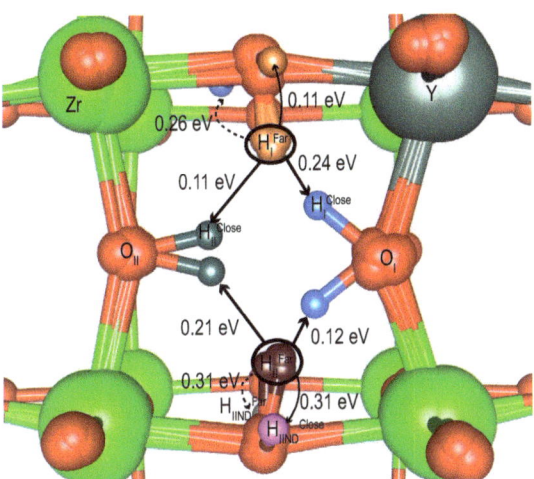

Figure 1. The lowest-energy proton–tuple in yttrium-doped barium zirconate is shown superimposed upon minima that are one proton move away from the lowest energy structure. The arrows highlight the move. The barrier to the move is listed by the arrows. The two protons in the lowest energy structure are each surrounded by an ellipse. Proton sites are labeled by color. The same notation as in our earlier work [10,21] is used.

Table 1 summarizes these moves and gives the energy of the final tuple minimum relative to the global minimum starting point. The minimum energies reveal forward barriers of nearly the same energy as the final site showing that the backward move barriers are much lower. In cases where the final tuple minimum has a populated site that is perpendicular to the plane shown in Figure 1, the symbol $\perp$ is added to more clearly specify the site and motion. The table groups minima including H_{II}^{Far} at the top and those including H_I^{Far} at the bottom. Within the groups, the minima are organized by relative energy to the global minimum (H_I^{Far}, H_{II}^{Far}).

Table 1. The energies of the minima superimposed in Figure 1 relative to the global minimum (H_I^{Far}, H_{II}^{Far}) are shown along with the single proton motion needed to get to the minimum from the global minimum (H_I^{Far}, H_{II}^{Far}). The top set are minima resulting from a move from the H_I^{Far} site, while the lower set results from a proton move from the H_{II}^{Far} site. Within each region, the sites are organized by energy. Proton sites perpendicular to the plane in Figure 1 have the $\perp$ symbol added to clarify their location.

Tuple	Relative Energy (eV)	Move From Global Minimum	Move Barrier (eV)
$(H_{II}^{Close}, H_{IL}^{Far})$	0.11	$T(H_I^{Far} \rightarrow H_{II}^{Close})$	0.11
$(H_I^{Far} \perp, H_{II}^{Far})$	0.11	$R(H_I^{Far} \rightarrow H_I^{Far} \perp)$	0.11
$(H_I^{Close} \perp, H_{II}^{Far})$	0.16	$R(H_I^{Far} \rightarrow H_I^{Close} \perp)$	0.26
$(H_I^{Close}, H_{II}^{Far})$	0.23	$T(H_I^{Far} \rightarrow H_I^{Close})$	0.24
(H_L^{Far}, H_L^{Close})	0.11	$T(H_{II}^{Far} \rightarrow H_L^{Close})$	0.12
$(H_I^{Far}, H_{II}^{Close})$	0.20	$T(H_{II}^{Far} \rightarrow H_{II}^{Close})$	0.21
$(H_L^{Far}, H_{UND}^{Close} \perp)$	0.31	$R(H_U^{Far} \rightarrow H_{UND}^{Close} \perp)$	0.31
$(H_I^{Far}, H_{IIND}^{Far} \perp)$	0.30	$R(H_{II}^{Far} \rightarrow H_{IIND}^{Far} \perp)$	0.31

2.2. Two Proton Moves Away from Global Minimum

All moves of two protons away from the global minimum tuple were considered. However, all but one nudge elastic band calculation showed sequential moves. The one exception was the concerted proton transfer shown in Figure 2a. Arrows highlight the proton moves as well as the moves of other ions. The concerted proton transfer has a barrier of 0.21 eV. Figure 2b shows related sequential moves with the same initial and final points. Solid arrows highlight the first steps while dashed arrows highlight the second step. In both Figure 2a,b, earlier ion positions are lightened. The sequential transfer moves have barriers of 0.12 eV for the first proton transfer from H_{II}^{Far} (brown) to H_I^{Close} (blue) and 0.09 eV for second proton transfer from H_I^{Far} (orange) to H_{II}^{Close} (teal). The order of sequential moves was determined by what was seen in a concerted attempted move calculation. Solid arrows highlight all the ion moves included in the first proton transfer, while dashed arrows highlight the corresponding moves for the second transfer. In both cases, significant oxygen ion rearrangement partially accompanies the proton move to stabilize the final proton tuple.

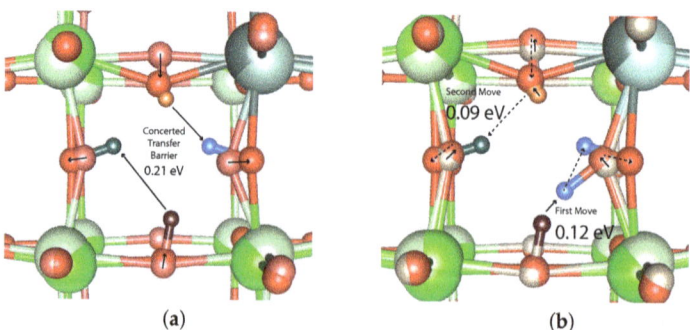

Figure 2. Two transfers from the original lowest-energy structure can happen in concert through a 0.21 eV barrier as seen in (**a**) or sequentially through 0.12 and 0.09 eV barriers as seen in (**b**). In (**b**), the motions in the first step are highlighted with solid arrows while the motions in the second step are highlighted with dashed arrows. While the barrier is shown by the arrow with the most significant movement in (**b**), many ion moves are part of each transfer. For each non-proton ion, the lighter colors are the earlier positions and the darkest the final positions.

While all possible moves were considered, our report focuses on the most important moves namely those barriers that limit long-range transport. Figure 1 suggests that the critical barrier for moving from close association with one yttrium ion trap to another is the rotational barriers at the bottom of the figure. Figure 3 shows the sequential moves that are most critical to escape a dopant trap. The moves involve a transfer of the H_I^{Far} proton (orange) followed by a rotation of the H_{II}^{Far} proton (brown). Solid and dashed arrows highlight the motion for the first and second moves, respectively. Two second move rotations are possible after the transfer. These are emphasized by short- and long-dashed arrows. The barrier for each move is written by the arrow showing the dominant motion. Comparing Figures 1 and 3 shows that the transfer move of the H_I^{Far} proton raises the H_{II}^{Far} proton rotational barrier.

Table 2 shows how barriers for the rotations and transfers in Figure 1 are affected by having the second proton in the tuple in different locations. Most useful are the last two sections, which show a range of barriers for the critical rotation to move a proton from being near one yttrium trap to another. The long-range proton motion limiting barrier is thus expected to be in the range of 0.31–0.41 eV. Single proton barriers for the same size system [10,21] and a smaller system [15] further highlight the effect of having a second proton in close proximity to the moving proton.

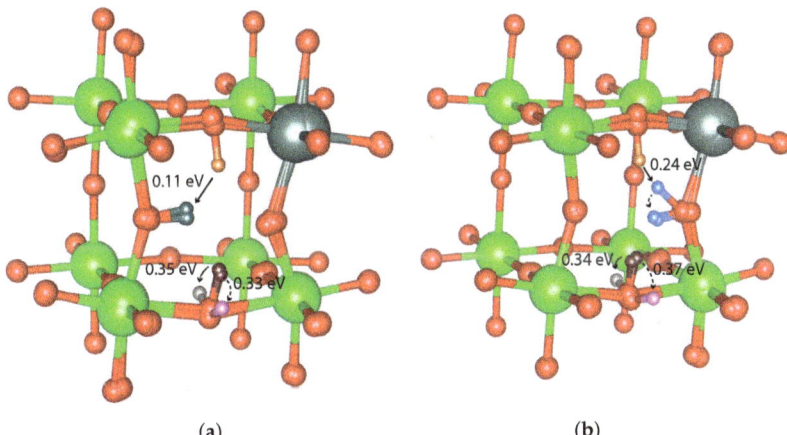

Figure 3. The sequential move most critical to escape one dopant trap is a transfer followed by a rotation and can occur in two ways shown in (**a**,**b**). In each case, the first transfer move is highlighted with a solid arrow, and the second two possible rotations are marked by short- and long-dashed arrows.

Table 2. Barriers are affected by nearby protons. This table summarizes the barrier for the motions in Figure 1 with different protons present. In addition, single proton movement barriers in the same size system [10,21] and a smaller system [15] are shown for comparison.

Move	Barrier (eV)	Nearby Proton	One Proton Barrier
$T(H_I^{Far} \rightarrow H_{II}^{Close})$	0.11 0.09	H_{II}^{Far} H_I^{Close}	0.14 [10], 0.32 [15]
$R(H_I^{Far} \rightarrow H_I^{Far} \perp)$	0.11 0.25	H_{II}^{Far} H_{II}^{Close}	0.02 [10], 0.12 [15]
$T(H_I^{Far} \rightarrow H_{II}^{Close})$	0.24	H_{II}^{Far}	0.24 [10], 0.25 [15]
$R(H_I^{Far} \rightarrow H_I^{Close} \perp)$	0.26	H_{II}^{Far}	0.10 [10], 0.22 [15]
$T(H_{II}^{Far} \rightarrow H_I^{Close})$	0.12 0.18 0.23	H_I^{Far} $H_I^{Far} \perp$ $H_I^{Close} \perp$	0.11 [10], 0.17 [15]
$T(H_{II}^{Far} \rightarrow H_{II}^{Close})$	0.21 0.26	H_I^{Far} $H_I^{Close} \perp$	0.15 [10], 0.43 [15]
$R(H_{II}^{Far} \rightarrow H_{IIND}^{Far} \perp)$	0.31 0.34 0.34 0.35 0.44	H_I^{Far} H_I^{Close} $H_I^{Far} \perp$ H_{II}^{Close} $H_I^{Close} \perp$	0.24 [10,21], 0.24 [15]
$R(H_{II}^{Far} \rightarrow H_{IIND}^{Close} \perp)$	0.31 0.33 0.37 0.41	H_I^{Far} H_{II}^{Close} H_I^{Close} $H_I^{Close} \perp$	0.25 [10,21] 0.27 [15]

3. Methods

Our earlier work on smaller systems [20,23] showed that the global energy minimum for two protons in 12.5 % yttrium-doped barium zirconate places the two protons within the same unit cell due to octahedral tilting bringing oxygen ions close to both protons. However, long-range octahedral tilting changes due to the presence of multiple nearby defects is assisted by periodic boundary conditions in small systems. In contrast, earlier work [21] with an optimized larger structure of 4 × 4 × 4 unit cells showed that a proton and an oxygen vacancy defect in the same unit cell did not disturb the long-range octahedral tilting arrangement. This study uses this same initial structure and finds the same tuple global minimum as our earlier studies in smaller systems [20,23]. All tuples with protons between 1.6 and 2.7 Å were considered, as this is the range of the second peak in the site–site radial distribution function. The closer proton sites characterizing the first peak are on the same oxygen and hence cannot be simultaneously populated. Some tuples with protons further than 2.7 Å were also optimized to confirm higher energy minima. From the global energy tuple minimum, moves to all other tuple arrangements within two moves are considered. The conjugate gradient method is used to find minima [33], while the nudged elastic band (NEB) method [27] is used to find transition states. All NEB calculations start with two images unless the path requires more. After NEB, the climbing NEB (cNEB) method [27] is used to converge the highest energy image to the saddle point. Optimization for minima and transition states is stopped when forces are less than 0.02 eV/Å.

The energy for all the above ionic relaxations is calculated using density functional theory (DFT) with the Perdew–Burke–Ernzerhof (PBE) functional in the Vienna ab initio simulation package (VASP) [28–32] with the same projector augmented wave (PAW) method as in our earlier work [21]. The valence states of Ba(5s,5p,6s,6d), Zr(4s,4p,5s,5d), Y(4s,4p,5s,4d), and O(2s,2p) are considered explicitly while inner core electrons are treated using PAW potentials. Projection operators are optimized automatically in real space. The preconditioned residual minimization method with direct inversion in the iterative subspace is used. The partial occupancy of orbitals is initialized with Gaussian smearing. An accurate level calculation is performed. The first ten steps are non-self consistent. Plane waves are generated using a single gamma point and a 600 eV cut off.

4. Discussion

The global minimum for two protons in $BaZr_{0.875}Y_{0.125}O_3$ has the pair in close proximity within a single unit cell even in a $4 \times 4 \times 4$ unit cell system. The result agrees with prior computational work [20,23] on smaller systems as well as neutron scattering work [20]. NMR and ab initio data on scandium-doped barium zirconate also show two protons in close proximity to each other [34] and the dopant [6]. NMR experiments and ab initio calculations with the yttrium dopant [35] reveal that the lowest two energy sites are the two sites populated in the global minimum tuple in this study namely the H_I^{Far} and H_{II}^{Far} sites. As seen in Figure 1, these sites are in close proximity to the dopant suggesting that the dopant stabilizes the proton tuple.

The $4 \times 4 \times 4$ unit cell system in this study allows a more systematic look at moves away from the global minimum without changing long-range octahedral tilting. Figure 1 and Table 1 show that the limiting barrier to long-range proton motion away from the global minimum is 0.31 eV. Considering two sequential proton moves as suggested by earlier work [20] shows an increase in this barrier as seen in Figure 3. Table 2 further suggests that the critical barrier to long-range motion is in the $0.31 - 0.41$ eV range depending on the location of the second proton. KMC simulations on smaller systems with a 1:1 proton:dopant ratio [13,23] rather than the 2:8 ratio in this work have shown long-range limiting barriers increasing from 0.39 to 0.45 eV with an increasing number of protons. The earlier simulations including multiple protons [23] used the ab initio binding site and transition state energies for single protons and included the effect of additional protons only through additional Coulombic interactions. The limiting barrier ranges of this work and our earlier work [23] are both in line with the experimental long-range barrier for 10% yttrium-doped barium zirconate of 0.44 eV [36,37]. However, the current work presents energies that allow backbone relaxation due to both protons in the tuple pair, which, as shown in Figure 2, can be significant.

While there is agreement in the overall limiting barrier, the different approaches have different individual barriers as seen by comparing the barrier and one proton barrier columns in Table 2. Within the one proton barrier column are barriers for larger 1:8 proton:dopant ratio and for smaller 1:1 proton:dopant ratio systems, respectively. The one proton smaller system tends to have larger barriers than the one proton larger system. This could be the effect of repulsions between a proton and its periodic image in the smaller simulation cell, but it may also be the effect of different methods. The larger systems simulations use a single gamma point, while the smaller system calculations use k-point meshes. The lattice size found in the larger simulations [10,21] was 4.29 Å, whereas the smaller simulations used a lattice size of 4.26 Å. The differences in the individual barriers led to a higher percentage of intraoctahedral transfer barriers in the small system kMC simulations [13,23] and a higher percentage of rotational limiting barriers in large system kMC simulations [10]. Hu et al. support a change in limiting barrier from intraoctahedral transfer to rotation with increasing the dopant concentration with rotation as the limiting barrier by 14.5% yttrium doping. The key result of the current work, however, is that the presence of a nearby proton raises the conduction limiting barrier of another despite the lattice motion of oxygen ions stabilizing close proton pair structures and the global tuple minimum having close proton–proton proximity.

While most of the tuple motion explored in this paper was sequential rather than concerted, sequential proton motion can be correlated, in that a move of one proton can promote the move of a nearby proton by preparing the lattice or by proton–proton repulsion. Quasi-elastic neutron studies extracted a proton–polaron effective mass of twice the proton mass suggesting the conductive species is a proton pair whose motion is correlated with lattice motion [20,24]. While this study does not calculate normal modes, prior computational studies considering single proton transfer show that there is significant lattice motion [22,38] at proton conduction temperatures in these systems. Through a simplified model, Geneste [39] also suggests that single proton transfer is facilitated by lattice reorganization and, in particular, octahedral tilting. Based on the proton tuple paths to escape the dopant found in this paper, long-range proton transport is expected to include both proton–proton correlation and proton–lattice correlation. The first proton can prepare the lattice for the next, and, at the same time, the latter proton can push the first one.

Author Contributions: Conceptualization, M.A.G.; methodology, M.A.G.; software, VASP and scripts written by Y.P., M.T.H., S.M. and M.A.G.; validation, Y.P., M.T.H., S.M. and M.A.G.; formal analysis, Y.P., M.T.H., S.M. and M.A.G.; investigation, Y.P., M.T.H., S.M. and M.A.G.; resources, M.A.G.; data curation, Y.P., M.T.H., S.M. and M.A.G.; writing—original draft preparation, Y.P., M.T.H. and M.A.G.; writing—review and editing, Y.P., M.T.H., S.M. and M.A.G.; visualization, Y.P. and M.A.G.; supervision, M.A.G.; project administration, M.A.G.; funding acquisition, M.A.G. All authors have read and agreed to the published version of the manuscript.

Funding: This research was supported by the National Science Foundation under grant DMR-1709975 and the Mount Holyoke College Department of Chemistry. Computational resources were provided in part by the MERCURY consortium under NSF grant CHE-1626238 and 2018427.

Data Availability Statement: The coordinates for all the figures are available at Gomez, Maria, 2023, Replication Data for "Exploring proton pair motion away from the global proton-tuple energy minimum in yttrium-doped barium zirconate", https://doi.org/10.7910/DVN/1ASCQG, accessed on 10 February 2023, Harvard Dataverse, V1.

Conflicts of Interest: The authors declare no conflict of interest.

Abbreviations

The following abbreviations are used in this manuscript:

kMC	Kinetic Monte Carlo
cNEB	Climbing Nudged Elastic Band
VASP	Vienna Ab Initio Simulations Package
NEB	Nudged Elastic Band
DFT	Density Functional Theory
PBE	Perdew–Burke–Ernzerhof

References

1. Regalado Vera, C.Y.; Hanping, D.; Peterson, D.; Gibbons, W.; Zhou, M.; Ding, D. A mini-review on proton conduction in BaZrO$_3$-based perovskite electrolytes. *J. Phys. Energy* **2021**, *3*, 032019. [CrossRef]
2. Hossain, M.K.; Biswas, M.C.; Chanda, R.K.; Rubel, M.H.K.; Khan, M.I.; Hashizume, K. A review on experimental and theoretical studies of perovskite barium zirconate proton conductors. *Emergent Mater.* **2021**, *4*, 999–1027. [CrossRef]
3. Vignesh, D.; Sonu, B.K.; Rout, E. Factors Constituting Proton Trapping in BaCeO$_3$ and BaZrO$_3$ Perovskite Proton Conductors in Fuel Cell Technology: A Review. *Energy Fuels* **2022**, *36*, 7219–7244. [CrossRef]
4. Toyoura, K.; Fujii, T.; Hatada, N.; Han, D.; Uda, T. Carrier–Carrier Interaction in Proton-Conducting Perovskites: Carrier Blocking vs Trap-Site Filling. *J. Phys. Chem.* **2019**, *123*, 26823–26830. [CrossRef]
5. Ding, J.; Balachandran, J.; Sang, X.; Guo, W.; Anchell, J.S.; Veith, G.M.; Bridges, C.A.; Cheng, Y.; Rouleau, C.M.; Poplawsky, J.D.; et al. The Influence of Local Distortions on Proton Mobility in Acceptor Doped Perovskites. *Chem. Mater.* **2018**, *30*, 4919–4925. [CrossRef]
6. Buannic, L.; Sperrin, L.; Dervisoglu, R.; Blanc, F.; Grey, C.P. Proton distribution in Sc-doped BaZrO$_3$: A solid state NMR and first principle calculations analysis. *Phys. Chem. Chem. Phys.* **2018**, *20*, 4317–4328. [CrossRef]
7. Yamazaki, Y.; Blanc, F.; Okuyama, Y.; Buannic, L.; Lucio-Vega, J.C.; Grey, C.P.; Haile, S.M. Proton trapping in yttrium-doped barium zirconate. *Nat. Mater.* **2013**, *12*, 647–651. [CrossRef]
8. Hu, H.; Zou, J.; Shan, L.; Jiang, X.; Ni, Y.; Li, X.; Qian, X.; Chen, W.; Zhou, Y.; Zhang, W.; et al. Conductivities in Yttrium-Doped Barium Zirconate: A First-Principles Study. *Crystals* **2023**, *13*, 401. [CrossRef]
9. Draber, F.M.; Denninger, J.R.; Müller, P.C.; Sommerfeld, I.K.; Martin, M. The Impact of Nanoscale Percolation in Yttrium-Doped BaZrO$_3$ on the Oxygen Ion and Proton Conductivities: A Density Functional Theory and Kinetic Monte Carlo Study. *Adv. Energy Sustain. Res.* **2022**, *3*, 2200007. [CrossRef]
10. Gomez, M.A.; Brooks-Randall, S.; Cai, G.; Glass-Klaiber, J.; Jiang, Y.; Jo, S.; Lin, Z.; Lin, S.; Marcellus, M.; Nguyen, H.A.; et al. Graph analysis of proton conduction pathways in scandium-doped barium zirconate. *J. Chem. Phys.* **2021**, *154*, 074711. [CrossRef]

11. Draber, F.M.; Ader, C.; Arnold, J.P.; Eisele, S.; Grieshammer, S.; Yamaguchi, S.; Martin, M. Nanoscale percolation in doped BaZrO3 for high proton mobility. *Nat. Mater.* **2020**, *19*, 338–346. [CrossRef]
12. Brieuc, F.; Dezanneau, G.; Hayoun, M.; Dammak, H. Proton diffusion mechanisms in the double perovskite cathode material GdBaCo$_2$O$_{5.5}$: A molecular dynamics study. *Solid State Ionics* **2017**, *309*, 187–191. [CrossRef]
13. Krueger, R.; Haibach, F.; Fry, D.; Gomez, M. Centrality measures highlight proton traps and access points to proton highways in kinetic Monte Carlo trajectories. *J. Chem. Phys.* **2015**, *142*, 154110. [CrossRef]
14. Hermet, J.; Torrent, M.; Bottin, F.; Dezanneau, G.; Geneste, G. Hydrogen diffusion in the protonic conductor BaCe$_{1-x}$Gd$_x$O$_{3-x/2}$ from density functional theory. *Phys. Rev. B* **2013**, *87*, 104303. [CrossRef]
15. Gomez, M.A.; Chunduru, M.; Chigweshe, L.; Foster, L.; Fensin, S.J.; Fletcher, K.M.; Fernandez, L.E. The effect of yttrium dopant on the proton conduction pathways of BaZrO$_3$, a cubic perovskite. *J. Chem. Phys.* **2010**, *132*, 214709. [CrossRef]
16. Merinov, B.; Goddard, W. Proton diffusion pathways and rates in Y-doped BaZrO$_3$ solid oxide electrolyte from quantum mechanics. *J. Chem. Phys.* **2009**, *130*, 194707. [CrossRef]
17. Björketun, M.E.; Sundell, P.G.; Wahnström, G. Effect of acceptor dopants on the proton mobility in BaZrO$_3$: A density functional investigation. *Phys. Rev. B* **2007**, *76*, 054307. [CrossRef]
18. Björketun, M.E.; Sundell, P.G.; Wahnström, G.; Engberg, D. A kinetic Monte Carlo study of proton diffusion in disordered perovskite structured lattices based on first-principles calculations. *Solid State Ionics* **2005**, *176*, 3035–3040. [CrossRef]
19. Munch, W.; Kreuer, K.D.; Seifert, G.; Maier, J. Proton diffusion in perovskites: Comparison between BaCeO$_3$, BaZrO$_3$, SrTiO$_3$, and CaTiO$_3$ using quantum molecular dynamics. *Solid State Ionics* **2000**, *136-7*, 183–189. [CrossRef]
20. Du, P.; Chen, Q.; Fan, Z.; Pan, H.; Haibach, F.G.; Gomez, M.A.; Braun, A. Cooperative origin of proton pair diffusivity in yttrium substituted barium zirconate. *Commun. Phys.* **2020**, *3*, 200. [CrossRef]
21. Ziqing, L.; Lin, S.; Tian, Y.; Van Bokkelen, A.; Valerio, M.; Gomez, M.A. Oxygen Vacancies Altering the Trapping in the Proton Conduction Landscape of Doped Barium Zirconate. *J. Phys. Chem.* **2020**, *124*, 27954–27964. [CrossRef]
22. Torayev, A.; Sperrin, L.; Gomez, M.A.; Kattirtzi, J.A.; Merlet, C.; Grey, C.P. Local Distortions and Dynamics in Hydrated Y-Doped BaZrO$_3$. *J. Phys. Chem. Nanomater. Interfaces* **2020**, *124*, 16689–16701. [CrossRef]
23. Gomez, M.; Fry, D.; Sweet, M. Effects on the proton conduction limiting barriers and trajectories in BaZr$_{0.875}$Y$_{0.125}$O$_3$ due to the presence of other protons. *J. Korean Ceram. Soc.* **2016**, *53*, 521–528. [CrossRef]
24. Braun, A.; Chen, Q. Experimental neutron scattering evidence for proton polaron in hydrated metal oxide proton conductors. *Nat. Commun.* **2017**, *8*, 15830. [CrossRef] [PubMed]
25. Fan, Z.; Li, N.; Du, P.; Yang, W.; Chen, Q. Influence of Lattice Dynamics on the Proton Transport in BaZrY-oxide perovskite under high pressure. *J. Phys. Chem.* **2020**, *124*, 22376–22382. [CrossRef]
26. Samgin, A.L. Lattice-assisted proton motion in perovskite oxides. *J. Phys. Chem. Solids* **2013**, *74*, 1661–1668. [CrossRef]
27. Henkelman, G.; Uberuaga, B.P.; Jónsson, H. A climbing image nudged elastic band method for finding saddle points and minimum energy paths. *J. Chem. Phys.* **2000**, *113*, 9901–9904. [CrossRef]
28. Kresse, G.; Furthmüller, J. Efficient iterative schemes for ab initio total-energy calculations using a plane-wave basis set. *Phys. Rev. B* **1996**, *54*, 11169–11186. [CrossRef]
29. Kresse, G.; Joubert, D. From ultrasoft pseudopotentials to the projector augmented-wave method. *Phys. Rev. B* **1999**, *59*, 1758–1775. [CrossRef]
30. Kresse, G. Ab Initio Molekular Dynamik für Flüssige Metalle. Ph.D. Thesis, Technische Universität at Wien, Vienna, Austria, 1993.
31. Kresse, G.; Hafner, J. Ab initio molecular-dynamics for liquid-metals. *Phys. Rev. B* **1993**, *47*, RC558. [CrossRef]
32. Kresse, G.; Furthmüller, J. Efficiency of ab-initio total energy calculations for metals and semiconductors using a plane-wave basis set. *Comput. Mater. Sci.* **1996**, *6*, 15–50. [CrossRef]
33. Payne, M.; Teter, M.P.; Ailan, D.C. Iterative minimization techniques for ab initio total energy calculation: molecular dynamics and conjugate gradients. *Rev. Mod. Phys.* **1992**, *64*, 1045–1097. [CrossRef]
34. Takahashi, H.; Oikawa, I.; Takamura, H. Atomistic Insight into the Correlation among Oxygen Vacancies, Protonic Defects, and the Acceptor Dopants in Sc-Doped BaZrO$_3$ Using First-Principles Calculations. *J. Phys. Chem.* **2018**, *122*, 6501–6507. [CrossRef]
35. Blanc, F.; Sperrin, L.; Lee, D.; Dervisoglu, R.; Yamazaki, Y.; Haile, S.M.; De Paëpe, G.; Grey, C.P. Dynamic nuclear polarization NMR of low-gamma nuclei: Structural Insights into hydrated yttrium-doped BaZrO$_3$. *J. Phys. Chem. Lett.* **2014**, *5*, 2431–2436. [CrossRef]
36. Kreuer, K.D.; Adams, S.; Munch, W.; Fuchs, A.; Klock, U.; Maier, J. Proton conducting alkaline earth zirconates and titanates for high drain electrochemical applications. *Solid Sate Ionics* **2001**, *145*, 295–306. [CrossRef]
37. Bohn, H.G.; Schober, T. Electrical conductivity of the high-temperature proton conductor BaZr$_{0.90}$Y$_{0.1}$O$_{2.95}$. *J. Am. Ceram. Soc.* **2000**, *83*, 768–72. [CrossRef]
38. Gomez, M.A.; Kwan, G.; Zhu, W.; Chelliah, M.; Zuo, X.; Eshun, A.; Blackmer, V.; Huynh, T.; Huynh, M. Ordered yttrium concentration effects on barium zirconate structure, proton binding sites and transition states. *Solid State Ionics* **2017**, *304*, 126–134. [CrossRef]
39. Geneste, G. Proton transfer in barium zirconate: Lattice reorganization, Landau-Zener curve-crossing approach. *Solid State Ionics* **2018**, *323*, 172–202. [CrossRef]

Disclaimer/Publisher's Note: The statements, opinions and data contained in all publications are solely those of the individual author(s) and contributor(s) and not of MDPI and/or the editor(s). MDPI and/or the editor(s) disclaim responsibility for any injury to people or property resulting from any ideas, methods, instructions or products referred to in the content.

Communication

Investigation on the Formation of Rare-Earth Metal Phenoxides via Metathesis

Jintao Wang [1,2], Qijun Pei [1], Yang Yu [1], Jirong Cui [1,2], Shangshang Wang [1,2], Khai Chen Tan [1,2], Jiaquan Guo [1,2], Teng He [1,2,*] and Ping Chen [1,2,3]

[1] Dalian Institute of Chemical Physics, Chinese Academy of Sciences, Dalian 116023, China
[2] Center of Materials Science and Optoelectronics Engineering, University of Chinese Academy of Sciences, Beijing 100049, China
[3] State Key Laboratory of Catalysis, Dalian Institute of Chemical Physics, Chinese Academy of Sciences, Dalian 116023, China
* Correspondence: heteng@dicp.ac.cn

Abstract: A number of alkali organometallic complexes with suitable thermodynamic properties and high capacity for hydrogen storage have been synthesized; however, few transition metal–organic complexes have been reported for hydrogen storage. Moreover, the synthetic processes of these transition metal–organic complexes via metathesis were not well characterized previously, leading to a lack of understanding of the metathesis reaction. In the present study, yttrium phenoxide and lanthanum phenoxide were synthesized via metathesis of sodium phenoxide with YCl_3 and $LaCl_3$, respectively. Quasi in situ NMR, UV-vis, and theoretical calculations were employed to characterize the synthetic processes and the final products. It is revealed that the electron densities of phenoxides in rare-earth phenoxides are lower than in sodium phenoxide due to the stronger Lewis acidity of Y^{3+} and La^{3+}. The synthetic process may follow a pathway of stepwise formation of dichloride, monochloride, and chloride-free species. Significant decreases in K-band and R-band absorption were observed in UV-vis, which may be due to the weakened conjugation effect between O and the aromatic ring after rare-earth metal substitution. Two molecular structures, i.e., planar and nonplanar, are identified by theoretical calculations for each rare-earth phenoxide. Since these two structures have very close single-point energies, they may coexist in the materials.

Keywords: rare-earth organometallic complexes; metathesis; yttrium phenoxide; lanthanum phenoxide; hydrogen storage

Citation: Wang, J.; Pei, Q.; Yu, Y.; Cui, J.; Wang, S.; Tan, K.C.; Guo, J.; He, T.; Chen, P. Investigation on the Formation of Rare-Earth Metal Phenoxides via Metathesis. *Inorganics* 2023, 11, 115. https://doi.org/10.3390/inorganics11030115

Academic Editors: Craig Buckley and Michael A. Beckett

Received: 8 February 2023
Revised: 1 March 2023
Accepted: 7 March 2023
Published: 10 March 2023

Copyright: © 2023 by the authors. Licensee MDPI, Basel, Switzerland. This article is an open access article distributed under the terms and conditions of the Creative Commons Attribution (CC BY) license (https://creativecommons.org/licenses/by/4.0/).

1. Introduction

Nowadays, oil and gas are used as the main sources of energy; however, the use of fossil sources leads to significant air pollution and global warming, which has become a huge challenge for the world [1]. To reduce energy consumption and CO_2 emissions from transportation and industries, the search for renewable energy sources is the foremost task [2]. Recently, some renewable energy sources have been used instead of traditional fossil sources; however, renewables such as solar and wind still have limitations, e.g., high cost, low efficiency, less reliability, and poor stability, due to the undesired environmental change. A suitable energy carrier is essential to solve these issues and controllably absorb or release energy. A promising energy carrier is hydrogen [3]. Hydrogen is abundant, combustible, and possesses high energy density compared to other energy sources. Hydrogen can be produced through several methods, including natural gas reforming [4], electrolysis [5], solar-driven [6], and biological processes [7]. Additionally, the hydrogen fuel cell not only transforms chemical energy to electricity with a high conversion (60% against 20% for combustion engines) but also produces water vapor through the chemical reaction between hydrogen and air in the present of a catalyst. Unfortunately, hydrogen storage is challenging due to the high flammability and low density of the hydrogen gas.

Lack of efficient hydrogen storage methods is one of the bottlenecks for the implementation of hydrogen energy [8–10]. Hydrogen may be absorbed by using material in which hydrogen bonds chemically or by physical absorption on solids, such as metal hydrides, alloys, carbon nanotubes, graphene, borohydrides, and ammonia borane [11]. It was widely accepted that storage of hydrogen in chemical compounds provided a safer and more efficient solution [12,13]. Diverse chemical materials have been developed, such as metal hydrides, complex hydrides, chemical hydrides, liquid organic hydrogen carriers, and so on. Very recently, organometallic complexes, a new family of hydrogen storage materials, were developed [14,15]. The dehydrogenation thermodynamics of H-rich organometallic complexes can be rationally tuned thanks to the electron-donating abilities of alkali metals, allowing reversible hydrogen uptake and release at ambient conditions from a thermodynamic point of view. A number of alkali organometallic complexes have been synthesized, including phenoxides [16], anilinides [17], pyrrolides [18], imidazolides [19], indolides [20], azaindolides [20], and carbazolides [19]. However, slow hydrogenation and dehydrogenation kinetics were observed even in the presence of transition metal catalysts at high temperatures, which may in part be due to the sluggish mass transfer in solid-state catalysis. Essentially, excellent contact between catalysts and substrates plays a vital role, where an ideal scenario is to fabricate a molecule that consists of transition metal and a hydrogen storage portion. Fortunately, organometallic complexes that have both metal cations and organic groups in one molecule provide a suitable platform to build such molecules. In this way, the hydrogenation and dehydrogenation of fabricated transition metal–organic complexes may be realized for hydrogen storage in the absence of catalysts. However, few transition metal–organic complexes for hydrogen storage have been reported.

Rare earth (RE) chemistry has gained increased attention in recent decades due to its application in catalysis, metallurgy, glass, hydrogen storage, superconductor, and so on [21,22]. Usually, RE elements are found in a variety of accessory minerals, such as phosphate, carbonates, fluorides, and silicates [23]. RE elements have been studied in the field of hydrogen storage in the last several decades. RE hydrides can reversibly absorb and desorb hydrogen. Although the RE hydrides are not good hydrogen storage materials due to their low hydrogen capacity and high dehydrogenation temperature, alloys consisting of RE metals and other metals (i.e., alkali metals, alkali earth metals, transition metals) are well known to exhibit high volumetric hydrogen capacity [24]. To enhance the hydrogen storage properties, a number of studies exploring new RE-based alloys and hydrides have been conducted using many methods, such as melting and ball milling. Typically, $LaNi_5$ can absorb hydrogen to form $LaNi_5H_6$ and release hydrogen upon heating in ambient conditions [25]. Moreover, thanks to the advances in anaerobic manipulation techniques and ligand designs, lanthanide metals (except radioactive promethium) and yttrium can form numerous molecular species with organic ligands [26]. Among them, the unique series of compounds of aliphatic alkoxides and aryloxides with the REs have come to the forefront for their unique properties and applications in the preparation of organometallics [22]. For instance, RE aryloxides can be employed as precursors to prepare new complexes, i.e., protonolysis reaction of 1-phenyl-3-N-(p-methoxyphenylimino)-1-butanone (HL) with RE aryloxides $[RE(OAr)_3(THF)_n]$ (ArO = 2,6-Bu^t_2-4-MeC_6H_2O) in THF can give mononuclear lanthanide aryloxides $L_2RE(OAr)(THF)_n$ (RE = Y, Nd) [27]. Another application of $RE(OAr)_3(THF)_n$ is to react with organolithium reagents to prepare new RE metal–organic complexes [28]. Additionally, RE cations, such as Eu^{3+} and Tb^{3+}, in which luminescence results from the 4f → 4f and 4f → 5d transitions, give rise to relatively broad absorption and high emission intensities, and RE aryloxides have become a new type of tunable and energy-efficient luminophores [29]. Usually, RE aryloxides are obtained from the metathesis of RE halides and aryloxides [30,31]. Through this method, diverse RE aryloxides have been developed; however, the synthetic processes are not well characterized, leading to a lack of understanding of the metathesis reaction.

Therefore, in the present study, two rare-earth organic complexes, i.e., yttrium phenoxide and lanthanum phenoxide, were synthesized via metathesis reaction. Quasi in situ nuclear magnetic resonance (NMR) and ultraviolet-visible (UV-vis) methods were employed to characterize the formation processes and electronic states of these two phenoxides. The molecular structures were predicated by theoretical calculations. Meanwhile, attempts at hydrogenation of yttrium and lanthanum phenoxides for hydrogen storage were conducted.

2. Results
2.1. Syntheses of Rare-Earth Metal Phenoxides

In our previous investigations, alkali organometallic complexes were synthesized via the reaction of alkali metal hydrides and organic compounds that have protic H, releasing one equivalent H_2 at the same time [14]. Accordingly, carbazole and indole were employed to react with rare-earth metal hydrides (YH_3 and LaH_3) in the present study; however, no reaction was observed. Even phenol, with more protic H than that of carbazole and indole, did not react with YH_3 and LaH_3 upon intensive ball milling. Therefore, a metathesis reaction was employed for the synthesis of yttrium phenoxide (Y(OPh)$_3$) and lanthanum phenoxide (La(OPh)$_3$). A schematic diagram of the preparation method is shown in Figure 1. First, sodium phenoxide (NaOPh) was prepared via the reaction of phenol and sodium hydride via ball mill, releasing one equivalent hydrogen. Then, sodium phenoxide reacted with corresponding chlorides (YCl_3 and $LaCl_3$) in ethanol, generating Y(OPh)$_3$ and La(OPh)$_3$, respectively, as shown in R1 and R2. Since NaCl has low solubility in ethanol, the products Y(OPh)$_3$ and La(OPh)$_3$ can be facilely separated through filtration. After evaporation of ethanol at 200 °C, target products without organic ligand were obtained.

$$3NaOPh + YCl_3 \rightarrow Y(OPh)_3 + 3NaCl \quad R1$$

$$3NaOPh + LaCl_3 \rightarrow La(OPh)_3 + 3NaCl \quad R2$$

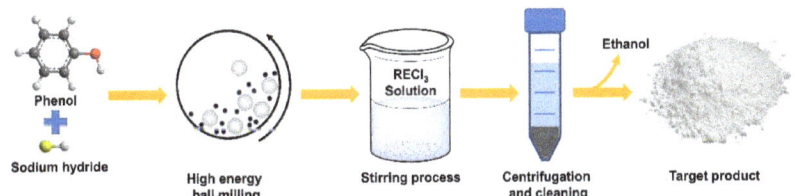

Figure 1. Schematic diagram of preparation method for rare-earth organic compounds.

The formation of NaCl from R1 and R2 in the precipitates was confirmed by X-ray diffraction (XRD), as shown in Figure 2a. The weights of precipitates from both reactions were consistent with the theoretical values of the NaCl produced, indicating almost all the Na and Cl were removed from the solution. Furthermore, the Na species could hardly be observed from ^{23}Na NMR solutions, further confirming that most of the NaCl was precipitated (Figure S1). However, both Y(OPh)$_3$ and La(OPh)$_3$ from the distillations exhibited a few broad diffraction peaks after removing the solvent (Figure 2b), demonstrating the amorphous state of the two samples, which is consistent with the transmission electron microscopy (TEM) images (Figure 2c). The absence of NaOPh in the XRD patterns also suggests its full conversion to La and Y phenoxides. Fourier-transform infrared (FT-IR) characterizations on these two samples demonstrated similar spectra as NaOPh, suggesting the retention of its phenol group (Figure S2). The appearance of vibrations at 583 and 566 cm^{-1} in Y(OPh)$_3$ and La(OPh)$_3$ indicate the formation of Y−O and La−O bonds and the replacement of Na by Y and La, respectively [32–34]. It was shown that all the ethanol solvent had been removed under vacuum at 200 °C in the solid products, as evidenced by ^{1}H NMR in DMSO-d_6 (Figure S3). Elemental analysis results for the amorphous Y(OPh)$_3$ showed that the ratio of Y:C was ca. 1:18, confirming the composition of Y(OPh)$_3$.

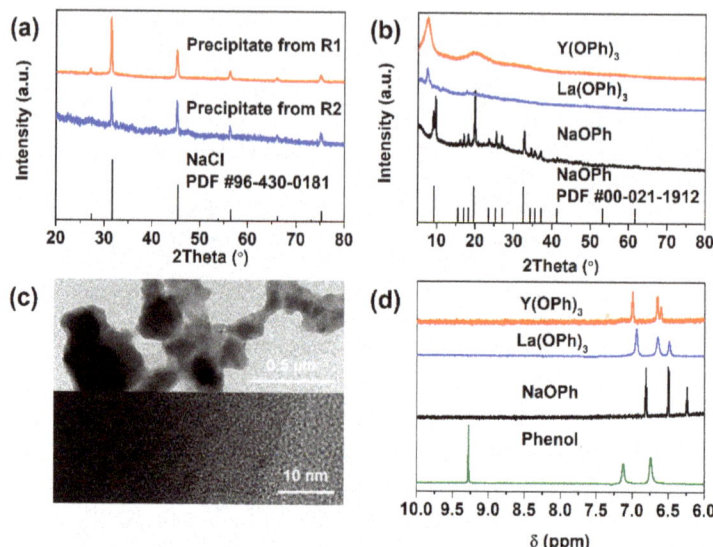

Figure 2. (a) The XRD patterns of the precipitates from R1 and R2. (b) The XRD patterns of Y(OPh)$_3$ and La(OPh)$_3$ compared with NaOPh. (c) Morphologies of Y(OPh)$_3$ observed by TEM. (d) ^{1}H NMR spectra of Y(OPh)$_3$ and La(OPh)$_3$ compared with NaOPh and phenol, respectively.

NMR was employed to characterize the two samples in comparison with phenol and sodium phenoxide. It was shown that signals of both Y(OPh)$_3$ and La(OPh)$_3$ in ^{1}H NMR shifted downfield compared with NaOPh, confirming the formation of new compounds. Their chemical shifts were different from those of phenol, excluding the hydrolysis/alcoholysis of sodium phenoxide to phenol. Compared with the signals of phenol, all signals of NaOPh, Y(OPh)$_3$, and La(OPh)$_3$ moved upfield and followed the order of NaOPh > La(OPh)$_3$ > Y(OPh)$_3$, which follows the Lewis acidity of the cations, i.e., Na$^+$ (0.159) < La^{3+} (0.343) < Y^{3+} (0.393) [35]. ^{13}C NMR spectra also revealed the impact of metathesis (Figure S4), i.e., the signals of C atoms at 1 and 2 sites in NaOPh moved to the high field and other signals to the low field upon replacing Na with Y cation. These results clearly demonstrated that rare-earth cations had successfully substituted the sodium cations in NaOPh, yielding Y(OPh)$_3$ and La(OPh)$_3$. However, no single crystal was obtained, leading to a lack of structural information about these two compounds. The crystal structure of ligand-free RE phenoxide has seldom been reported in the literature [21,36].

2.2. Characterizations

Although the synthesis of Y and La phenoxides has been reported previously, the synthetic processes are not well characterized, leading to a lack of understanding of the metathesis reaction. To get an understanding of the transformation processes from NaOPh to Y(OPh)$_3$ and La(OPh)$_3$, quasi in situ NMR was performed, where NaOPh was gradually added to pristine YCl$_3$ and LaCl$_3$ solutions. For pristine NaOPh in Figure 3a, chemical shifts at 6.81, 6.50, and 6.24 ppm were assigned to β-H, α-H, and γ-H in the phenoxide ring, respectively. When one equivalent NaOPh was added into YCl$_3$, chemical shifts at 6.95, 6.60, and 6.55 ppm were observed, which were different from those of pristine NaOPh, indicating the occurrence of cation exchange. When the molar ratio of NaOPh to YCl$_3$ increased from 1 to 1.2, 1.5, 2, and 3, all the signals shifted slightly downfield, suggesting the gradual formation of dichloride, monochloride, and chloride-free YCl$_n$(OPh)$_{3-n}$ (0 ≤ n ≤ 3) species. The elemental analysis results on the products of 1–1, 1–2 and 1–3 showed that the ratios of Y:C were 1:6, 1:12, and 1:18, respectively, confirming the formation of dichloride, monochloride, and chloride-free species. The slight downfield shift may be due to Y$^{\delta+}$

3. Galyametdinov, Y.; Athanassopoulou, M.A.; Griesar, K.; Kharitonova, O.; Soto Bustamante, E.A.; Tinchurina, L.; Ovchinnikov, I.; Haase, W. Synthesis and Magnetic Investigations on Rare-Earth-Containing Liquid Crystals with Large Magnetic Anisotropy. *Chem. Mater.* **1996**, *8*, 922–926. [CrossRef]
4. Luo, Y.; Han, Y.; Lin, J. Synthesis and Luminescent Properties of Europium (III) Schiff Base Complexes Covalently Bonded to Silica Xerogels. *J. Lumin.* **2007**, *122–123*, 83–86. [CrossRef]
5. Gagne, O.C.; Hawthorne, F.C. Empirical Lewis Acid Strengths for 135 Cations Bonded to Oxygen. *Acta Crystallogr. Sect. B* **2017**, *73*, 956–961. [CrossRef] [PubMed]
6. Moehring, S.A.; Miehlich, M.; Hoerger, C.J.; Meyer, K.; Ziller, J.W.; Evans, W.J. A Room-Temperature Stable Y(II) Aryloxide: Using Steric Saturation to Kinetically Stabilize Y(II) Complexes. *Inorg. Chem.* **2020**, *59*, 3207–3214. [CrossRef] [PubMed]
7. Spinner, E.; Late, A.B. The Electronic Spectra of Acetophenones Substituted in the Methyl Group. *Spectrochim. Acta* **1961**, *17*, 558–567. [CrossRef]
8. Rowe, W.F.; Marginean, I.; Carnes, S.; Lurie, I.S. The Role of Diode Array Ultraviolet Detection for the Identification of Synthetic Cathinones. *Drug Test. Anal.* **2017**, *9*, 1512–1521. [CrossRef]
9. Berger, J.; Staretz, M.E.; Wood, M.; Brettell, T.A. Ultraviolet Absorption Properties of Synthetic Cathinones. *Forensic Chem.* **2020**, *21*, 100286. [CrossRef]
10. Becke, A.D. Density-Functional Thermochemistry. III. The Role of Exact Exchange. *J. Chem. Phys.* **1993**, *98*, 5648–5652. [CrossRef]
11. Tomasi, J.; Mennucci, B.; Cammi, R. Quantum Mechanical Continuum Solvation Models. *Chem. Rev.* **2005**, *105*, 2999–3094. [CrossRef] [PubMed]
12. Wu, Y.; Yu, H.; Guo, Y.; Zhang, Y.; Jiang, X.; Sun, B.; Fu, K.; Chen, J.; Qi, Y.; Zheng, J.; et al. Promoting Hydrogen Absorption of Liquid Organic Hydrogen Carriers by Solid Metal Hydrides. *J. Mater. Chem. A* **2019**, *7*, 16677–16684. [CrossRef]
13. Zheng, X.; Li, G.; Guo, Y.; Yu, H.; Li, S.; Xiao, R.; Zheng, J.; Li, X. Yttrium Trihydride Enhanced Lithium Storage in Carbon Materials. *Carbon* **2020**, *164*, 317–323. [CrossRef]
14. Frisch, M.J.; Trucks, G.; Schlegel, H.B.; Scuseria, G.E.; Robb, M.A.; Cheeseman, J.; Scalmani, G.; Barone, V.; Mennucci, B.; Petersson, G.A.; et al. *Gaussian 09 Revision A.1*; Gaussian Inc.: Wallingford, CT, USA, 2009.
15. Reed, A.E.; Curtiss, L.A.; Weinhold, F. Intermolecular Interactions from a Natural Bond Orbital, Donor-Acceptor Viewpoint. *Chem. Rev.* **1988**, *88*, 899–926. [CrossRef]

Disclaimer/Publisher's Note: The statements, opinions and data contained in all publications are solely those of the individual author(s) and contributor(s) and not of MDPI and/or the editor(s). MDPI and/or the editor(s) disclaim responsibility for any injury to people or property resulting from any ideas, methods, instructions or products referred to in the content.

MDPI
St. Alban-Anlage 66
4052 Basel
Switzerland
www.mdpi.com

Inorganics Editorial Office
E-mail: inorganics@mdpi.com
www.mdpi.com/journal/inorganics

Disclaimer/Publisher's Note: The statements, opinions and data contained in all publications are solely those of the individual author(s) and contributor(s) and not of MDPI and/or the editor(s). MDPI and/or the editor(s) disclaim responsibility for any injury to people or property resulting from any ideas, methods, instructions or products referred to in the content.

www.ingramcontent.com/pod-product-compliance
Lightning Source LLC
LaVergne TN
LVHW070701100526
838202LV00013B/1008